BRIEF CALCULUS

FOR BUSINESS, SOCIAL SCIENCES, AND LIFE SCIENCES

BRIEF CALCULUS
FOR BUSINESS, SOCIAL SCIENCES, AND LIFE SCIENCES

Produced by the Consortium based at Harvard which was formed under a National Science Foundation Grant. All proceeds from the sale of this work are used to support the work of the Consortium

Deborah Hughes-Hallett
Harvard University

William G. McCallum
University of Arizona

Andrew M. Gleason
Harvard University

Brad G. Osgood
Stanford University

Patti Frazer Lock
St. Lawrence University

Andrew Pasquale
Chelmsford High School

Daniel E. Flath
University of South Alabama

Jeff Tecosky-Feldman
Haverford College

Sheldon P. Gordon
Suffolk County Community College

Joe B. Thrash
University of Southern Mississippi

David O. Lomen
University of Arizona

Karen R. Thrash
University of Southern Mississippi

David Lovelock
University of Arizona

Thomas W. Tucker
Colgate University

with the assistance of
Otto K. Bretscher
Harvard University

John Wiley & Sons, Inc.

New York　　　Chichester　　　Brisbane　　　Toronto　　　Singapore　　　Weinheim

**This project was supported, in part,
by the**

National Science Foundation

**Opinions expressed are those of the authors
and not necessarily those of the Foundation**

Recognizing the importance of preserving what has been written, it is a policy of John Wiley & Sons, Inc. to have books of enduring value published in the United States printed on acid-free paper, and we exert our best efforts to that end.

This book was set in Times Roman by the Consortium based at Harvard using TeX, Mathematica, and the package *AsTeX*, which was written by Alex Kasman. Special thanks to S.Alex Mallozzi and Mike Esposito for managing the process. It was printed and bound by R.R. Donnelley & Sons, Company. The cover was printed by The Lehigh Press, Inc.

Photo Credits: Greg Pease

Problems from *Calculus: The Analysis of Functions*, by Peter D. Taylor (Toronto: Wall & Emerson, Inc. 1992). Reprinted with permission of the publisher.

ISBN 0-471-17646-X

Printed in the United States of America

10 9 8 7 6 5 4 3 2 1

Dedicated to Robin, Kari, Eric, and Dennis,
and
Laura, Hannah, Ben, Natty, Isaac, and Matthias.

PREFACE

Calculus is one of the greatest achievements of the human intellect. Inspired by problems in astronomy, Newton and Leibniz developed the ideas of calculus 300 years ago. Since then, each century has demonstrated the power of calculus to illuminate questions in mathematics, the physical sciences, engineering, and the social and biological sciences.

Calculus has been so successful because of its extraordinary power to reduce complicated problems to simple rules and procedures. Therein lies the danger in teaching calculus: it is possible to teach the subject as nothing but the rules and procedures – thereby losing sight of both the mathematics and of its practical value. With the generous support of the National Science Foundation, our consortium set out to create a new calculus curriculum that would restore that insight. This book brings this new calculus curriculum to the applied calculus course.

Basic Principles

Two principles guided our efforts. The first is our prescription for restoring the mathematical content to calculus:

> **The Rule of Four:** *Every topic should be presented geometrically, numerically, algebraically, and verbally.*

We continually encourage students to think about the geometrical and numerical meaning of what they are doing. It is not our intention to undermine the purely algebraic aspect of calculus, but rather to reinforce it by giving meaning to the symbols. In the homework problems dealing with applications, we continually ask students to explain verbally what their answers mean in practical terms.

The second principle, inspired by Archimedes, is our prescription for restoring practical understanding:

> **The Way of Archimedes:** *Formal definitions and procedures evolve from the investigation of practical problems.*

Archimedes believed that insight into mathematical problems is gained by first considering them from a mechanical or physical point of view.[1] For the same reason, our text is problem driven. Whenever possible, we start with a practical problem and derive the general results from it. By practical problems we usually, but not always, mean real world applications. These two principles have led to a dramatically new curriculum – more so than a cursory glance at the table of contents might indicate.

Technology

We take advantage of computers and graphing calculators to help students learn to think mathematically. For example, using a graphing calculator to zoom in on functions is an excellent way of seeing local linearity. The

---

[1] . . . I thought fit to write out for you and explain in detail . . . the peculiarity of a certain method, by which it will be possible for you to get a start to enable you to investigate some of the problems in mathematics by means of mechanics. This procedure is, I am persuaded, no less useful even for the proof of the theorems themselves; for certain things first became clear to me by a mechanical method, although they had to be demonstrated by geometry afterwards because their investigation by the said method did not furnish an actual demonstration. But it is of course easier, when we have previously acquired, by the method, some knowledge of the questions, to supply the proof than it is to find it without any previous knowledge. From *The Method*, in *The Works of Archimedes* edited and translated by Sir Thomas L. Heath (Dover, NY).

ability to use technology effectively as a tool is important. Students are expected to use their own judgement to determine where technology is a useful tool.

However, the book does not require any specific software or technology. Test sites have used the materials with graphing calculators, graphing software, and computer algebra systems. Any technology with the ability to graph functions and perform numerical integration will suffice.

What Student Background is Expected?

This book is intended for students in business, the social sciences, and the life sciences. We have found the material to be thought-provoking for well-prepared students while still accessible to students with weak algebra backgrounds. Providing numerical and graphical approaches as well as the algebraic gives students several ways of mastering the material. This approach encourages students to persist, thereby lowering failure rates.

Content

We began work on this book by talking to faculty in business, economics, biology, and a wide range of other fields, as well as to many mathematicians who teach applied calculus. As a result of these discussions we included some new topics, and omitted some traditional topics whose inclusion we could not justify. In the process, we also changed the focus of certain topics. In order to meet individual needs or course requirements, topics can easily be added or deleted, or the order changed.

Chapter 1: Measuring Change

Chapter 1 introduces the concept of a function and the idea of change, including the distinction between total change and rate of change. Linear functions, exponential functions, and power functions are discussed. Although the functions are probably familiar, the graphical, numerical, and modeling approach to them is fresh. Our purpose is to acquaint the student with each function's individuality: the shape of its graph, characteristic properties, comparative growth rates, and general uses. We expect to give the student the skill to read graphs and think graphically, to read tables and think numerically, and to apply these skills, along with their algebraic skills, to modeling the real world. We introduce exponential functions at the earliest possible stage, since they are fundamental to the understanding of real-world processes. Further attention is given to using these functions to model real data through an understanding of regression analysis.

We encourage you to cover this chapter thoroughly, as the time spent on it will pay off when you get to the calculus.

Chapter 2: Rate of Change: The Derivative

Chapter 2 presents the key concept of the derivative according to the Rule of Four. The purpose of this chapter is to give the student a practical understanding of the meaning of the derivative and its interpretation as an instantaneous rate of change without complicating the discussion with differentiation rules. After finishing this chapter, a student will be able to find derivatives numerically (by taking arbitrarily fine difference quotients), visualize derivatives graphically as the slope of the graph, and interpret the meaning of first and second derivatives in various applications. The student will also understand the concept of marginality and recognize the derivative as a function in its own right.

Chapter 3: Accumulated Change: The Definite Integral

Chapter 3 presents the key concept of the definite integral, along the same lines as Chapter 2. Chapter 3 (and Section 5.7 on antiderivatives) can be delayed until after Chapter 5 without difficulty.

The purpose of this chapter is to give the student a practical understanding of the definite integral as a limit of Riemann sums, and to bring out the connection between the derivative and the definite integral in the Fundamental

Theorem of Calculus. We use the same method as in Chapter 2, introducing the fundamental concept in depth without going into technique. The motivating problem is computing the total distance traveled from the velocity function. The student will finish the chapter with a good grasp of the definite integral as a limit of Riemann sums, with the ability to compute it numerically, and with an understanding of how to interpret the definite integral in various contexts.

Chapter 4: A Library of Functions

Chapter 4 extends the library of functions begun in Chapter 1. Exponential functions with base e, logarithmic functions, polynomials, periodic functions, logistic functions, and surge functions are all introduced as families of functions used to model real world phenomena. The emphasis is on understanding the behavior of the functions and the effect of parameters on this behavior. Further attention is given to constructing new functions from old – how to shift, flip and stretch the graph of any basic function to give the graph of a new related function.

Chapter 5: Short-Cuts to Differentiation

Chapter 5 presents the symbolic approach to differentiation. The title is intended to remind the student that the basic methods of differentiation are not to be regarded as the definition of the derivative. The derivatives of all the basic functions are introduced as well as the rules for differentiating combinations of functions. Antiderivatives are introduced in Section 5.7 The student will finish this chapter with basic proficiency in differentiation and an understanding of why the various rules are true.

Chapter 6: Applications

Chapter 6 presents applications of the derivative and the definite integral. Our aim in this chapter is to enable the student to use calculus in solving problems, rather than to learn a catalogue of application templates. It is not meant to be comprehensive, and you do not need to cover all the sections. The student will finish this chapter with the experience of having successfully tackled a few problems that required sustained thought over more than one session.

Chapter 7: Functions of Many Variables

Chapter 7 introduces functions of two variables from several points of view, using contour diagrams, formulas, and tables. It gives students the skills to read contour diagrams and think graphically, to read tables and think numerically, and to apply these skills, along with their algebraic skills, to modeling the real world. The idea of the partial derivative is introduced from graphical, numerical, and analytical viewpoints. Partial derivatives are then applied to optimization problems, ending with a discussion of Lagrange multipliers. Students will finish this chapter with a solid understanding of functions of two variables.

Appendices

There are two appendices: one on roots and accuracy and one on compound interest.

What is the Relationship Between This Book and the Calculus Books by the Same Consortium?

Much of this book is based on Chapters 1–8 of the text *Calculus* (first edition, 1994) and Chapters 11, 13, and 14 of *Multivariable Calculus* (first edition, 1997), both by the same consortium. However, the content of this book was thought out from scratch with substantial input from faculty in business, social, and life sciences. For example, symbolic antidifferentiation plays a much smaller role in this text; the emphasis is instead on when and how to use a definite integral. Since a firm understanding of graphs and tabular data is especially important for this audience, this book emphasizes the graphical and numerical aspects to an even greater extent than in the original texts.

What is the Relationship Between This Book and the Two-Semester Applied Calculus Book by the Same Consortium?

This book is designed for a one-semester applied calculus course. The book contains material from the text *Applied Calculus* (Preliminary Edition, 1996). We have limited the amount of material included in this text to ensure that students have time to develop a solid understanding of the key ideas. We have increased the emphasis on the concept of change and the distinction between rate of change and accumulated change. In addition, we have divided the first chapter of the *Applied Calculus* book into two chapters in this book (Chapters 1 and 4). This book starts by emphasizing conceptual ideas in the first three chapters: functions and change, the rate of change, and accumulated change.

Supplementary Materials

- **Instructor's Manual containing** containing teaching tips, calculator programs, and some overhead transparency masters.
- **Instructor's Solution Manual** with complete solutions to all problems.
- **Answer Manual** with brief answers to all odd-numbered problems.
- **Student's Solution Manual** with complete solutions to half the odd-numbered problems.
- **Student Workbook** with study guides and supplementary materials.

Acknowledgements

First and foremost, we want to express our appreciation to the National Science Foundation for their faith in our ability to produce a revitalized calculus curriculum and, in particular, to Louise Raphael, John Kenelly, John Bradley, Bill Haver, and James Lightbourne. We also want to thank the members of our Advisory Board, Benita Albert, Lida Barrett, Bob Davis, Lovenia DeConge-Watson, John Dossey, Ron Douglas, Don Lewis, Seymour Parter, John Prados, and Steve Rodi for their ongoing guidance and advice.

In addition, we want to thank all the people across the country who encouraged us to write this book and who offered so many helpful comments. We would like to thank the following people, for all that they have done to help our project succeed: Wayne Anderson, Leonid Andreev, David Arias, Ruth Baruth, Graeme Bird, J.Curtis Chipman, David Chua, Eric Connally, Bob Condon, Josh Cowley, Larry Crone, Gene Crossley, Jie Cui, Jane Devoe, Mike Esposito, Gail Small Ferrell, Joe Fiedler, Holland Filgo, Sally Fischbeck, Hermann Flaschka, David Flath, Ron Frazer, Lynn Garner, David Graser, David Grenda, David Harris, John Hennessey, David Hornung, Richard Iltis, Adrian Iovita, Jerry Johnson, Donna Krawczyk, Theodore Laetsch, Sylvain Laroche, Kurt Lemmert, Suzanne Lenhart, Tom Lucas, Alex Mallozzi, Alfred Manaster, Elliot Marks, Georgia Mederer, Kurt Mederer, David Meredith, Jean Morris, Saadat Moussavi, Jim Osterburg, Edmund Park, Greg Peters, Rick Porter, Rebecca Rapoport, Harry Row, Virginia Stallings, Brian Stanley, Virginia Stover, "Suds" Sudholz, Noah Syroid, John.S. Thomas, Tom Timchek, J.Jerry Uhl, Tilaka Vijithakumara.

Deborah Hughes-Hallett	David O. Lomen	Jeff Tecosky-Feldman
Andrew M. Gleason	David Lovelock	Joe B. Thrash
Patti Frazer Lock	William G. McCallum	Karen R. Thrash
Daniel E. Flath	Brad G. Osgood	Thomas W. Tucker
Sheldon P. Gordon	Andrew Pasquale	

To Students: How to Learn from this Book

- This book may be different from other math textbooks that you have used, so it may be helpful to know about some of the differences in advance. At every stage, this book emphasizes the *meaning* (in practical, graphical or numerical terms) of the symbols you are using. There is much less emphasis on "plug-and-chug" and using formulas, and much more emphasis on the interpretation of these formulas than you may expect. You will often be asked to explain your ideas in words or to explain an answer using graphs.

- The book contains the main ideas of calculus in plain English. Success in using this book will depend on reading, questioning, and thinking hard about the ideas presented. It will be helpful to read the text in detail, not just the worked examples.

- There are few examples in the text that are exactly like the homework problems, so homework problems can't be done by searching for similar–looking "worked out" examples. Success with the homework will come by grappling with the ideas of calculus.

- Many of the problems in the book are open-ended. This means that there is more than one correct approach and more than one correct solution. Sometimes, solving a problem relies on common sense ideas that are not stated in the problem explicitly but which you know from everyday life.

- This book assumes that you have access to a calculator or computer that can graph functions, find (approximate) roots of equations, and compute integrals numerically. There are many situations where you may not be able to find an exact solution to a problem, but can use a calculator or computer to get a reasonable approximation. An answer obtained this way is usually just as useful as an exact one. However, the problem does not always state that a calculator is required, so use your own judgement.

  If you mistrust technology, listen to this student, who started out the same way:

  > Using computers is strange, but surprisingly beneficial, and in my opinion is what leads to success in this class. I have difficulty visualizing graphs in my head, and this has always led to my downfall in calculus. With the assistance of the computers, that stress was no longer a factor, and I was able to concentrate on the concepts behind the shapes of the graphs, and since these became gradually more clear, I got increasingly better at picturing what the graphs should look like. It's the old story of not being able to get a job without previous experience, but not being able to get experience without a job. Relying on the computer to help me avoid graphing, I was tricked into focusing on what the graphs meant instead of how to make them look right, and what graphs symbolize is the fundamental basis of this class. By being able to see what I was trying to describe and learn from, I could understand a lot more about the concepts, because I could change the conditions and see the results. For the first time, I was able to see how everything works together . . . .

  That was a student at the University of Arizona who took calculus in Fall 1990, the first time we used some of the material in this text. She was terrified of calculus, got a C on her first test, but finished with an A for the course.

- This book attempts to give equal weight to three methods for describing functions: graphical (a picture), numerical (a table of values) and algebraic (a formula). Sometimes it's easier to translate a problem given in one form into another. For example, you might replace the graph of a parabola with its equation, or plot a table of values to see its behavior. It is important to be flexible about your approach: if one way of looking at a problem doesn't work, try another.

- Students using this book have found discussing these problems in small groups helpful. There are a great many problems which are not cut-and-dried; it can help to attack them with the other perspectives your colleagues can provide. If group work is not feasible, see if your instructor can organize a discussion session in which additional problems can be worked on.

- You are probably wondering what you'll get from the book. The answer is, if you put in a solid effort, you will get a real understanding of one of the most important accomplishments of the millennium – calculus –

as well as a real sense of the power of mathematics in the age of technology.

Deborah Hughes-Hallett   David O. Lomen   Jeff Tecosky-Feldman

Andrew M. Gleason   David Lovelock   Joe B. Thrash

Patti Frazer Lock   William G. McCallum   Karen R. Thrash

Daniel E. Flath   Brad G. Osgood   Thomas W. Tucker

Sheldon P. Gordon   Andrew Pasquale

Table of Contents

4 A LIBRARY OF FUNCTIONS   197

5 SHORT-CUTS TO DIFFERENTIATION   265

6 APPLICATIONS   313

7 FUNCTIONS OF MANY VARIABLES   371

APPENDIX   437

CHAPTER ONE

MEASURING CHANGE

Calculus is the study of change. We investigate the total change in a quantity and the average rate of change of one quantity with respect to another. The idea of changing one quantity to investigate the resulting change in another quantity leads us naturally to the concept of a function. Functions are truly fundamental to mathematics, and the study of calculus begins with the study of functions.

This chapter will lay the foundation for studying calculus by surveying the behavior of linear functions, exponential functions, and power functions. We will explore ways of handling the graphs, tables, and formulas that represent these functions. We will consider the different ways in which these functions change, and we will investigate some applications and how to use these functions in modeling data.

1.1   HOW DO WE MEASURE CHANGE?

Change is all around us. The temperature outside, the population of your town, the price of a stock, the size of a cancerous tumor, or the velocity of a baseball are all examples of quantities that are changing. Calculus is the study of change.

The Height of a Child

Kari was born in May of 1982, and her height (in inches) each year on her birthday is given in Table 1.1.

TABLE 1.1   *The growth of a child*

Age (yrs.)	Birth	1	2	3	4	5	6	7	8	9	10	11	12	13	14
Height (in.)	19	28	33	36	39	42	44	47	49	51	54	56	59	61	63

Kari is growing, and so her height is changing. What is the change in her height during the first four years of her life? We see that

$$\text{Total change in height between birth and age 4} = 39 \text{ inches} - 19 \text{ inches} = 20 \text{ inches}.$$

What is the change in her height between age 4 and age 14?

$$\text{Total change in height between age 4 and age 14} = 63 \text{ inches} - 39 \text{ inches} = 24 \text{ inches}.$$

Total Change and Rate of Change

Was Kari growing faster during the first four years of her life or the following ten years? We saw above that she grew 20 inches during the first 4 year period and she grew 24 inches during the following 10 years. However, the numbers 20 and 24 are not very helpful in answering the question of when she was growing fastest. To answer this question, we need a *rate of change*. While the change in height is measured in inches, the rate of change is measured in inches *per year*.

Since Kari grew 20 inches during her first 4 years, the average rate of change during this period is 20 inches divided by 4 years, or 5 inches per year. We see that

$$\text{Average rate of change between age 0 and age 4} = \frac{\text{change in height}}{\text{change in age}} = \frac{39 - 19}{4 - 0} = \frac{20}{4} = 5 \text{ in/yr.}$$

$$\text{Average rate of change between age 4 and age 14} = \frac{\text{change in height}}{\text{change in age}} = \frac{63 - 39}{14 - 4} = \frac{24}{10} = 2.4 \text{ in/yr.}$$

Kari grew at an average rate of 5 inches per year between birth and age 4, and at an average rate of 2.4 inches per year between age 4 and age 14. Kari was growing faster during the first 4 years of her life.

The **change** or **total change** in a quantity between time a and time b = the value of the quantity at time b − the value of the quantity at time a

The **average rate of change** of a quantity between time a and time b = $\dfrac{\text{the change in quantity}}{\text{the change in time}}$

Public Debt of the United States

Table 1.2 gives the public debt, D, of the United States for the years 1980 to 1993.[1] The total change in the public debt during this 13-year period was $4351.2 - 907.7 = 3443.5$ billion dollars. If we want to know the rate at which the public debt has been increasing, in billions of dollars per year, we use the average rate of change.

$$\begin{aligned}
\text{Average rate of change} \atop \substack{\text{of the public debt} \\ \text{between 1980 and 1993}} &= \frac{\text{the change in the public debt}}{\text{the change in time}} \\
&= \frac{4351.2 - 907.7}{1993 - 1980} \\
&= \frac{3443.5 \text{ billion dollars}}{13 \text{ years}} \\
&= 264.88 \text{ billion dollars per year.}
\end{aligned}$$

Notice that the units for the average rate of change are units of public debt over units of time, or billions of dollars per year. Between 1980 and 1993, the public debt of the United States increased at an average rate of 264.88 billion dollars per year. (This represents an increase of $725,700,000, over 700 million dollars, every day!)

The Δ Notation

To find the change in the public debt, D, we subtracted one value of D from another. The notation ΔD stands for the change in D and is a difference of two values of D. Likewise, Δt stands for the change in t and is a difference of two values of t.

ΔD = Change in D. The units of ΔD are the same as the units of D.

$$\text{Average rate of} \atop \text{change of } D = \frac{\text{change in } D}{\text{change in } t} = \frac{\Delta D}{\Delta t}. \text{ The units of } \frac{\Delta D}{\Delta t} \text{ are the } D \text{ units over the } t \text{ units.}$$

TABLE 1.2   *Public Debt of the United States*

Year	Debt (billions of dollars)	Year	Debt (billions of dollars)
1980	907.7	1987	2350.3
1981	997.9	1988	2602.3
1982	1142.0	1989	2857.4
1983	1377.2	1990	3233.3
1984	1572.3	1991	3665.3
1985	1823.1	1992	4064.6
1986	2125.3	1993	4351.2

---

[1]*The World Almanac 1995.*

Example 1   Find the average rate of change of the US public debt between 1980 and 1985, and between 1985 and 1993.

Solution   Between 1980 and 1985

$$\text{Average rate of change } = \frac{\Delta D}{\Delta t} = \frac{1823.1 - 907.7}{1985 - 1980} = \frac{915.4}{5} = 183.1 \text{ billion dollars per year.}$$

Between 1985 and 1993

$$\text{Average rate of change } = \frac{\Delta D}{\Delta t} = \frac{4351.2 - 1823.1}{1993 - 1985} = \frac{2528.1}{8} = 316.0 \text{ billion dollars per year.}$$

Farmland in the United States

Example 2   Figure 1.1 shows a graph of the number of farms (in millions) in the United States between 1940 and 1993.[2] Estimate the average rate at which the number of farms is changing between 1950 and 1970. Interpret your answer.

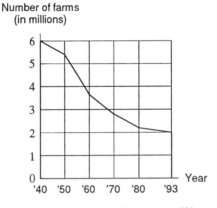

Figure 1.1: Number of farms in the US

Solution   We see from Figure 1.1 that the number of farms in the US is approximately 5.4 million in 1950 and approximately 2.8 million in 1970. We have

$$\text{Average rate of change } = \frac{2.8 - 5.4}{1970 - 1950} = -0.13 \text{ million farms per year.}$$

The average rate of change is negative since the number of farms is decreasing. We see that during this 20 year period, the number of farms in the US went down at an average rate of 0.13 million farms per year, or an average decrease of 130,000 farms per year.

PCB's and Pelicans

We have looked at the change in a child's height, the change in the public debt, and the change in the number of farms. All of these quantities are changing over time. In this example, we look at a quantity that is changing for a reason other than the passage of time.

---

[2]*The World Almanac 1995.*

High levels of PCB (polychlorinated biphenyl, an industrial pollutant) in the environment affect many animal populations. In Table 1.3 we look at the effect of PCB on the thickness of pelican eggs. The concentration of PCB in the eggshell is given in parts per million, and the thickness of the shell is given in millimeters. We see in Table 1.3 that as the concentration of PCB goes up, the thickness of the eggshell goes down, which is bad for pelicans.[3]

TABLE 1.3

Concentration (in ppm)	87	147	204	289	356	452
Thickness (in mm)	0.44	0.39	0.28	0.23	0.22	0.14

Example 3   Find the average rate of change in the thickness of the shell as the PCB concentration changes from 87 ppm to 452 ppm. Give units with your answer. What does the fact that your answer is negative tell you about PCB and pelican eggs?

Solution   Since we are looking for the average rate of change of thickness, with respect to change in PCB concentration rather than change in time, we have

$$\text{Average rate of change of thickness} = \frac{\text{change in the thickness}}{\text{change in the PCB level}}$$
$$= \frac{0.14 - 0.44}{452 - 87}$$
$$= \frac{-0.30}{365}$$
$$= -0.00082 \frac{\text{mm}}{\text{ppm}}$$

The units are thickness units (mm) over PCB concentration units (ppm), or millimeters over parts per million. The average rate of change is negative because the thickness of the eggshell goes *down* as the PCB concentration goes up. We see that the thickness of pelican eggs goes down by an average of 0.00082 mm for every additional part per million of PCB in the eggshell.

Distance and Velocity

Consider the motion of a grapefruit thrown up in the air. The height of the grapefruit above ground is changing: the grapefruit will go up, turn around, fall down, and then "splat". (See Figure 1.2.) The height of the grapefruit is increasing and then decreasing. Table 1.4 gives the height, y, of the grapefruit above ground t seconds after it is thrown.

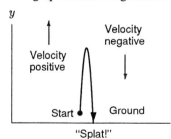

TABLE 1.4   *Height of the grapefruit above the ground*

t (sec)	0	1	2	3	4	5	6
y (feet)	6	90	142	162	150	106	30

Figure 1.2: The grapefruit's path

[3]Risebrough, R.W., "Effects of environmental pollutants upon animals other than man." *Proceedings of the 6th Berkeley Symposium on Mathematics and Statistics, VI,* (Berkeley: University of California Press, 1972) 443-463.

Example 4   Find the change and average rate of change of the height of the grapefruit during the first 3 seconds. Give units with your answers and interpret your answers.

Solution   The change in height during the first 3 seconds $= 162 - 6 = 156$ ft. The grapefruit goes up a total of 156 feet during the first 3 seconds. The average rate of change during this 3 second interval is $156/3 = 52$ ft/sec. During the first 3 seconds, the grapefruit is rising at an average rate of 52 ft/sec.

Notice that the units for the average rate of change of the height of the grapefruit are feet per second. You may recognize these as units of *velocity*. When the quantity we are measuring is a distance against time, then the units of the average rate of change are distance units divided by time units, which are the units of velocity. The average rate of change of distance with respect to time is velocity. (This is why we use the word "rate" to mean "velocity".)

$$\text{Average velocity} = \frac{\text{change in distance}}{\text{change in time}}$$
$$= \text{the average rate of change of distance with respect to time}$$

Velocity versus Speed

From now on, we will make a distinction between velocity and speed. Suppose an object moves along a line. If we pick one direction to be positive, the *velocity* is positive if it is in the positive direction and negative if it is in the opposite direction. For the grapefruit, upward is positive and downward is negative. (See Figure 1.2.) *Speed* is the absolute value of the velocity and so is always positive or zero.

Example 5   Compute the average velocity of the grapefruit over the interval $t = 4$ to $t = 6$. What is the significance of the sign of your answer?

Solution

$$\text{Average velocity} = \frac{\text{change in distance}}{\text{change in time}}$$
$$= \frac{30 - 150}{6 - 4}$$
$$= -60 \text{ ft/sec}$$

The negative sign means the height is decreasing and the grapefruit is moving downward.

Problems for Section 1.1 ━━━━━━━━━━━━━━━━━━━━━━━━━━━━━━━

1. In 1993, there were 108 million bicycles produced worldwide, which is almost three times more than the number of automobiles produced. Table 1.5 shows world bicycle production (in millions of bicycles) for selected years between 1950 and 1993.[4]

   (a) Find the change in bicycle production between 1950 and 1990. Give units with your answer.

   (b) Find the average rate of change in bicycle production between 1950 and 1990. Give units with your answer and interpret it in terms of bicycle production.

   TABLE 1.5   *World bicycle production*

Year	1950	1960	1970	1980	1990	1993
Production (millions)	11	20	36	62	90	108

2. The graph of the total value of world exports (internationally traded goods), in billions of dollars, is shown in Figure 1.3 for the years 1950 to 1993.[5]

   (a) Is the value of the exports higher in 1990 or in 1960? Approximately how much higher?

   (b) Estimate the average rate of change between 1960 and 1990. Give units with your answer and interpret it in terms of the value of world exports.

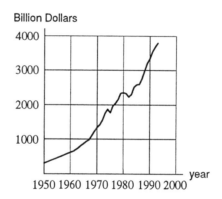

Figure 1.3: World Exports

3. Table 1.6 shows the number of NCAA Division I men's basketball games for selected years between 1948 and 1990.[6]

   (a) Find the average rate of change between 1948 and 1990 and give units with your answer.

   (b) Find the annual increase in the number of games for each year from 1983 to 1989. (Your answer should be six numbers.)

   (c) Find the average rate of change from 1983 to 1989.

   (d) Show that the average rate of change found in part (c) is the average of the six yearly changes found in part (b).

---

[4]Lester R. Brown, et. al., *Vital Signs 1994*, (New York: W.W. Norton & Company, 1994) 87.
[5]Lester R. Brown, et. al., *Vital Signs 1994*, (New York: W.W. Norton & Company, 1994) 77.
[6]*The World Almanac and Book of Facts 1992*, (New York: Pharos Books, 1991) 863.

TABLE 1.6 *NCAA Division I Men's Basketball Games*

Year	Games	Year	Games	Year	Games
1948	3945	1965	4520	1983	7957
1950	3659	1967	4602	1984	8029
1952	4009	1969	4883	1985	8269
1953	3754	1971	5232	1986	8360
1955	3829	1973	5582	1987	8580
1958	4153	1975	6147	1988	8587
1960	4295	1979	7131	1989	8677
1963	4180	1981	7407	1990	8646

4. Would you expect the average rate of change (in units per year) of each of the following to be positive or negative? Explain your reasoning in each case.

   (a) Number of acres of rain forest in the world.
   (b) Population of the world.
   (c) Number of polio cases each year in the United States, since 1950.
   (d) Height of a sand dune that is being eroded.
   (e) Cost of living in the United States.

5. Table 1.7 shows the total amount (in billions of dollars) spent by consumers on tobacco products in the US between 1987 and 1993.

   (a) What is the average rate of change in the amount spent on tobacco between 1987 and 1993? Give units with your answer and interpret it in terms of money spent on tobacco.
   (b) During this six-year period, is there any interval during which the average rate of change is negative? If so, when?

TABLE 1.7

Year	1987	1988	1989	1990	1991	1992	1993
Billions of dollars	35.6	36.2	40.5	43.4	45.4	50.9	50.5

6. Table 1.8 shows the total labor force (in thousands of workers) in the US between 1930 and 1990.[7] Find the average rate of change between 1930 and 1990. Between 1930 and 1950. Between 1950 and 1970. Give units with your answers and interpret your answers in terms of the labor force. Is the average rate of change increasing or decreasing over time?

TABLE 1.8

Year	1930	1940	1950	1960	1970	1980	1990
Labor force (in thousands)	29,424	32,376	45,222	54,234	70,920	90,564	103,905

[7]*The World Almanac and Book of Facts 1995*, (New Jersey: Funk & Wagnalls Corporation, 1994) 154.

7. The position, s, of a car is given in Table 1.9.
   (a) Find the average velocity of the car between $t = 0$ and $t = 15$. Give units with your answer.
   (b) Find the average velocity of the car between $t = 10$ and $t = 30$. Give units with your answer.
   (c) Find the distance traveled by the car between $t = 10$ and $t = 30$. Give units with your answer.

TABLE 1.9

t (sec)	0	5	10	15	20	25	30
s (ft)	0	30	55	105	180	260	410

8. Some scientists suspect that synthetic chemicals common in such things as pesticides, plastics, detergents, and toiletries are interfering with the human hormone system.[8] A controversial 1992 Danish study in the *British Medical Journal* reported that the average male sperm count had decreased from 113 million per milliliter in 1940 to 66 million per milliliter in 1990.
   (a) Find the average rate of change of the sperm count.
   (b) A man's fertility is affected if his sperm count drops below about 20 million per milliliter. If the average rate of change stays the same as that found in the Danish study, in what year will the average sperm count go below 20 million per milliliter?

1.2 WHAT'S A FUNCTION?

Let's look at an example. In the summer of 1990, the temperatures in Arizona reached an all-time high (so high, in fact, that some airlines decided it might be unsafe to land their planes there). The daily high temperatures in Phoenix for June 19–29 are given in Table 1.10.

TABLE 1.10   *Temperature in Phoenix, Arizona, June 1990*

Date: June (1990)	19	20	21	22	23	24	25	26	27	28	29
Temperature (°F)	109	113	114	113	113	113	120	122	118	118	108

Although you may not have thought of something so unpredictable as temperature as being a function, the temperature *is* a function of date, because each day gives rise to one and only one high temperature. There is no formula for temperature (otherwise we would not need the weather bureau), but nevertheless the temperature does satisfy the definition of a function: Each date, t, has a unique high temperature, H, associated with it.

We define a function as follows:

> One quantity, H, is a **function** of another, t, if each value of t has a unique value of H associated with it. We say H is the *value* of the function or the *dependent variable*, and t is the *argument* or *independent variable*. Alternatively, think of t as the *input* and H as the *output*. We write $H = f(t)$, where f is the name of the function.

---
[8] Adapted from "Investigating the next 'Silent Spring'", *U.S. News & World Report* (March 11, 1996) 50-52.

In the temperature example above, the independent variable is the date, and the dependent variable is the temperature. The function assigns temperatures to dates. For example, we see in Table 1.10 that $f(25) = 120$.

Functions play an important role in many fields. Frequently, one observes that one quantity is a function of another and then tries to find a reasonable formula to express this function. For example, before about 1590 there was no quantitative idea of temperature. Of course, people understood relative notions like warmer and cooler, and some absolute notions like boiling hot, freezing cold, or body temperature, but there was no numerical measure of temperature. It took the genius of Galileo to realize that the expansion of fluids as they warmed was the key to the measurement of temperature. He was the first to think of temperature as a function of fluid volume.

Finding a function which represents a given situation is called making a *mathematical model*. Such a model can throw light on the relationship between the variables and can thereby help us make predictions.

Representation of Functions: Tables, Graphs, and Formulas

Functions can be represented in at least four different ways: by tables of values, by graphs, by formulas, and in words. For example, the function giving the temperatures in Phoenix, Arizona, as a function of time can be represented by the graphs in Figure 1.4 as well as by Table 1.10.

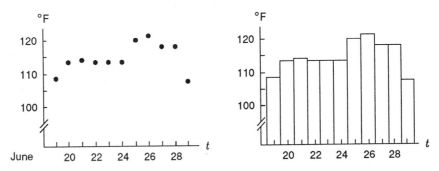

Figure 1.4: Phoenix temperatures, June 1990

Other functions arise naturally as graphs. Figure 1.5 contains electrocardiogram (EKG) pictures showing the heartbeat patterns of two patients, one normal and one not. Although it is possible to construct a formula to approximate an EKG function, this is seldom done. The pattern of repetitions is what a doctor needs to know, and these are much more easily seen from a graph than from a formula. However, each EKG represents a function showing electrical activity as a function of time.

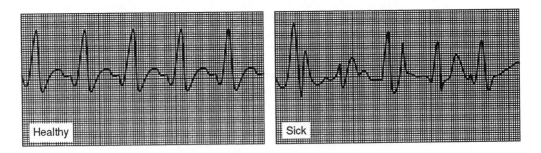

Figure 1.5: EKG readings on two patients

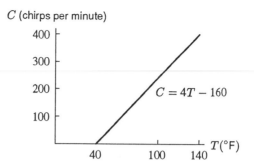

Figure 1.6: Cricket chirp rate versus temperature

As another example of a function, consider the snow tree cricket. Surprisingly enough, all such crickets chirp at essentially the same rate if they are at the same temperature. That means that the chirp rate is a function of temperature. In other words, if we know the temperature, we can determine the chirp rate. Even more surprisingly, the chirp rate, C, in chirps per minute, increases steadily with the temperature, T, in degrees Fahrenheit, and to a high degree of accuracy can be computed by the formula

$$C = 4T - 160.$$

The formula for C is written $C = f(T)$ to express the fact that we are thinking of C as a function of T. The graph of this function is in Figure 1.6.

Change and Rate of Change of Functions

The notation $y = f(t)$ tells us that y is a function of t. The independent variable is t and the dependent variable is y. We learned how to find the change and average rate of change in Section 1.1, and we can now extend this idea to functions.

If $y = f(t)$ is a function, then

$$\text{change in } y \text{ between } t = a \text{ and } t = b \quad = \quad \Delta y \quad = f(b) - f(a).$$

The units of change in a function are the units of the dependent variable.

If $y = f(t)$ is a function, then

$$\text{average rate of change of } y \text{ between } t = a \text{ and } t = b \quad = \quad \frac{\Delta y}{\Delta t} \quad = \frac{f(b) - f(a)}{b - a}.$$

The units of average rate of change of a function are $\dfrac{\text{units of the dependent variable}}{\text{units of the independent variable}}$.

The average rate of change tells the average change in the dependent variable given a unit increase in the independent variable.

Example 1   Table 1.10 gives the daily high temperatures in Phoenix, Arizona during a heat wave. Find the change and the average rate of change of the high temperature between June 20 and June 26, 1990. Give units with your answers.

Solution   High temperature is a function of date, and so temperature H is the dependent variable and date t is the independent variable. The change in the high temperature between June 20 and June 26 is $122 - 113 = 9°$. The temperature rose $9°$ during this period.

$$\text{Average rate of change} = \frac{\Delta H}{\Delta t} = \frac{f(26) - f(20)}{26 - 20} = \frac{122 - 113}{26 - 20} = \frac{9}{6} = 1.5 \text{ degrees per day.}$$

The high temperature was going up, on the average, at the rate of $1.5°$ per day.

Example 2   We saw above that, for the snow tree cricket, chirps per minute, C, as a function of temperature, T, is given by
$$C = f(T) = 4T - 160.$$
Find the average rate of change of chirps per minute between temperatures of $60°$ F and $70°$ F.

Solution   Since $f(60) = 4(60) - 160 = 80$, and $f(70) = 4(70) - 160 = 120$, we have

$$\text{average rate of change} = \frac{\Delta C}{\Delta T} = \frac{f(70) - f(60)}{70 - 60} = \frac{120 - 80}{70 - 60} = \frac{40}{10} = 4\frac{\text{chirps/minute}}{\text{degree}}.$$

The chirping rate goes up by 4 chirps per minute for every $1°$ increase in the temperature.

So far we have used the temperature to predict the chirp rate and thought of the temperature as the *independent variable* and the chirp rate as the *dependent variable*. However, we could do this backward, calculating the temperature from the chirp rate. From this point of view, the temperature would be dependent on the chirp rate; thus, which variable is dependent and which is independent may depend on your viewpoint.

Thinking of temperature as a function of chirp rate would enable us (in theory, at least) to use the chirp rate instead of a thermometer to measure temperature. The way we actually do measure temperature is based on another function: the relation between the height of the liquid in a thermometer and temperature. The height of the mercury is certainly a function of temperature; however, we always use this the other way around, and determine the temperature from the height of the mercury, as suggested by Galileo.

Example 3   Figure 1.7[9] shows the birth rate (number of births per year per 1000 of the population) in developed countries as a function of year, between 1775 and 1977. If B is the birth rate and y is the year, we can write $B = f(y)$.
(a)   Describe what this graph tells you about how the birth rate has changed since 1775.
(b)   Approximately what is $f(1900)$? What information does this give you?
(c)   Estimate the average rate of change of the birth rate between 1875 and 1975. Interpret your answer.

[9]Elaine Murphy, "Food and Population: A Global Concern", (Washington DC: Population Reference Bureau Inc., 1984)

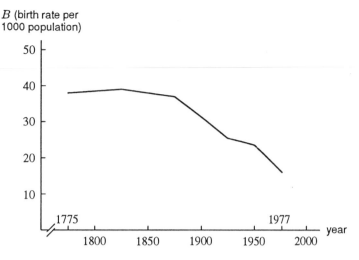

Figure 1.7: Birth rate in developed countries

Solution (a) We can see from the graph in Figure 1.7 that the value of the function stays about the same until about 1875, and then it begins to decrease. Since the value of the function corresponds to the birth rate, we see that the birth rate stayed about the same from 1775 until 1875, and that it has been decreasing since 1875.

(b) $f(1900)$ is approximately 32. This means that in 1900, there were approximately 32 births for every 1000 people.

(c) Using Figure 1.7, we estimate the birth rate in 1875 to be about 38 and in 1975 to be about 16. The average rate of change of the birth rate over this time interval is

$$\text{average rate of change} = \frac{f(1975) - f(1875)}{1975 - 1875} = \frac{16 - 38}{100} = -0.22.$$

The average rate of change is negative because the birth rate is going down, as we see in Figure 1.7. The average rate of change tells us that, between 1875 and 1975, the birth rate decreased at an average rate of 0.22 births per thousand every year.

Supply and Demand Curves

Economists are interested in how the quantity, q, of an item which is manufactured and sold, depends on its price, p. Since manufacturers and consumers react differently to changes in price, there are two functions relating p and q. For a given item, the *supply curve* represents how the quantity of an item that manufacturers are willing to make depends on the price for which the item can be sold. The *demand curve* represents how the quantity of an item demanded by consumers depends on the price of the item. It is usually assumed that as the price increases, the manufacturers will be willing to supply more of the product, so the quantity supplied will go up as price goes up. On the other hand, as price increases, it is assumed that consumer demand will fall, and so we expect the quantity demanded to go down as prices go up.

Economists think of the quantities supplied and demanded as functions of price. However, for historical reasons, the economists put price (the independent variable) on the vertical axis and quantity (the dependent variable) on the horizontal axis. (The reason for this state of affairs is that economists originally took price to be the dependent variable and put it on the vertical axis. Unfortunately when the point of view changed, the axes did not.) Thus, typical supply and demand curves look like those shown in Figure 1.8.

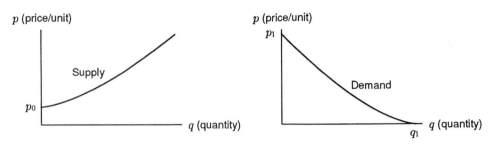

Figure 1.8: Supply and demand curves

> The **supply function**, $S(p)$, gives the quantity of some good that producers will supply to the market if its unit price is p.
>
> The **demand function**, $D(p)$, gives the quantity of some good that consumers will demand if its unit price is p.

Example 4   What is the economic meaning of the prices p_0 and p_1 and the quantity q_1 in Figure 1.8?

Solution   The vertical axis corresponds to a quantity of zero; since the price p_0 is the vertical intercept on the supply curve, p_0 is the price at which the quantity supplied is zero. In other words, unless the price is above p_0, the suppliers will not produce anything. The price p_1 is also on the vertical axis, so it corresponds to the price on the demand curve where the quantity demanded is zero. In other words, unless the price is below p_1, consumers won't buy any of the product.

The horizontal axis corresponds to a price of zero, so the quantity q_1 on the demand curve is the quantity that would be demanded if the price were zero—or the quantity which could be given away if the item were free.

If we plot the supply and demand curves on the same axes, as shown in Figure 1.9, there will be exactly one point where the graphs cross. This point (q^*, p^*) is called the *equilibrium point*.

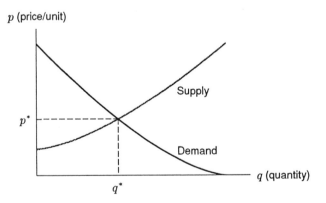

Figure 1.9: The equilibrium point

The values p^* and q^* at this point are called the *equilibrium price* and *equilibrium quantity*, respectively. At this equilibrium point, a quantity q^* of an item is produced and sold for a price of p^*

per unit. It is assumed that the market will naturally settle to this equilibrium point. (See Problems 14 and 15.)

The **equilibrium price** and **equilibrium quantity** are at the point where the supply and demand curves cross. At the equilibrium price

$$\text{Demand} = \text{Supply}.$$

Proportionality

A common functional relationship occurs when one quantity is *proportional* to another. For example, if apples are 60 cents a pound, we say the price you pay, p cents, is proportional to the weight you buy, w pounds, because

$$p = f(w) = 60w.$$

As another example, the area, A, of a circle is proportional to the square of the radius, r:

$$A = f(r) = \pi r^2.$$

In general, y is (directly) **proportional** to x if there is a constant k, called the *constant of proportionality*, such that

$$y = kx.$$

We also say that one quantity is *inversely proportional* to another if one is proportional to the reciprocal of the other. For example, the speed, v, at which you make a 50-mile trip is inversely proportional to the time, t, taken, because v is proportional to $1/t$:

$$v = \frac{50}{t} = 50\left(\frac{1}{t}\right).$$

Notice that if y is directly proportional to x, then the magnitude of one variable increases when the magnitude of the other variable increases, while the magnitude of one variable decreases if the magnitude of the other variable decreases. If, however, y is inversely proportional to x, then the magnitude of one variable increases when the value of the other decreases, and vice-versa.

Example 5   Table 1.11 gives values for a function $p = f(t)$. Is p proportional to t?

TABLE 1.11

t	0	10	20	30	40	50
p	0	25	60	100	140	200

Solution   If p is proportional to t, then $p = kt$ for some fixed constant k. From the first point $(t = 0, p = 0)$, we have $0 = k(0)$, so k can be anything. From the second point $(t = 10, p = 25)$, we have $25 = k(10)$, so we see that $k = 2.5$. To see if p is proportional to t, we must see if $p = 2.5t$ gives all the values in Table 1.11. However, when we check the point $t = 20, p = 60$, we see that $60 \neq 2.5(20)$, and thus p is not proportional to t.

Problems for Section 1.2

1. Which of the graphs in Figure 1.10 best matches the following three stories?[10] Write a story for the remaining graph.

    (a) I had just left home when I realized I had forgotten my books, and so I went back to pick them up.

    (b) Things went fine until I had a flat tire.

    (c) I started out calmly but sped up when I realized I was going to be late.

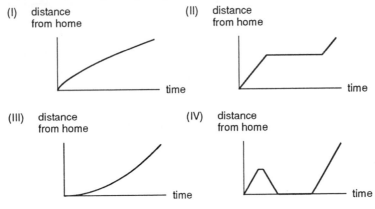

Figure 1.10

2. It warmed up throughout the morning, and then suddenly got much cooler around noon, when a storm came through. After the storm, it warmed up before cooling off at sunset. Sketch a possible graph of this day's temperature as a function of time.

3. Right after a certain drug is administered to a patient with a rapid heart rate, the heart rate plunges dramatically and then slowly rises again as the drug wears off. Sketch a possible graph of the heart rate against time from the moment the drug is administered.

4. Generally, the more fertilizer that is used, the better the yield of the crop. However, if too much fertilizer is applied, the crops become poisoned, and the yield goes down rapidly. Sketch a possible graph showing the yield of the crop as a function of the amount of fertilizer applied.

5. Describe what Figure 1.11 tells you about an assembly line whose productivity is represented as a function of the number of workers on the line.

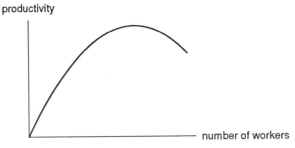

Figure 1.11

---

[10] Adapted from Jan Terwel, "Real Maths in Cooperative Groups in Secondary Education." *Cooperative Learning in Mathematics*, ed. Neal Davidson (Reading: Addison Wesley, 1990) 234.

6. A flight from Dulles Airport in Washington, D.C. to LaGuardia Airport in New York City has to circle LaGuardia several times before being allowed to land. Plot a graph of the distance of the plane from Washington against time, from the moment of takeoff until landing.

7. In her *Guide to Excruciatingly Correct Behavior*, Miss Manners states that

> There are three possible parts to a date of which at least two must be offered: entertainment, food and affection. It is customary to begin a series of dates with a great deal of entertainment, a moderate amount of food and the merest suggestion of affection. As the amount of affection increases, the entertainment can be reduced proportionately. When the affection has replaced the entertainment, we no longer call it dating. Under no circumstances can the food be omitted.

Based on this statement, sketch a graph showing entertainment as a function of affection, assuming the amount of food to be constant. Mark the point on the graph at which the relationship starts, as well as the point at which the relationship ceases to be called dating.

8. Recall that functions can be given by a formula, a table of values, or a graph. Five different functions are given below. In each case, find $f(5)$.

(a) $f(x) = 2x + 3$

(b)

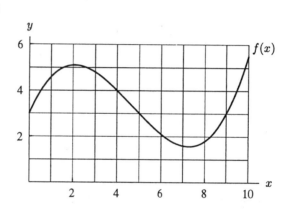

Figure 1.12

(c)

x	1	2	3	4	5	6	7	8
$f(x)$	2.3	2.8	3.2	3.7	4.1	4.9	5.6	6.2

(d)

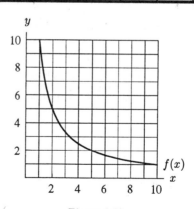

Figure 1.13

(e) $f(x) = 10x - x^2$

9. Find the average rate of change, between $x = 2$ and $x = 10$, of the function $y = f(x)$ whose graph is given in Figure 1.13. Is your answer positive or negative?

10. The graph of $r = f(p)$ is given in Figure 1.14.

    (a) What is the value of r when p is zero?
    (b) What is the value of r when p is three?
    (c) What is $f(2)$?

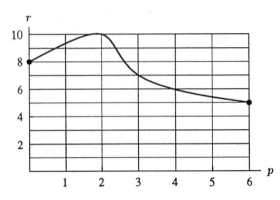

Figure 1.14

11. Let $y = f(x) = x^2 + 2$.
    (a) Find the value of y when x is zero.
    (b) What is $f(3)$?
    (c) What values of x give y a value of 11?
    (d) Are there any values of x that give y a value of 1?

12. Find the average rate of change of $f(x) = 2x^2$ between $x = 1$ and $x = 3$.

13. One of the graphs in Figure 1.15 is a supply curve, and the other is a demand curve. Which is which? Explain how you made your decision using what you know about the effect of price on supply and demand.

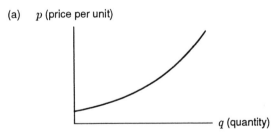 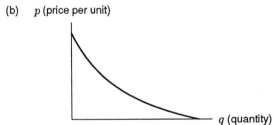

Figure 1.15

14. Figure 1.16 shows the supply and demand curves for a particular product.
    (a) What is the equilibrium price for this product? At this price, what quantity will be produced?

(b)   Choose a price above the equilibrium price—for example, $p = 12$. At this price, how many items will suppliers be willing to produce? How many items will consumers want to buy? Use your answers to these questions to explain why, if prices are above the equilibrium price, the market tends to push prices lower (towards the equilibrium).

(c)   Now choose a price below the equilibrium price—for example, $p = 8$. At this price, how many items will suppliers be willing to produce? How many items will consumers want to buy? Use your answers to these questions to explain why, if prices are below the equilibrium price, the market tends to push prices higher (towards the equilibrium).

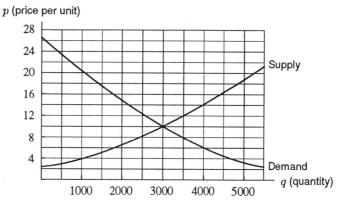

Figure 1.16

15.   Figure 1.17 shows the supply and demand curves for a particular product.

(a)   What is the equilibrium price for this product? At this price, what quantity will be produced?

(b)   Choose a price above the equilibrium price—for example, $p = 300$. At this price, how many items will suppliers be willing to produce? How many items will consumers want to buy? Use your answers to these questions to explain why, if prices are above the equilibrium price, the market tends to push prices lower (towards the equilibrium).

(c)   Now choose a price below the equilibrium price—for example, $p = 200$. At this price, how many items will suppliers be willing to produce? How many items will consumers want to buy? Use your answers to these questions to explain why, if prices are below the equilibrium price, the market tends to push prices higher (towards the equilibrium).

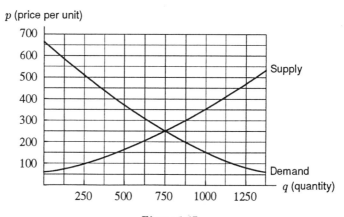

Figure 1.17

16. Based upon its market research, a corporate office provides the demand curve in Figure 1.18 to its ice cream shop franchises. A franchise owner has determined that 240 scoops per day can be sold at a price of $1.00 per scoop.

    (a) According to the corporate demand curve, estimate how many scoops could be sold per day during a half-price sale. (i.e., when the price per scoop is 50¢).
    (b) The owner is considering an increase in the price per scoop after the half-price sale. Using the corporate demand curve, estimate how many scoops per day could be sold at a price of $1.50 per scoop.

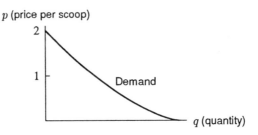

Figure 1.18: Demand curve for ice cream

17. Tables 1.12 and 1.13 are given below. One represents a supply curve and the other represents a demand curve.

    TABLE 1.12

p (price/unit)	182	167	153	143	133	125	118
q (quantity)	5	10	15	20	25	30	35

    TABLE 1.13

p (price/unit)	6	35	66	110	166	235	316
q (quantity)	5	10	15	20	25	30	35

    (a) Which table represents which curve? Why?
    (b) At a price of $155, approximately how many items would consumers purchase?
    (c) At a price of $155, approximately how many items would manufacturers supply?
    (d) Will the market push prices higher or lower than $155?
    (e) What would the price have to be if you wanted consumers to buy at least 20 items?
    (f) What would the price have to be if you wanted manufacturers to supply at least 20 items?

18. The US production, Q, of copper in metric tons and the value, P, in thousands of dollars per metric ton are given for the years 1984 through 1989 in Table 1.14[11]. Plot these data points with Q, the production, on the horizontal axis and P, the value, on the vertical axis. Sketch a possible supply curve which fits these data reasonably well.

    TABLE 1.14

Year	1984	1985	1986	1987	1988	1989
Q	1103	1105	1144	1244	1417	1497
P	1473	1476	1456	1818	2656	2888

    ---
    [11]"US Copper, Lead, and Zinc Production", *The World Almanac 1992*, 688.

19. Examine the demand curve shown in Figure 1.19.

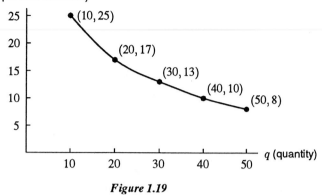

Figure 1.19

(a) If the price is \$17 per item, how many items do consumers purchase?
(b) If the price is \$8 per item, how many items do consumers purchase?
(c) At what price do consumers purchase 30 items?
(d) At what price do consumers purchase 10 items?

20. Examine the supply curve shown in Figure 1.20.

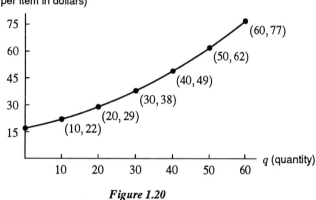

Figure 1.20

(a) If the price is \$49 per item, how many items do manufacturers supply?
(b) If the price is \$77 per item, how many items do manufacturers supply?
(c) At what price are manufacturers willing to supply 20 items?
(d) At what price are manufacturers willing to supply 50 items?

21. When Galileo was formulating the laws of motion, he considered the motion of a body starting from rest and falling under gravity. He originally thought that the velocity of such a falling body was proportional to the distance it had fallen. What light do the data in Table 1.15 shed on Galileo's hypothesis? What alternative hypothesis is suggested by the two sets of data in Table 1.15 and Table 1.16?

TABLE 1.15

Distance (ft)	0	1	2	3	4
Velocity (ft/sec)	0	8	11.3	13.9	16

TABLE 1.16

Time (sec)	0	1	2	3	4
Velocity (ft/sec)	0	32	64	96	128

22. According to the National Association of Realtors[12], the minimum annual gross income, m, in thousands of dollars needed to obtain a 30-year home loan of A-thousand dollars at 9% is given in Table 1.17.

TABLE 1.17

A	50	75	100	150	200
m	17.242	25.863	34.484	51.726	68.968

The minimum annual gross income, m, in thousands of dollars needed for a home loan of $100,000 at various interest rates, r, is given in Table 1.18.

TABLE 1.18

r	8	9	10	11	12
m	31.447	34.484	37.611	40.814	44.084

(a) Is the size of the loan, A, proportional to the minimum annual gross income, m?
(b) Is the percentage rate, r, proportional to the minimum annual gross income, m?

1.3 LINEAR FUNCTIONS

Probably the most commonly used functions are the *linear functions*. These are functions that represent a steady increase or a steady decrease. A function is linear if any change, or increment, in the independent variable causes a proportional change, or increment, in the dependent variable. The graph of a linear function is a line.

The Olympic Pole Vault

During the early years of the Olympics, the height of the winning pole vault increased approximately as shown in Table 1.19. Since the winning height increased regularly by 8 inches every four years, the height is a linear function of time over the period from 1900 to 1912. The height starts at 130 inches and increases by the equivalent of 2 inches every year, so if y is the height in inches and t is the number of years since 1900, we can write

$$y = f(t) = 130 + 2t.$$

The coefficient 2 tells us the rate at which the height increases and is the *slope* of the line $f(t) = 130 + 2t$.

TABLE 1.19   *Olympic pole vault records (approximate)*

Year	1900	1904	1908	1912
Height (inches)	130	138	146	154

You can visualize the slope in Figure 1.21 as the ratio

$$\text{Slope} = \frac{\text{Rise}}{\text{Run}} = \frac{8}{4} = 2.$$

---

[12]"Income needed to get a Mortgage", *The World Almanac 1992*, 720

Calculating the slope (rise/run) using any other two points on the line gives the same value. It is this fact—that the slope, or rate of change, is the same everywhere—that makes a line straight. For a function that is not linear, the rate of change will vary from point to point. Since $y = f(t)$ increases with t, we say that f is *an increasing function*. What about the constant 130? This represents the initial height in 1900, when $t = 0$. Geometrically, the 130 is the *intercept* on the vertical axis. You may wonder whether the linear trend continues beyond 1912. Not surprisingly, it doesn't exactly. The formula $y = 130 + 2t$ predicts that the height in the 1996 Olympics would be 322 inches or 26 feet 10 inches, which is considerably higher than the actual value of 19 feet 5 inches. In fact, the height does increase at almost every session of the Olympics, but not at a constant rate. Thus, there is clearly a danger in *extrapolating* too far from the given data. You should also observe that the data in Table 1.19 is *discrete*, because it is given only at specific points (every four years). However, we have treated the variable t as though it were *continuous*, because the function $y = 130 + 2t$ makes sense for all values of t. The graph in Figure 1.21 is of the continuous function because it is a solid line, rather than four separate points representing the years in which the Olympics were held.

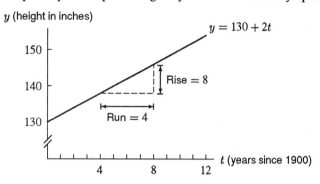

Figure 1.21: Olympic pole vault records

Linear Functions in General

A **linear function** has the form

$$y = f(x) = b + mx$$

Its graph is a line such that
- m is the slope, or rate of change of y with respect to x
- b is the vertical intercept, or value of y when x is zero.

Notice that if the slope is zero, $m = 0$, we have $y = b$, a horizontal line.

To recognize that a function $y = f(x)$ given by a table of values is linear, look for differences in y-values that are constant for equal differences in x.

The slope of a linear function can be calculated from values of the function at two points, given by $x = a$ and $x = c$, using the formula

$$m = \frac{\text{Rise}}{\text{Run}} = \frac{f(c) - f(a)}{c - a}.$$

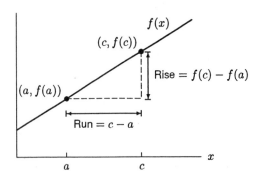

Figure 1.22: Difference quotient $= \dfrac{f(c) - f(a)}{c - a}$

The quantity $(f(c) - f(a))/(c - a)$ is called a *difference quotient* because it is the quotient of two differences. (See Figure 1.22). This difference quotient should look familiar; it is the same formula we saw earlier for the average rate of change.

If a function is linear, the average rate of change is the same on all intervals. Furthermore, the average rate of change on any interval equals the slope m.

The Success of Search and Rescue Teams

Consider the problem of the "search and rescue" teams working to find lost hikers in remote areas in the West. To search for an individual, members of the search team separate and walk parallel to one another through the area to be searched. Experience has shown that the team's chance of finding a lost individual is related to the distance, d, by which team members are separated. The percentages found[13] for various separations are recorded in Table 1.20.

From the data in the table, you can see that as the separation distance decreases, a larger percentage of the lost hikers is found, which makes sense. Since $P = f(d)$ decreases as d increases, we say that P is a *decreasing function* of d. You can also see that for the data given, each 20-foot increase in distance causes the percentage found to drop by 10. This constant decrease in P as d increases by a fixed amount is a clear indication that the graph of P against d is a line. (See Figure 1.23.) Notice that, since the function is linear, we can compute the slope using any interval.

TABLE 1.20 *Separation of searchers versus success rate*

Separation distance d (ft)	Percent found, P
20	90
40	80
60	70
80	60
100	50

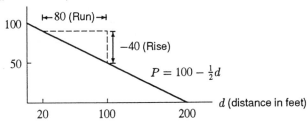

Figure 1.23: Separation of searchers versus success rate

[13] From *An Experimental Analysis of Grid Sweep Searching*, by J. Wartes (Explorer Search and Rescue, Western Region, 1974).

Using the interval $d = 20$ to $d = 100$, we have

$$\text{slope } m = \frac{\Delta P}{\Delta d} = \frac{50 - 90}{100 - 20} = \frac{-40}{80} = -\frac{1}{2}\frac{\text{percent}}{\text{foot}}.$$

The negative sign shows that P decreases as d increases. The slope of $-\frac{1}{2}$ tells that the percent found, P, decreases by one half a percent for each one foot increase in the separation distance.

What about the vertical intercept? If $d = 0$, the searchers are walking shoulder to shoulder and you'd expect everyone to be found, so $P = 100$. This is exactly what you get if the line is continued to the vertical axis (a decrease of 20 in d causes an increase of 10 in P). Therefore, the equation of the line is

$$P = f(d) = 100 - \frac{1}{2}d.$$

What about the horizontal intercept? When $P = 0$, or $0 = 100 - \frac{1}{2}d$, then $d = 200$. The value $d = 200$ represents the separation distance at which, according to the model, no one is found. This is unreasonable, because even when the searchers are far apart, the search will sometimes be successful. This suggests that somewhere outside the given data, the linear relationship ceases to hold. As in the pole vault example, extrapolating too far beyond the given data may not give accurate answers.

Increasing versus Decreasing Functions

Let's summarize what we know about increasing and decreasing functions:

A function f is **increasing** if the values of $y = f(x)$ increase as x increases.
A function f is **decreasing** if the values of $y = f(x)$ decrease as x increases.

The graph of an *increasing* function *climbs* as you move from left to right.
The graph of a *decreasing* function *descends* as you move from left to right.

Example 1   Which of the following tables of values could represent linear functions?

(a)

x	y
0	25
1	30
2	35
3	40
4	45

(b)

x	y
0	10
2	16
4	26
6	40
8	62

(c)

t	s
20	2.4
30	2.2
40	2.0
50	1.8
60	1.6

Solution   (a)   Since y goes up by 5 for every increase of 1 in x, this could be the table of values of a linear function (with a slope of $5/1 = 5$).
    (b)   Between $x = 0$ and $x = 2$, y goes up by 6 as x goes up by 2. Between $x = 2$ and $x = 4$, y goes up by 10 as x goes up by 2. Since the slope is not constant, this is not the table of values of a linear function.
    (c)   Since s goes down by 0.2 for every increase of 10 in t, this could be the table of values of a linear function (with a slope of $-0.2/10 = -0.02$).

Example 2   Table 1.21 gives data for the demand function for a certain product, where p is the price of the product in dollars and q is the quantity sold, in tons, at that price. Notice that the demand function is linear.

(a)   Write q as a linear function of p, and interpret the slope in terms of demand for this product.

(b)   Write p as a linear function of q, and interpret the slope in terms of demand for this product.

TABLE 1.21

p dollars	16	18	20	22	24
q tons	500	460	420	380	340

Solution   (a)   If we think of q as a linear function of p, then q is the dependent variable and p is the independent variable, and we have slope $m = \Delta q/\Delta p$. We can use any interval to find the slope. In the computations below, we are using the first two points.

$$\text{slope } m = \frac{\Delta q}{\Delta p} = \frac{460 - 500}{18 - 16} = \frac{-40}{2} = -20.$$

The units are the units of q over the units of p, or tons per dollar. The slope tells us that, for every dollar increase in price, the number of tons sold will go down by 20.

To finish writing q as a linear function of p, we need to find the vertical intercept, b. Since q is a linear function of p, we have $q = b + mp$. We know that $m = -20$ and we can use any of the points given in Table 1.21 to find b.

$$q = b + mp$$
$$500 = b + (-20)(16)$$
$$500 = b - 320$$
$$820 = b.$$

Therefore, the vertical intercept is 820 and the equation of the line is

$$q = 820 - 20p.$$

(b)   If we now consider p as a linear function of q, we have

$$\text{slope } m = \frac{\Delta p}{\Delta q} = \frac{18 - 16}{460 - 500} = \frac{2}{-40} = -\frac{1}{20} = -0.05.$$

The units of the slope are dollars per ton. The slope tells us that, if we want to sell one more ton of the product, we should reduce the price by $0.05.

Since p is a linear function of q, we have $p = b + mq$. Since $m = -0.05$, the equation becomes $p = b - 0.05q$. To find b, we substitute any point from Table 1.21 into this equation.

$$p = b + mq$$
$$16 = b + (-0.05)(500)$$
$$16 = b - 25$$
$$41 = b.$$

The equation of our line is $p = 41 - 0.05q$. Notice that we would have arrived at the same answer by taking our answer to part (a), $q = 820 - 20p$, and solving the equation for p.

Example 3   The solid waste generated each year in the cities of the United States is increasing. The solid waste generated, in millions of tons, was 82.3 in 1960 and 139.1 in 1980.[14]

(a)   Construct a mathematical model of the amount of municipal solid waste generated in the US by finding the equation of the line through these two points.

(b)   Use this model to predict the amount of municipal solid waste generated in the US, in millions of tons, in the year 2000.

Solution   (a)   We are looking at the amount of municipal solid waste, W, as a function of year, t, and the two points are $(1960, 82.3)$ and $(1980, 139.1)$. For the model, we assume that the quantity of solid waste is a linear function of year. The slope of the line is

$$m = \frac{139.1 - 82.3}{1980 - 1960} = \frac{56.8}{20} = 2.84 \frac{\text{millions of tons}}{\text{year}}.$$

This slope tells us that the amount of solid waste generated in the cities of the US has been going up at a rate of 2.84 million tons per year. To find the equation of the line, we must find the vertical intercept. We substitute the point $(1960, 82.3)$ and the slope $m = 2.84$ into the equation $W = b + mt$:

$$W = b + mt$$
$$82.3 = b + (2.84)(1960)$$
$$82.3 = b + 5566.4$$
$$-5484.1 = b.$$

The equation of the line is $W = -5484.1 + 2.84t$, where W is the amount of municipal solid waste in the US in millions of tons, and t is the year.

(b)   How much solid waste does this model predict in the year 2000? We can graph the line and find the vertical coordinate when $t = 2000$, or we can substitute $t = 2000$ into the equation of the line, and solve for W:

$$W = -5484.1 + 2.84t$$
$$W = -5484.1 + (2.84)(2000)$$
$$W = -5484.1 + 5680$$
$$W = 195.9.$$

The model predicts that in the year 2000, the solid waste generated by cities in the US will be 195.9 million tons.

Families of Linear Functions

Formulas such as $f(x) = mx$, and $f(x) = b + mx$, containing constants such as m and b which can take on various values, are said to define a *family of functions*. The constants m and b are called *parameters*. Each of the functions in this section belongs to the family $f(x) = b + mx$.

---

[14]*Statistical Abstracts of the US*, 1988, p. 193, Table 333.

Grouping functions into families which share important features is particularly useful for mathematical modeling. We often choose a family to represent a given situation on theoretical grounds and then use data to determine the particular values of the parameters. The meaning of the parameters m and b in the family $f(x) = b + mx$ is shown in Figures 1.24 and 1.25.

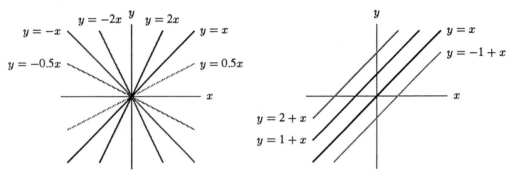

Figure 1.24: The family $y = mx$
(with $b = 0$)

Figure 1.25: The family $y = b + x$
(with $m = 1$)

Visualizing Change and Rate of Change

The change in a function $y = f(t)$ between $t = a$ and $t = b$ is $\Delta y = f(b) - f(a)$. How do we visualize this on a graph? Since it is a difference of two y-values, it represents a vertical distance. See Figure 1.26.

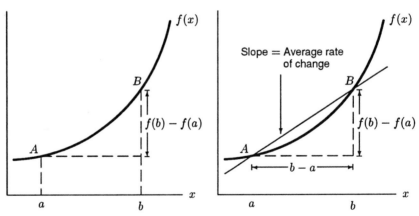

Figure 1.26: Visualizing the change in a
function

Figure 1.27: Visualizing average rate of
change

The average rate of change between $t = a$ and $t = b$ is the same as the slope of the line between the points $(a, f(a))$ and $(b, f(b))$. The average rate of change is represented graphically by the slope of this line. See Figure 1.27.

> The average rate of change of a function $f(t)$ over any interval is the slope of the line joining the points on the graph of $f(t)$ corresponding to the endpoints of the interval.

Example 4   A graph of world cigarette production is shown in Figure 1.28.[15]

(a)   Did cigarette production increase more during the 1950s or during the 1980s?

(b)   Over which decade (1950-60, 1960-70, 1970-80, or 1980-90) is the average rate of change of cigarette production the greatest? Justify your answers graphically.

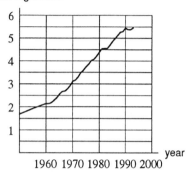

Trillion Cigarettes

Figure 1.28: World Cigarette Production

Solution   (a)   The change in cigarette production was greater between 1980 and 1990 since the vertical rise is greater between 1980-90 than between 1950-60. See Figure 1.29.

(b)   Since the secant line between 1970 and 1980 is the steepest, it has the greatest slope. We see that the average rate of change was greatest during the 1970s. See Figure 1.30.

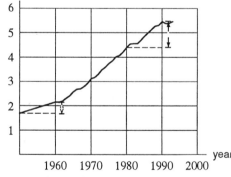

Trillion Cigarettes

Figure 1.29: Change in cigarette production is greater in the 1980s than in the 1950s

Trillion Cigarettes

Figure 1.30: Average rate of change is greatest in the 1970s

Example 5   (a)   Find the average rate of change of $f(x) = 7x - x^2$ between $x = 1$ and $x = 4$.

(b)   Sketch a graph of $f(x)$ and represent this average rate of change graphically as the slope of a line.

Solution   (a)   Since $f(x) = 7x - x^2$, we see that the average rate of change between $x = 1$ and $x = 4$ is

$$\frac{\Delta f}{\Delta x} = \frac{f(4) - f(1)}{4 - 1} = \frac{12 - 6}{3} = 2.$$

[15]Lester R. Brown, et. al., *Vital Signs 1994*, (New York: W.W. Norton & Company, 1994), p.101.

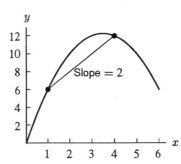

Figure 1.31: Average rate of change =
slope of secant line

(b)   A graph of $f(x) = 7x - x^2$ is given in Figure 1.31. The average rate of change is represented graphically as the slope of the secant line between $x = 1$ and $x = 4$.

Example 6   A car travels away from home in a straight line and its distance from home at time t is shown in Figure 1.32. Is the average velocity of the car greater during the first hour or the second hour?

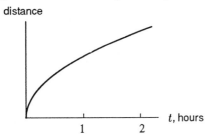

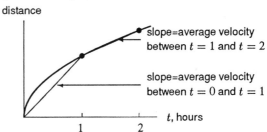

Figure 1.32: Distance of car from home   *Figure 1.33:* Visualizing average velocity of a car

Solution   Average velocity is represented graphically by the slope of a secant line. We see that the secant line between $t = 0$ and $t = 1$ is much steeper than the secant line between $t = 1$ and $t = 2$. (See Figure 1.33.) Thus, the average velocity is greater during the first hour.

Problems for Section 1.3

1.   Match the graphs in Figure 1.34 with the equations below.
   (a) $y = x - 5$ (c) $5 = y$ (e) $y = x + 6$
   (b) $-3x + 4 = y$ (d) $y = -4x - 5$ (f) $y = x/2$

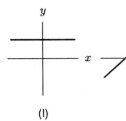

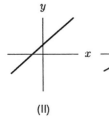

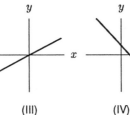

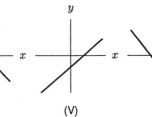

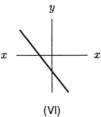

(I)   (II)   (III)   (IV)   (V)   (VI)

Figure 1.34

2. Match the graphs in Figure 1.35 with the equations below.

(a) $y = -2.72x$ (c) $y = 27.9 - 0.1x$ (e) $y = -5.7 - 200x$

(b) $y = 0.01 + 0.001x$ (d) $y = 0.1x - 27.9$ (f) $y = x/3.14$

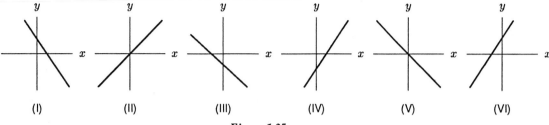

Figure 1.35

3. Find the slope and vertical intercept of the line whose equation is $2y + 5x - 8 = 0$.

4. Find the equation of the line through the points $(-1, 0)$ and $(2, 6)$.

5. Find the equation of the line with slope m through the point (a, c).

6. Estimate the slope of the line shown in Figure 1.36 and use the slope to find an equation for that line. (Note that the x and y scales are different.)

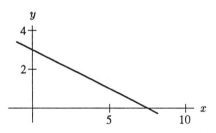

Figure 1.36

7. Which of the following tables of values could correspond to linear functions?

(a)

x	y
0	27
1	25
2	23
3	21

(b)

t	s
15	62
20	72
25	82
30	92

(c)

u	w
1	5
2	10
3	18
4	28

8. For each of the tables in Problem 7 which could correspond to a linear function, find a formula for that linear function.

9. Corresponding values of p and q are given in the table below.

(a) Find q as a linear function of p.

(b) Find p as a linear function of q.

p	1	2	3	4
q	950	900	850	800

10. A linear equation was used to generate the values in the table below. Find that equation.

x	5.2	5.3	5.4	5.5	5.6
y	27.8	29.2	30.6	32.0	33.4

11. An equation of a line is $3x + 4y = -12$.
    (a) Find the x and y intercepts.
    (b) Find the length of the portion of the line that lies between the x and y intercepts of the equation.

12. Is the demand function for a product an increasing or decreasing function? Is the supply function for a product an increasing or decreasing function?

13. A company's pricing schedule is designed to encourage large orders. A table of the price per dozen, p, of a certain item versus the size of the order in gross, q, is given. (A gross is equal to 12 dozen.)
    (a) Find a formula for q as a linear function of p.
    (b) Find a formula for p as a linear function of q.

q	3	4	5	6
p	15	12	9	6

14. The graph of a linear function is given in Figure 1.37. Estimate $f(0)$, $f(1)$, and $f(3)$.

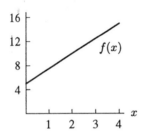

Figure 1.37

15. A linear equation was used to generate the following table.
    (a) Find the equation.   (b) Sketch the graph of $f(t)$.

t	0	1	2	3	4
$f(t)$	19.72	18.48	17.24	16.00	14.76

16. The Canadian gold reserve, Q, in millions of fine troy ounces is given for the years 1986 through 1990 in the following table.[16] Find a linear formula which approximates this data reasonably well and gives the gold reserve as a function of time measured since 1986.

year	1986	1987	1988	1989	1990
Q	19.72	18.52	17.14	16.10	14.76

---

[16]"Gold Reserves of Central Banks and Governments," *The World Almanac 1992*, 158

17. According to a 1979 *Build and Blood Pressure Study*[17] by the Society of Actuaries, the average weight, w, in pounds, of American men in their sixties for various heights, h, in inches, is given in the following table. Find a linear function which gives a reasonable approximation to the average weight as a function of height for men in their sixties. What is the slope of your line? What are the units for this slope? Interpret the slope in terms of height and weight.

h	68	69	70	71	72	73	74	75
w	167	172	176	181	186	191	196	200

18. A demand curve is given by $75p + 50q = 300$, where p is the price of the product, in dollars, and q is the quantity demanded at that price. Find the vertical and horizontal intercepts and interpret them in terms of consumer's demand for the product. (Put price p on the vertical axis and quantity q on the horizontal axis.)

19. Refer to the graph of the function k in Figure 1.38:
    (a) Between which pair of consecutive points is the average rate of change of k greatest?
    (b) Between which pair of consecutive points is the average rate of change of k closest to zero?
    (c) Between which two pairs of consecutive points are the average rates of change of k closest?

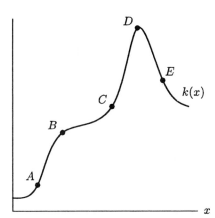

Figure 1.38

20. Figure 1.39 shows the position of an object at time t.
    (a) Is average velocity greater between $t = 0$ and $t = 3$ or between $t = 3$ and $t = 6$?
    (b) Is average velocity positive or negative between $t = 6$ and $t = 9$?
    (c) On a sketch similar to Figure 1.39, represent the average velocity between $t = 2$ and $t = 8$ as the slope of a line.

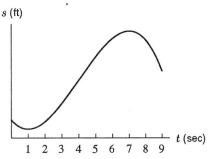

Figure 1.39

21. Draw a graph of distance against time satisfying the following properties: the average velocity is positive for all intervals and the average velocity for the first half of the trip is less than the average velocity for the second half of the trip.

---

[17]"Average Weight of Americans by Height and Age," *The World Almanac 1992*, 956

22. Total sales for a company are shown in Figure 1.40.

   (a) Is the average rate of change of sales positive or negative between $t = 0$ and $t = 5$?
   (b) Is the average rate of change of sales positive or negative between $t = 0$ and $t = 10$?
   (c) Is the average rate of change of sales positive or negative between $t = 0$ and $t = 15$?
   (d) Is the average rate of change of sales positive or negative between $t = 0$ and $t = 20$?
   (e) During which time interval is the average rate of change larger: $0 \leq t \leq 5$ or $0 \leq t \leq 10$?
   (f) During which interval is the average rate of change larger: $0 \leq t \leq 10$ or $0 \leq t \leq 20$?
   (g) Estimate the average rate of change between $t = 0$ and $t = 10$ and interpret your answer in terms of sales.

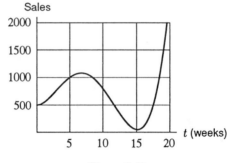

Figure 1.40

23. Find the average rate of change of $f(x) = 3x^2 + 4$, from $x = -2$ to $x = 1$, and illustrate your answer graphically.

24. A tour boat operator found that when the price charged for the scenic boat tour was $25, the average number of customers per week was 500. When the price was reduced to $20, the average number of customers per week went up to 650. Find the formula for the demand function, assuming that it is linear.

25. A car rental company offers cars at $40 a day and 15 cents a mile. Its competitor's cars are $50 a day and 10 cents a mile.

   (a) For each company, write a formula giving the cost of renting a car for a day as a function of the distance traveled.
   (b) On the same axes, sketch graphs of both functions.
   (c) How should you decide which company is cheaper?

26. The graph of Fahrenheit temperature, °F, against Celsius temperature, °C, is a line. You know that 212°F and 100°C both represent the temperature at which water boils. Similarly, 32°F and 0°C both represent water's freezing point.

   (a) What is the slope of the graph?
   (b) What is the equation of the line?
   (c) Use the equation to find what Fahrenheit temperature corresponds to 20°C.
   (d) What temperature is the same number of degrees in both Celsius and Fahrenheit?

27. Suppose you are driving at a constant speed from Chicago to Detroit, about 275 miles away. About 120 miles from Chicago you pass through Kalamazoo, Michigan. Sketch a graph of your distance from Kalamazoo as a function of time.

28. When a cold yam is put into a hot oven to bake, the temperature of the yam rises. The rate, R (in degrees per minute), at which the temperature of the yam rises is governed by Newton's

Law of Heating, which says that the rate is proportional to the temperature difference between the yam and the oven. If the oven is at 350°F and the temperature of the yam is H°F,

(a)   Write a formula giving R as a function of H.
(b)   Sketch the graph of R against H.

29.   When a cup of hot coffee sits on the kitchen table, its temperature falls. The rate, R, at which its temperature changes is governed by Newton's Law of Cooling, which says that the rate is proportional to the temperature difference between the coffee and the surrounding air. Let's think of the rate, R, as a negative quantity because the temperature of the coffee is falling. If the temperature of the coffee is H°C and the temperature of the room is 20°C,

(a)   Write a formula giving R as a function of H.
(b)   Sketch a graph of R against H.

1.4 APPLICATIONS OF FUNCTIONS TO ECONOMICS

Management decisions within a particular firm or industry usually depend on many functions. In this section we will look at some of these functions. One of the most important of these is the cost of production.

The Cost Function

The **cost function**, $C(q)$, gives the total cost of producing a quantity q of some good.

What sort of function do you expect C to be? The more goods that are made, the higher the cost, so C is an increasing function. For most goods, such as cars or cases of soda, q can only be an integer, so the graph of C might look like that in Figure 1.41.

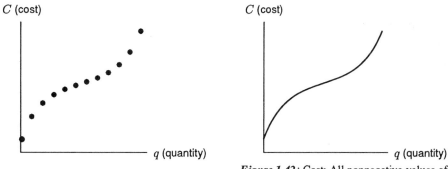

Figure 1.41: Cost: Nonnegative integer values of q **Figure 1.42:** Cost: All nonnegative values of q

However, economists usually imagine the graph of C as the smooth curve drawn through these points, as in Figure 1.42. This is equivalent to assuming that $C(q)$ is defined for all nonnegative values of q, not just for integers.

Costs of production can be separated into two parts: the *fixed cost,* which is incurred even if nothing is produced, and the *variable cost* which varies depending on how many units are produced.

A Soccer Ball Example

Let's consider a company that makes soccer balls. The factory and machinery needed to begin production are fixed costs, since these costs are incurred even if no soccer balls are made. The costs of labor and raw materials are variable costs since these quantities depend on how many balls are made. Suppose that the fixed costs for this company are $24,000 and the variable costs are $7 per soccer ball. The total costs for the company are seven times the number of soccer balls produced, plus the fixed cost of $24,000. We have

$$C(q) = 24,000 + 7q,$$

where q is the number of soccer balls produced. Notice that this is the equation of a line, with slope equal to the variable costs per ball (7) and vertical intercept equal to the fixed costs (24,000). Our cost function for soccer balls is a linear function of q. (We will see examples of cost functions that are not linear in Chapter 2.)

Example 1   Sketch the graph of the cost function $C(q) = 24,000 + 7q$. On the graph, label the fixed costs and variable cost per unit.

Solution   The graph is in Figure 1.43. The graph of the cost function is a line with a vertical intercept of 24,000 and a slope of 7. The fixed costs are represented graphically by the vertical intercept. The variable cost per unit is represented graphically by the slope of the line, or equivalently, by the "rise" (the change in cost) corresponding to a horizontal change of one unit.

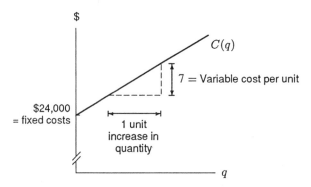

Figure 1.43: Cost function for the soccer ball company

Example 2   In each case, draw a graph of a linear cost function satisfying the given conditions:
(a)   Fixed costs are large but variable cost per unit is small.
(b)   There are no fixed costs but variable cost per unit is high.

Solution   (a)   A solution is the graph of any line with a large vertical intercept and a small slope, such as that shown in Figure 1.44.
(b)   A solution is the graph of any line with a vertical intercept of zero (so the line goes through the origin) and a large positive slope. See Figure 1.45.

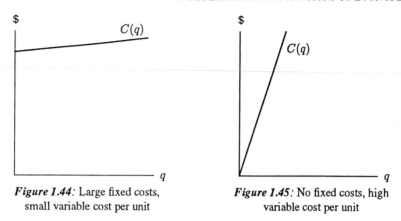

Figure 1.44: Large fixed costs, small variable cost per unit

Figure 1.45: No fixed costs, high variable cost per unit

The Revenue Function

> The **revenue function**, $R(q)$, gives the total revenue received by a firm from selling a quantity q of some good.

The soccer ball company makes money by selling soccer balls. The revenue for this company depends on the price of the soccer balls and the quantity sold. If 10 soccer balls are sold at $15 each, the total revenue is $15 \times 10 = \$150$.

If the price is p and the quantity sold is q, then

$$\text{Revenue} = \text{Price} \times \text{Quantity}, \quad \text{so} \quad R = pq.$$

If the unit price does not depend on the quantity sold, the graph of the revenue as a function of q is a line through the origin, with slope equal to the price p. (We will consider nonlinear revenue functions in Chapter 2.)

Example 3   If the company sells soccer balls for $15 each, sketch a graph of the company's revenue function and mark on it the price of a soccer ball.

Solution   Since $R(q) = pq = 15q$, the revenue graph is a line through the origin with a slope of 15. See Figure 1.46. The price is the slope of the line and represents the increase in revenue corresponding to a unit increase in quantity sold.

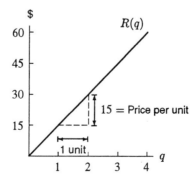

Figure 1.46: Revenue function for the soccer ball company

Example 4   Sketch graphs of the cost function $C(q) = 24,000 + 7q$ and the revenue function $R(q) = 15q$ on the same axes. For what values of q does the firm make money? Explain your answer graphically.

Solution   The firm makes money whenever revenues are greater than costs, so we want to find the values of q for which the graph of $R(q)$ lies above the graph of $C(q)$. See Figure 1.47.

In Figure 1.47, the graph of $R(q)$ is above the graph of $C(q)$ for all values of q to the right of the value q_0, where the graphs of $R(q)$ and $C(q)$ cross. What is this point of intersection q_0? It is the point where revenue and cost are equal. We see on the graph in Figure 1.47 that the graphs of revenue and cost cross at $q_0 = 3,000$, or we can find q_0 analytically as follows:

$$\text{Revenue} = \text{Cost}$$
$$15q = 24,000 + 7q$$
$$8q = 24,000$$
$$q = 3000.$$

Thus, the company makes a profit if it produces and sells more than 3000 soccer balls.

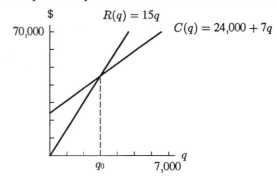

Figure 1.47: Cost and revenue functions for the soccer ball company

The Profit Function

Decisions are often made by considering the profit, usually written as π (to distinguish it from the price, p; this π has nothing to do with the area of a circle, and merely stands for the Greek equivalent of the letter "p").

$$\boxed{\text{Profit} = \text{Revenue} - \text{Cost} \quad \text{so} \quad \pi = R - C.}$$

The *break-even point* for a company is the point where the profit is zero, or, equivalently, the point where revenue equals cost. We saw in Example 4 that the break-even point for the soccer ball company is $q_0 = 3000$. If the company produces fewer than 3000 balls, it loses money. If it produces exactly 3000 balls, it breaks even. If it produces more than 3000 balls, it makes a profit.

Example 5   A company producing jigsaw puzzles has fixed costs of $6000 and variable costs of $2 per puzzle. The company sells the puzzles for $5 each.
(a) Find a formula for the cost function.
(b) Find a formula for the revenue function.
(c) Find a formula for the profit function.
(d) Sketch a graph of $R(q)$ and $C(q)$ on the same axes.
(e) What is the break-even point for the company?

Solution (a) $C(q) = 6000 + 2q$

(b) $R(q) = 5q$

(c) $\pi(q) = R(q) - C(q) = 5q - (6000 + 2q) = 3q - 6000$

(d) See Figure 1.48.

(e) The break-even point is labeled q_0 in Figure 1.48. We find q_0 by setting the revenue equal to the cost and solving for q:

$$\text{Revenue} = \text{Cost}$$
$$5q = 6000 + 2q$$
$$3q = 6000$$
$$q = 2000.$$

Notice that this is the same answer we get if we set the profit function equal to zero. The break-even point is $q_0 = 2000$ puzzles.

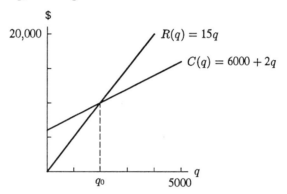

***Figure 1.48*:** Cost and revenue functions for the jigsaw puzzle company

---

Example 6 A company estimates that its revenue and cost at different quantities are those given in Table 1.22.

(a) Estimate the break-even point for the company, and explain what it means.

(b) Find the company's profit if 1000 units are produced.

(c) What price is the company charging for its product?

TABLE 1.22 *Cost and Revenue Function*

q	500	600	700	800	900	1000	1100
$C(q)$	5000	5500	6000	6500	7000	7500	8000
$R(q)$	4000	4800	5600	6400	7200	8000	8200

Solution (a) The break-even point is the value of q for which revenue equals cost. Since revenue is below cost at $q = 800$ and revenue is greater than cost at $q = 900$, the break-even point is between 800 and 900. A good estimate is about 833. If the company produces more than 833 units, it will make a profit, and if it produces fewer, it will not make a profit.

(b) If the company produces 1000 units, the cost will be $7500 and the revenue will be $8000, and so the profit is $8000 - 7500 = \$500$.

(c) Since $R(q) = 8q$, the company is selling the product for $8 each.

The Depreciation Function

Suppose that the soccer ball company has a big machine that costs $20,000. The managers of the company plan to keep the machine for ten years and then sell it for $3,000. They expect that the value of their machine will *depreciate* from $20,000 today to a resale value of $3000 in ten years. For tax purposes, they can depreciate the value of this machine over the ten-year period. The depreciation formula gives the value, $V(t)$, of the machine as a function of the number of years, t, since the machine was purchased. One widely-used depreciation method is known as *straight-line depreciation*, where we assume that the value of the machine depreciates linearly. The corresponding depreciation formula will be a linear function.

Since the value of the machine when it is new (that is, when $t = 0$) is $20,000, we have $V(0) = 20,000$, and since the resale value at time $t = 10$ is $3,000, we have $V(10) = 3,000$. The slope of the graph of $V(t)$ is:

$$m = \frac{3,000 - 20,000}{10 - 0} = \frac{-17,000}{10} = -1,700.$$

This slope tells us that the value of the machine is decreasing at a rate of $1,700 per year. Since $V(0) = 20,000$, the vertical intercept is 20,000, so the depreciation function for this machine is $V(t) = 20,000 - 1,700t$. It makes sense that this function is decreasing since the value of the machine is going down as time passes.

Revisiting the Supply and Demand Curves

In Section 1.2, we learned that a demand function tells us what quantity will be demanded at a given price, and that a supply function tells us what quantity will be produced at a given price. The market tends to settle at the equilibrium price and quantity, which is the point at which demand equals supply.

Example 7   Suppose that the market demand and supply functions are given by

$$D(p) = 100 - 2p \quad \text{and} \quad S(p) = 3p - 50.$$

Find the equilibrium price and quantity.

Solution   To find the equilibrium price and quantity, we find the point where supply equals demand. We can do this algebraically or graphically. Algebraically, we have

$$\text{Demand} = \text{Supply}$$
$$100 - 2p = 3p - 50$$
$$150 = 5p$$
$$p = 30.$$

The equilibrium price is $30. To find the equilibrium quantity, we can use either the demand function or the supply function. At a price of $30, the quantity produced is $100 - 2(30) = 100 - 60 = 40$ items. The equilibrium quantity is 40 items.

Alternatively, if we graphed the demand and supply curves on the same axes, we would see that the point of intersection is at a price of $30 and a quantity of 40 items.

The Effect of Taxes on Equilibrium[9]

Suppose that the demand and supply functions for a product are

$$D(p) = 100 - 2p \quad \text{and} \quad S(p) = 3p - 50.$$

What effect do taxes have on the equilibrium price and quantity for this product? And who (the producer or the consumer) ends up paying for the tax? We will consider two types of taxes. A *specific tax* is a tax that collects a fixed amount per unit of a product sold regardless of the actual selling price. This is the case with such items as gasoline, alcohol, and cigarettes. A *sales tax* is a tax that collects a fixed percentage of the product price. Many cities and states collect sales tax on a wide variety of items. We consider the case of a specific tax here. The effect of a sales tax is considered in Problems 20 and 21.

Suppose a specific tax of $5 per unit is imposed upon suppliers. This means that a selling price of p dollars will not bring forth the same quantity supplied, since suppliers will only receive $p - 5$ dollars. The amount supplied will depend on $p - 5$, while the amount demanded will still depend on p (the price the consumers have to pay). We have

$$D(p) = 100 - 2p \quad \text{and} \quad \begin{aligned} S(p) &= 3(p - 5) - 50 \\ &= 3p - 15 - 50 \\ &= 3p - 65. \end{aligned}$$

What are the equilibrium price and quantity in this situation? At the equilibrium price, we have

$$\begin{aligned} \text{Demand} &= \text{Supply} \\ 100 - 2p &= 3p - 65 \\ 165 &= 5p \\ p &= 33. \end{aligned}$$

The equilibrium price when there is a specific tax is $33. We saw in Example 7 that the equilibrium price in a purely competitive market is $30, so the equilibrium price increases by $3 as a result of the tax. Notice that this is less than the amount of the tax. The consumer ends up paying $3 more than he or she would have if the tax did not exist. However the government must receive $5 per item. Thus the producer pays the other $2 of the tax and is able to retain $28 of the amount paid per item.

The equilibrium quantity at a price of $33 is 34 units. Not surprisingly, the tax has reduced the number of items sold. The government, when imposing a $5 per unit tax on an industry selling 40 units, should not expect to receive $(\$5)(40) = \$200$ in tax revenue. In this example, tax revenue would be only $(\$5)(34) = \$170$.

A Budget Constraint

An ongoing debate in the federal government concerns the allocation of money between defense and social programs. In general, the more that is spent on defense, the less that is available for social programs, and vice versa. Let's simplify the example to guns and butter. Assuming a constant budget, we will show that the relationship between the number of guns and the quantity of butter is linear. Suppose that there is $12,000 to be spent and that it is to be divided between guns, costing $400 each, and butter, costing $2000 a ton. Suppose also that the number of guns bought is g, and

---

[9] Adapted from Barry Bressler, *A Unified Approach to Mathematical Economics* (New York: Harper & Row, 1975) 81-88

that the number of tons of butter is b. Then the amount of money spent on guns is $\$400g$ (because each one is $\$400$), and the amount spent on butter is $\$2000b$. Assuming all the money is spent,

$$\text{Amount spent on guns} + \text{Amount spent on butter} = \$12,000$$

or

$$400g + 2000b = 12,000.$$

Thus,

$$g + 5b = 30.$$

This equation is the budget constraint. Its graph is the line shown in Figure 1.49, which can be found by plotting points. We will calculate the points at which the graph crosses the axes. If $b = 0$, then

$$g + 5(0) = 30 \quad \text{so} \quad g = 30.$$

If $g = 0$, then

$$0 + 5b = 30 \quad \text{so} \quad b = 6.$$

Since the number of guns bought determines the amount of butter bought (because all the money which doesn't go to guns goes to butter), b is a function of g. Similarly, the amount of butter bought determines the number of guns, so g is a function of b. The budget constraint represents an *implicitly defined function*, because neither quantity is given explicitly in terms of the other. If we solve for g, giving

$$g = 30 - 5b$$

we have an explicit formula for g in terms of b. Similarly,

$$b = \frac{30 - g}{5} \quad \text{or} \quad b = 6 - 0.2g$$

gives b as an *explicit function* of g. Since the explicit functions

$$g = 30 - 5b \quad \text{and} \quad b = 6 - 0.2g$$

are linear, the graph of the budget constraint must be a line.

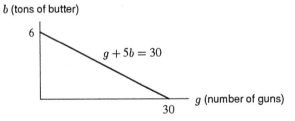

Figure 1.49: Budget constraint

Problems for Section 1.4

1. An amusement park charges an admission fee of $\$7$ as well as an additional $\$1.50$ for each ride.
   (a) Find the total cost $C(n)$ as a function of the number of rides, n.
   (b) Find $C(2)$ and $C(8)$ and interpret your answers in terms of amusement park fees.

2. The cost function for a certain company is $C(q) = 4000 + 2q$, and the revenue function is $R(q) = 10q$.

    (a) What are the fixed costs for the company?
    (b) What is the variable cost per unit?
    (c) What price is the company charging for its product?
    (d) Graph $C(q)$ and $R(q)$ on the same axes and label the break-even point, q_0. Explain graphically how you know the company will make a profit if the quantity produced is greater than q_0.
    (e) Find the break-even point q_0.

3. A table of values for a linear cost function is given below. What are the fixed costs and the variable cost per unit? Find a formula for the cost function.

q	0	5	10	15	20
$C(q)$	5000	5020	5040	5060	5080

4. A company has a cost function of $C(q) = 6000 + 10q$ and a revenue function of $R(q) = 12q$.

    (a) Find the revenue and cost if the company is producing 500 units. Does the company make a profit?
    (b) Find the revenue and cost if the company is producing 5000 units. Does the company make a profit?
    (c) Find the break-even point and illustrate it graphically.

5. A company that makes Adirondack chairs has fixed costs of $5000 and variable costs of $30 per chair. The company plans to sell the chairs for $50 each.

    (a) Find formulas for the cost function and the revenue function.
    (b) Graph the cost and the revenue functions on the same axes.
    (c) Find the break-even point.

6. The Quick-Food company wants to provide students with an alternative to the college food service plan. Quick-Food has fixed costs of $350,000 per term that are incurred regardless of how many students are served. In addition, the cost of providing and serving the food amounts to $400 per student served. Quick-Food plans to charge $800 per student. How many students must sign up with the Quick-Food plan in order for the company to make a profit?

7. A company manufactures and sells a new type of running shoe. Total production costs to the company consist of fixed overhead of $650,000 plus production costs of $20 per pair of shoes. Each pair of shoes produced is sold for $70.

    (a) Find the total cost, $C(q)$, as a function of the number of pairs of shoes produced, q.
    (b) Find the total revenue, $R(q)$, as a function of the number of pairs of shoes produced, q.
    (c) Find the total profit, $\pi(q)$, as a function of the number of pairs of shoes produced, q.
    (d) How many pairs of shoes must be produced and sold in order for the company to make a profit?

8. A graph of the cost function for a certain company is shown in Figure 1.50.

    (a) Estimate the fixed costs for this company.
    (b) Estimate the variable cost per unit.
    (c) Estimate $C(10)$ and interpret it in terms of cost.

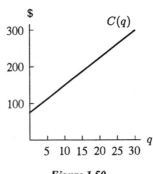

Figure 1.50

9. A graph of the cost function for a certain company is shown in Figure 1.51.

    (a) What are the fixed costs for this company?
    (b) What is the variable cost per unit?
    (c) We see from the graph that $C(100) = 2500$. Explain in words what this is telling you about costs.

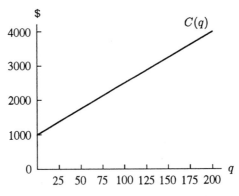

Figure 1.51

10. (a) Give an example of a possible company where the fixed costs would be zero (or very small).
    (b) Give an example of a possible company where the variable cost per unit would be zero (or very small).

11. Graphs of the cost and revenue functions for a certain company are given in Figure 1.52.

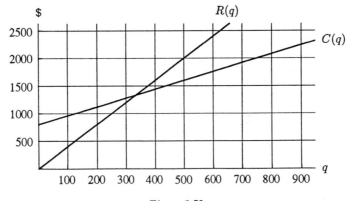

Figure 1.52

(a) Approximately what quantity of goods does this company have to produce in order to make a profit?

(b) Estimate the profit if the company produces 600 units.

12. For large orders, a photocopying company allows customers to choose from two different price lists. The first price list is $100 plus 3 cents per copy, whereas the second price list is $200 plus 2 cents per copy.

(a) For each price list, find the total cost as a function of the number of copies needed.

(b) Determine which price list is cheaper if you need 5000 copies.

(c) For what number of copies do both price lists charge you the same amount?

13. A $15,000 robot depreciates completely (down to a value of zero) in 10 years. Assume that this depreciation is linear.

(a) Find a formula for its value as a function of time.

(b) How much is the robot worth three years after it is purchased?

14. A $50,000 tractor has a resale value of $10,000 twenty years after it was purchased. Assume that the value of the tractor depreciates linearly and that time is measured from the time of purchase.

(a) Find a formula for the value of the tractor as a function of the time since it was purchased.

(b) Sketch a graph of the value of the tractor against time.

(c) Find the horizontal and vertical intercepts, give units for them, and interpret them.

15. Suppose that you have a budget of $1000 for the school year to cover your books and social outings, and suppose that books cost (on average) $40 each and social outings cost (on average) $10 each. Let b denote the number of books purchased per year and s denote the number of social outings in a year.

(a) What is the equation of your budget constraint?

(b) Graph the budget constraint. (It doesn't matter which variable you put on which axis.)

(c) Find the vertical and horizontal intercepts and give a financial interpretation for each one.

16. A company has a total budget of $500,000 and spends this budget on raw materials and personnel. The company uses m units of raw materials, at a cost of $100 per unit, and hires r employees, at a cost of $25,000 each.

(a) What is the equation of the company's budget constraint?

(b) Solve for m, the quantity of raw materials the company can buy, as a function of r, the number of employees.

(c) Solve for r, the number of employees the company can hire, as a function of m, the quantity of raw materials used.

17. Hot peppers have been rated according to Scoville units, with a maximum human tolerance level of 14,000 Scovilles per dish. The West Coast Restaurant, known for spicy dishes, promises a daily special to satisfy the most avid spicy-dish fans. The restaurant imports Indian peppers rated at 1200 Scovilles each and Mexican peppers with a Scoville rating of 900 each.

(a) Determine the Scoville constraint equation relating the maximum number of Indian and Mexican peppers the restaurant should use for their speciality dish.

(b) Solve the equation from part (a) to show explicitly the number of Indian peppers needed in the hottest dishes as a function of the number of Mexican peppers.

18. You have a fixed budget of k to spend on soda and suntan oil, which cost p_1 per liter and

p_2 per liter respectively.

(a) Write an equation expressing the relationship between the number of liters of soda and the number of liters of suntan oil that you can buy if you exhaust your budget. This is your *budget constraint*.

(b) Graph the budget constraint, assuming that you can buy fractions of a liter. Label the intercepts.

(c) Suppose your budget is suddenly doubled. Graph the new budget constraint on the same axes.

(d) With a budget of k, the price of suntan oil suddenly doubles. Sketch the new budget constraint on the same axes.

19. The demand and supply functions for a certain product are given by

$$D(p) = 2500 - 20p \quad \text{and} \quad S(p) = 10p - 500.$$

(a) Find the equilibrium price and quantity.

(b) If a specific tax of $6 is imposed on this product, find the new equilibrium price and quantity.

(c) How much of the $6 tax is paid by consumers, and how much by producers?

(d) What is the total tax revenue received by the government?

20. In Example 7, we saw that when the demand and supply functions are given by $D(p) = 100 - 2p$ and $S(p) = 3p - 50$, the equilibrium price is $30 and the equilibrium quantity is 40 units. Suppose that a sales tax of 5% is imposed on the supplier, so that the supplier only keeps $p - 0.05p$ of the price.

(a) Find the new equilibrium price and quantity.

(b) How much is paid in taxes on each unit? How much of this is paid by the consumer, and how much by the producer?

21. Suppose that the sales tax in Problem 20 is imposed on the consumer instead of the producer, so that the consumer's price is $p + 0.05p$, while the producer's price is p. Answer the questions in Problem 20 under these circumstances, and compare your answers with those for Problem 20.

22. Linear supply and demand curves are shown in Figure 1.53, with price on the vertical axis.

(a) Label the equilibrium price p_0 and the equilibrium quantity q_0 on the axes.

(b) Explain the effect on equilibrium price and quantity if the slope of the supply curve increases. Illustrate your answer graphically.

(c) Explain the effect on equilibrium price and quantity if the slope of the demand curve becomes more negative. Illustrate your answer graphically.

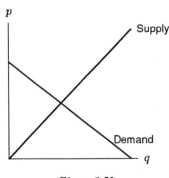

Figure 1.53

1.5 EXPONENTIAL FUNCTIONS

Population Growth

Consider the data for the population of Mexico in the early 1980s in Table 1.23. To see how the population is growing, you might look at the increase in population from one year to the next, as shown in the third column. If the population had been growing linearly, all the numbers in the third column would be the same. But populations usually grow faster as they get bigger, because there are more people to have babies. So you shouldn't be surprised to see the numbers in the third column increasing.

TABLE 1.23 *Population of Mexico (estimated), 1980–1986*

Year	Population (millions)	Change in population (millions)
1980	67.38	
		1.75
1981	69.13	
		1.80
1982	70.93	
		1.84
1983	72.77	
		1.89
1984	74.66	
		1.94
1985	76.60	
		1.99
1986	78.59	

Suppose we divide each year's population by the previous year's population. We get, approximately,

$$\frac{\text{Population in 1981}}{\text{Population in 1980}} = \frac{69.13 \text{ million}}{67.38 \text{ million}} = 1.026$$

$$\frac{\text{Population in 1982}}{\text{Population in 1981}} = \frac{70.93 \text{ million}}{69.13 \text{ million}} = 1.026.$$

The fact that both calculations give 1.026 shows the population grew by about 2.6% between 1980 and 1981 *and* between 1981 and 1982. If you do similar calculations for other years, you will find that the population grew by a factor of about 1.026, or 2.6%, every year. Whenever you have a constant *growth factor* (here 1.026), you have *exponential growth*. If t is the number of years since 1980,

$$\text{When } t = 0, \quad \text{population} = 67.38 = 67.38(1.026)^0$$
$$\text{When } t = 1, \quad \text{population} = 69.13 = 67.38(1.026)^1$$
$$\text{When } t = 2, \quad \text{population} = 70.93 = 69.13(1.026) = 67.38(1.026)^2$$
$$\text{When } t = 3, \quad \text{population} = 72.77 = 70.93(1.026) = 67.38(1.026)^3$$

and so t years after 1980, the population is given by

$$P = 67.38(1.026)^t.$$

This is an *exponential function* with *base* 1.026. It is called exponential because the variable, t, is in the exponent. The base represents the factor by which the population grows each year.

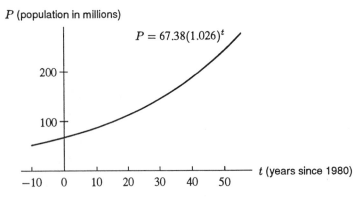

Figure 1.54: Population of Mexico (estimated): Exponential growth

If we assume that the same formula will hold for the next 50 years or so, the population will have the shape shown in Figure 1.54. Since the population is growing, the function is increasing. Notice also that the population grows faster and faster as time goes on. This behavior is typical of an exponential function. You should compare this with the behavior of a linear function, which climbs at the same rate everywhere and so has a straight-line graph. Because this graph is bending upward, we say it is *concave up*. Even exponential functions which climb slowly at first, such as this one, climb extremely quickly eventually. That is why exponential population growth is such a threat to the world.

Even if it represents reliable data, the smooth graph in Figure 1.54 is actually only an approximation to the true graph of the population of Mexico. Since we can't have fractions of people, the graph should really be jagged, jumping up or down by one each time someone is born or dies. However, with a population in the millions, the jumps are so small as to be invisible at the scale we are using. Therefore, the smooth graph is an extremely good approximation.

Example 1   Predict the population of Mexico in the year
(a)   2007 (when $t = 27$).   (b)   2034 (when $t = 54$).   (c)   2061 (when $t = 81$).

Solution   Extrapolating so far into the future is very risky because it assumes that the population continues to grow exponentially with the same constant growth factor. (There could, for example, be a medical breakthrough that would increase the growth factor, or an epidemic that would decrease it.) Writing "$\approx$" to represent approximately equal, the model we are using predicts
(a) $P = 67.38(1.026)^{27} \approx 67.38(2) = 134.76 \approx 135$ million.
(b) $P = 67.38(1.026)^{54} \approx 67.38(4) = 269.52 \approx 270$ million.
(c) $P = 67.38(1.026)^{81} \approx 67.38(8) = 539.04 \approx 539$ million.

Musical Pitch

The pitch of a musical note is determined by the frequency of the vibration which causes it. Middle C on the piano, for example, corresponds to a vibration of 263 hertz (cycles per second). A note one octave above middle C vibrates at 526 hertz, and a note two octaves above middle C vibrates at 1052 hertz. (See Table 1.24.)

TABLE 1.24

Pitch of notes above middle C

Number, n, of octaves above middle C	Number of hertz $V = f(n)$
0	263
1	526
2	1052
3	2104
4	4208

TABLE 1.25

Pitch of notes below middle C

n	$V = 263 \cdot 2^n$
-3	$263 \cdot 2^{-3} = 263(1/2^3) = 32.875$
-2	$263 \cdot 2^{-2} = 263(1/2^2) = 65.75$
-1	$263 \cdot 2^{-1} = 263(1/2) = 131.5$
0	$263 \cdot 2^0 = 263$

Notice that

$$\frac{526}{263} = 2 \quad \text{and} \quad \frac{1052}{526} = 2 \quad \text{and} \quad \frac{2104}{1052} = 2$$

and so on. In other words, each value of V is twice the value before, so

$$f(1) = 526 = 263 \cdot 2 = 263 \cdot 2^1$$
$$f(2) = 1052 = 526 \cdot 2 = 263 \cdot 2^2$$
$$f(3) = 2104 = 1052 \cdot 2 = 263 \cdot 2^3.$$

In general

$$V = f(n) = 263 \cdot 2^n,$$

where n is the number of octaves above middle C. The base 2 represents the fact that as we go up an octave, the frequency of vibrations doubles. Indeed, our ears hear a note as one octave higher than another precisely because it vibrates twice as fast. For the negative values of n in Table 1.25, this function represents the octaves below middle C. The notes on a piano are represented by values of n between -3 and 4, and the human ear finds values of n between -4 and 7 audible.

Although $V = f(n) = 263 \cdot 2^n$ makes sense in musical terms only for certain values of n, values of the function $f(x) = 263 \cdot 2^x$ can be calculated for all real x, and its graph has the typical exponential shape, as can be seen in Figure 1.55. It is concave up, climbing faster and faster as x increases.

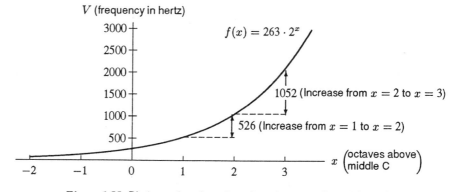

Figure 1.55: Pitch as a function of number of octaves above middle C

Removal of Pollutants from Jet Fuel

Now we will look at an example in which a quantity is decreasing instead of increasing. Before kerosene can be used as jet fuel, federal regulations require that the pollutants in it be removed by passing the kerosene through clay. We will suppose the clay is in a pipe and that each foot of the pipe removes 20% of the pollutants that enter it. Therefore each foot leaves 80% of the pollution. If P_0 is the initial quantity of pollutant and $P = f(n)$ is the quantity left after n feet of pipe, then

$$f(0) = P_0$$
$$f(1) = (0.8)P_0$$
$$f(2) = (0.8)(0.8)P_0 = (0.8)^2 P_0$$
$$f(3) = (0.8)(0.8)^2 P_0 = (0.8)^3 P_0,$$

and so, after n feet,

$$P = f(n) = P_0(0.8)^n.$$

In this example, n must be non-negative. However, the *exponential decay function*

$$P = f(x) = P_0(0.8)^x$$

makes sense for any real x. We'll plot it with $P_0 = 1$ in Figure 1.56; some values of the function are in Table 1.26.

Notice the way the function in Figure 1.56 is decreasing: each downward step is smaller than the one before. This is because as the kerosene gets cleaner, there's less dirt to remove, and so each foot of clay takes out less pollutant than the previous one. Compare this to the exponential growth in Figures 1.54 and 1.55 on pages 48 and 49, where each step upward is larger than the one before. Notice, however, that all three graphs mentioned are concave up.

TABLE 1.26 *Values of decay function*

x	$P = (0.8)^x$
-2	1.56
-1	1.25
0	1
1	0.8
2	0.64
3	0.51
4	0.41

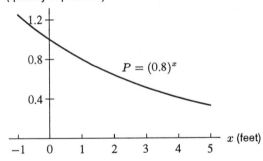

Figure 1.56: Pollutant removal: Exponential decay

The General Exponential Function

P is an **exponential function** of t with base a if

$$P = P_0 a^t$$

where P_0 is the initial quantity (when $t = 0$) and a is the factor by which P changes when t increases by 1.
If $a > 1$, we have exponential growth; if $0 < a < 1$, we have exponential decay.

The largest possible domain for the exponential function is all real numbers, provided $a > 0$. (Why do we not want $a \leq 0$?)

To recognize that a function $P = f(t)$ given by a table of values is exponential, look for ratios of P values that are constant for equally spaced t values.

Example 2   Determine whether each of the tables of values in Table 1.27 could correspond to an exponential function, a linear function, or neither. For those which could correspond to an exponential or linear function, find a formula for the function.

TABLE 1.27

(a)

x	$f(x)$
0	16
1	24
2	36
3	54
4	81

(b)

x	$g(x)$
0	14
1	20
2	24
3	29
4	35

(c)

x	$h(x)$
0	5.3
1	6.5
2	7.7
3	8.9
4	10.1

Solution   (a)   We see that $f(x)$ cannot be a linear function, since $f(x)$ increases by different amounts ($24 - 16 = 8$ and $36 - 24 = 12$) as x goes up by one. Could it be an exponential function? We look at the ratios of $f(x)$ values:

$$\frac{24}{16} = 1.5 \qquad \frac{36}{24} = 1.5 \qquad \frac{54}{36} = 1.5 \qquad \frac{81}{54} = 1.5.$$

The ratios of $f(x)$ values are all equal to 1.5, and so this table of values could correspond to an exponential function with a base of 1.5. Since $f(0) = 16$, a formula for $f(x)$ is

$$f(x) = 16(1.5)^x.$$

A good way to check this answer is by substituting $x = 0$, 1, 2, 3, 4 into this function to check that it produces the function values given for $f(x)$ in Table 1.27a.

(b)   Now we look at the table of values for $g(x)$. As x goes up by one, g goes up by 6 (from 14 to 20), then 4 (from 20 to 24), and so $g(x)$ is not linear. Now we check to see if $g(x)$ could be an exponential function: $\frac{20}{14} = 1.43$ and $\frac{24}{20} = 1.2$. Since these ratios (1.43 and 1.2) are different, $g(x)$ is not an exponential function either.

(c)   How about $h(x)$? As x increases by one, the value of $h(x)$ increases by 1.2 each time, so $h(x)$ could be a linear function with a slope of 1.2. Since $h(0) = 5.3$, a formula for $h(x)$ is

$$h(x) = 5.3 + 1.2x$$

Doubling Time and Half-Life

In Example 1, you were asked to predict the population of Mexico at $t = 27, t = 54$, and $t = 81$. If you look back at the answers to Example 1, you will see something that may surprise you. After 27 years the population has doubled; after another 27 years (at $t = 54$), it has doubled again. Another 27 years later (when $t = 81$), the population has doubled again. As a result, we say that the *doubling time* of the population of Mexico is 27 years.

Every exponentially growing population has a fixed doubling time. The world's population currently has a doubling time of about 38 years. Notice what this means: If you live to be 76, the world's population is expected to quadruple in your lifetime.

For functions decaying exponentially, the equivalent notion is the *half-life*. The half-life is the time period for half the mass to decay. The most common way to express the rate of decay for radioactive substances such as uranium is to give the half-life. The important thing to remember about the radioactive decay is that two half-lives do not make a whole life! Rather, in the time period of two half-lives, the substance will decay to $(1/2)(1/2) = 1/4$ of its original mass.

The **doubling time** of an exponentially increasing quantity is the time for it to double.
The **half-life** of an exponentially decaying quantity is the time for it to be reduced by half.

One of the most well-known radioactive substances is carbon-14, which is used to date organic objects. When the object, such as a piece of wood or bone, was part of a living organism, it accumulated small amounts of radioactive carbon-14, so that a certain proportion of the carbon in the object was carbon-14. Once the organism dies, it no longer picks up carbon-14 through interaction with its environment (for example, through respiration). By measuring the proportion of carbon-14 in the object and comparing that to the proportion in living material, we can estimate how much of the original carbon-14 has decayed.

Example 3   The half-life of carbon-14 is roughly 5000 years. Sketch a graph of the percent of carbon-14 left at time t, in years since the organism died.

Solution   At time $t = 0$, the organism has 100% of the original carbon-14. At time $t = 5000$ (the half-life), the amount has been reduced by half, so 50% remains. After another half-life (at $t = 10,000$), the amount is again reduced by half, to 25% remaining. A graph is shown in Figure 1.57.

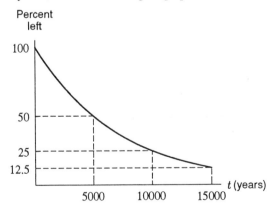

Figure 1.57: Decay of carbon-14

Concavity

We have used the term *concave up* to describe the graphs shown earlier in this section. These graphs all bend upward. We call a graph *concave down* if it bends downward.

> The graph of any function is **concave up** if it bends upward, and it is **concave down** if it bends downward. A line is neither concave up nor concave down.

Figure 1.58 might represent the graph of a growing population, and is increasing and concave up. Exponential growth looks like Figure 1.58. Figure 1.59 is a reasonable graph of the amount of pollution left in a lake as the pollutants are cleaned out. It is decreasing and concave up. Exponential decay looks like Figure 1.59. Figure 1.60 could be the temperature of an object in the oven, and is increasing and concave down. Figure 1.61 might represent the confidence of an ill-prepared student taking an exam. It is decreasing and concave down.

Of course, the graph of a function may be concave up on one section and concave down on another, as we will see in the next example.

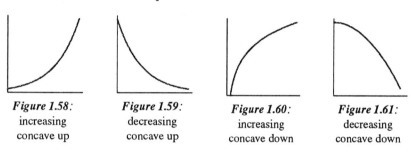

Figure 1.58:
increasing
concave up

Figure 1.59:
decreasing
concave up

Figure 1.60:
increasing
concave down

Figure 1.61:
decreasing
concave down

Example 4   The graph of a function $y = f(x)$ is shown in Figure 1.62.
  (a)   Over what intervals is the function increasing? decreasing?
  (b)   Over what intervals is the graph concave up? concave down?

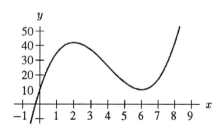

Figure 1.62: Discuss the concavity of this
graph

Solution (a)   By looking at the graph, we can see that the function is increasing for $x < 2$ and for $x > 6$. It is decreasing for $2 < x < 6$.
  (b)   We see that the graph is concave down on the left and concave up on the right. It is difficult to tell exactly where the graph changes concavity, although it appears to be close to $x = 4$. Roughly, the graph is concave down for $x < 4$ and concave up for $x > 4$.

Example 5   A table of values for a function $y = f(t)$ is given in Table 1.28. Does this function appear to be increasing or decreasing? Is its graph concave up or concave down?

TABLE 1.28

t	0	5	10	15	20	25	30
y	12.6	13.1	14.1	16.2	20.0	29.6	42.7

Solution   As t increases, we see that y is going up, so y appears to be an increasing function of t. Is the graph concave up or concave down? As we read from left to right, we see that the change in y as t increases by 5 starts small and gets larger and larger as t increases, so the graph is climbing faster and faster. The graph must be concave up. (Another way to see this is to plot the points and notice that a curve through these points bends up.)

The Family of Exponential Functions

The formula $P = P_0 a^t$ gives a family of exponential functions with parameters P_0 (the initial quantity) and a (the base). The base is as important for an exponential function as the slope is for a linear function. We assume $a > 0$ and $a \neq 1$. The base tells you whether the exponential function is increasing ($a > 1$) or decreasing ($0 < a < 1$). Since a is the factor by which P changes when t is increased by 1, large values of a mean fast growth; values of a near 0 mean fast decay. (See Figures 1.63 and 1.64.)

Alternative Formula for the Exponential Function

Exponential growth is often described in terms of percentages. For example, the population of Mexico is growing at a rate of 2.6% per year; in other words, the growth factor is $a = 1.026$. Similarly, each foot of clay removes 20% of the pollution from jet fuel, so the decay factor is $a = 1 - 0.20 = 0.8$. In general, the following formulas apply.

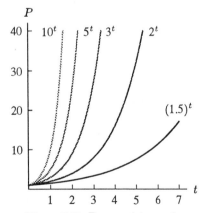

Figure 1.63: Exponential growth: $P = a^t, a > 1$

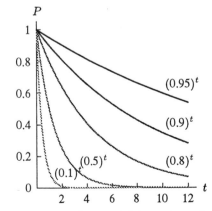

Figure 1.64: Exponential decay: $P = a^t$, $0 < a < 1$

If r is the *growth* rate, then $a = 1 + r$, and

$$P = P_0 a^t = P_0 (1 + r)^t.$$

If r is the *decay* rate, then $a = 1 - r$, and

$$P = P_0 a^t = P_0 (1 - r)^t.$$

Note, for example, that $r = 0.20$ when the percentage decay rate is 20%.

Absolute Rate of Change versus Relative Rate of Change

The population, P, in millions, of Nicaragua was 3.6 million in 1990 and growing at a rate of 3.4% per year. The 3.4% is a rate of change but it is not the same as the average rate of change we discussed earlier. The average rate of change is in absolute terms and is sometimes called the absolute rate of change to emphasize this fact. It tells how many people the population increased by during a certain time interval. The 3.4% is a relative rate of change: it tells the *percent* increase in the population. The percent increase is the absolute increase divided by the population.

If $P = f(t)$, then we have

$$\text{Average (absolute) rate of change of } P \text{ between } t = a \text{ and } t = b = \frac{\Delta P}{\Delta t} = \frac{f(b) - f(a)}{b - a}$$

$$\text{Relative rate of change of } P \text{ between } t = a \text{ and } t = b = \frac{1}{P}\frac{\Delta P}{\Delta t} = \frac{1}{f(a)}\left(\frac{f(b) - f(a)}{b - a}\right)$$

The absolute and relative rates of change for a time interval are sometimes referred to as the absolute and relative growth rates for the interval. For an exponential growth function, the relative rate of change is constant for all time intervals of the same length, but the absolute rate of change increases as the population increases.

Example 6   The population, P, in millions, of Nicaragua was 3.6 million in 1990 and growing at a rate of 3.4% per year.

(a) Write a formula for P as a function of t, where t is years since 1990.
(b) Use your formula to find the average rate of change (or absolute growth rate) in Nicaragua between 1990 and 1991, and between 1991 and 1992. Explain why your answers are different.
(c) Use your answers to part (b) to find the relative rate of change (or relative growth rate) over both time intervals. The relative growth rate should be 3.4% both times.

Solution   (a) Population growth is exponential and we use $P = P_0 a^t$. In 1990 (at $t = 0$), we know $P = 3.6$, so $P_0 = 3.6$. Nicaragua is growing at a rate of 3.4%, so $a = 1.034$. We have

$$P = 3.6(1.034)^t.$$

(b)   We first find the average rate of change between $t = 0$ (1990) and $t = 1$ (1991). At $t = 0$, we know $P = 3.6$ million. At $t = 1$, we find that $P = 3.6(1.034)^t = 3.7224$ million. We have

$$\text{Average rate of change} = \frac{f(1) - f(0)}{1 - 0} = \frac{3.7224 - 3.6}{1} = 0.1224.$$

Between 1990 and 1991, the population of Nicaragua grew by 0.1224 million people per year, or 122,400 people per year.

When $t = 2$ (1992), we have $P = 3.6(1.034)^2 = 3.8490$. Between 1991 and 1992, we have

$$\text{Average rate of change} = \frac{f(2) - f(1)}{2 - 1} = \frac{3.8490 - 3.7224}{1} = 0.1266.$$

Between 1990 and 1991, the population of Nicaragua grew by 0.1266 million people per year, or 126,600 people per year. Notice that the population increase was greater between 1991 and 1992. This is because the population was greater in 1991 than in 1990 and so there were more people to have babies.

(c)   The relative rate of change between 1990 and 1991 is the absolute change divided by the population in 1990, which is $0.1224/3.6 = 0.034$. The relative rate of change is 3.4%, as we expected. Between 1991 and 1992, the relative rate of change is $0.1266/3.7224 = 0.034$. Over small intervals, the relative rate of change will always be approximately 3.4%.

Example 7   A population has size 100 at time $t = 0$, where t is measured in years.

(a)   If the population has a constant absolute growth rate of 10 people per year, find a formula for the size of the population at time t.

(b)   If the population has a constant relative growth rate of 10% per year for intervals of 1 year, find a formula for the size of the population at time t.

(c)   Graph both functions on the same axes.

Solution   (a)   Since the absolute growth rate for one year periods is constant, the function is linear, and the formula is

$$P = 100 + 10t.$$

(b)   Since the relative growth rate for one year periods is constant, the function is exponential, and the formula is

$$P = 100(1.10)^t.$$

(c)   See Figure 1.65.

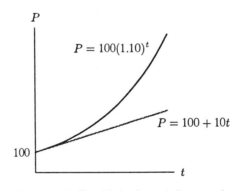

Figure 1.65: Two kinds of population growth

Problems for Section 1.5

1. The number of cancer cells in a tumor grows slowly at first but then grows with increasing rapidity. Draw a possible graph of the number of cancer cells against time.

2. Each year the world's annual consumption of electricity rises. In addition, each year the increase in annual consumption also rises. Sketch a possible graph of the annual world consumption of electricity as a function of time.

3. A drug is injected into a patient's bloodstream over a five-minute interval. During this time, the quantity in the blood increases linearly. After five minutes the injection is discontinued, and the quantity then decays exponentially. Sketch a graph of the quantity versus time.

4. When there are no other steroid hormones (for example, estrogen) in a cell, the rate at which steroid hormones diffuse into the cell is fast. The rate slows down as the amount in the cell builds up. Sketch a possible graph of the quantity of steroid hormone in the cell against time, assuming that initially there are no steroid hormones in the cell.

5. Each of the functions in Table 1.29 is increasing, but each increases in a different way. Which of the graphs in Figure 1.66 below best fits each function?

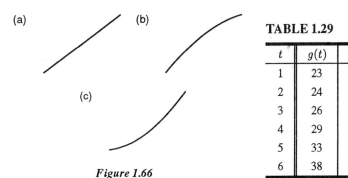

Figure 1.66

TABLE 1.29

t	$g(t)$	$h(t)$	$k(t)$
1	23	10	2.2
2	24	20	2.5
3	26	29	2.8
4	29	37	3.1
5	33	44	3.4
6	38	50	3.7

6. Each of the functions in Table 1.30 decreases, but each decreases in a different way. Which of the graphs in Figure 1.67 below best fits each function?

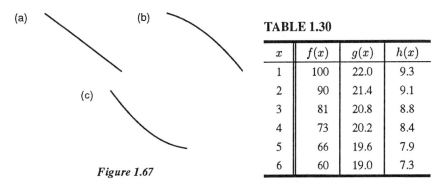

Figure 1.67

TABLE 1.30

x	$f(x)$	$g(x)$	$h(x)$
1	100	22.0	9.3
2	90	21.4	9.1
3	81	20.8	8.8
4	73	20.2	8.4
5	66	19.6	7.9
6	60	19.0	7.3

7. Table 1.31 gives some values of a function $w = f(t)$. Is this function increasing or decreasing? Is the graph of this function concave up or concave down?

TABLE 1.31

t	0	4	8	12	16	20	24
w	100	58	32	24	20	18	17

8. Match up the functions $h(s)$, $f(s)$, and $g(s)$, whose values are in Table 1.32, with the formulas

   $$y = a(1.1)^s \ , \ y = b(1.05)^s \ , \ y = c(1.03)^s,$$

   assuming a, b, and c are constants. Note that the function values have been rounded to two decimal places.

   TABLE 1.32

s	$h(s)$	s	$f(s)$	s	$g(s)$
2	1.06	1	2.20	3	3.47
3	1.09	2	2.42	4	3.65
4	1.13	3	2.66	5	3.83
5	1.16	4	2.93	6	4.02
6	1.19	5	3.22	7	4.22

For Problems 9–10, find a possible formula for the function represented by the data.

9.

x	0	1	2	3
$f(x)$	4.30	6.02	8.43	11.80

10.

t	0	1	2	3
$g(t)$	5.50	4.40	3.52	2.82

11. Determine whether each of the following tables of values could correspond to a linear function, an exponential function, or neither. For each table of values that could correspond to a linear or an exponential function, find a formula for the function.

(a)

x	$f(x)$
0	10.5
1	12.7
2	18.9
3	36.7

(b)

t	$s(t)$
−1	50.2
0	30.12
1	18.072
2	10.8432

(c)

u	$g(u)$
0	27
2	24
4	21
6	18

12. Total per capita health expenditures in the United States for the years 1970 to 1982 are given in Table 1.33.[10]

   (a) Check to see that the increase in health expenditures is approximately exponential. Show your work.

   (b) Estimate the doubling time of US per capita health expenditures.

   TABLE 1.33

Year	Yearly Health Expenditures ($ per capita)
1970	349
1972	428
1974	521
1976	665
1978	822
1980	1054
1982	1348

[10]*Statistical Abstracts of the US 1988*, 86, Table 129.

13. A population is currently 200 and is growing at 5% per year.

    (a) Write a formula for the population, P, as a function of time, t, in years.
    (b) Sketch a graph of P against t.
    (c) Use your graph to estimate the population 10 years from now.
    (d) Use your graph to estimate the doubling time of the population.

14. The half-life of nicotine in the blood is about 2 hours. After smoking a typical cigarette, a person will have absorbed about 0.4 mg of nicotine into the bloodstream. Fill in Table 1.34 to show the nicotine concentration remaining in the blood after t hours. Use your table to estimate the length of time until the nicotine concentration is reduced to about 0.04 mg.

    TABLE 1.34

t (hours)	0	2	4	6	8	10
Nicotine (mg)	0.4					

15. One of the main contaminants of a nuclear accident, such as that at Chernobyl, is strontium-90, which decays exponentially at a rate of approximately 2.5% per year.

    (a) Write the percent of strontium-90 remaining, P, as a function of years, t, since the nuclear accident. (Notice that at time $t = 0$, there is 100% of the contaminant remaining.)
    (b) Sketch the graph of P against t.
    (c) Use your graph to estimate the half-life of strontium-90.
    (d) Preliminary estimates after the Chernobyl disaster suggested that it would be about 100 years before the region would again be safe for human habitation. Use your graph to estimate the percent of original strontium-90 remaining at this time.

16. A population is known to be growing exponentially. Estimate the doubling time of the population shown by the graph in Figure 1.68, and verify graphically that the doubling time is independent of where you start on the graph.

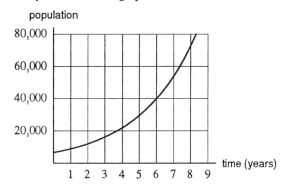

Figure 1.68

17. A certain region has a population of 10,000,000 and an annual growth rate of 2%. Estimate the doubling time by trial and error.

18. Assume that the median price, P, of a home rose from $50,000 in 1970 to $100,000 in 1990. Let t be the number of years since 1970.

    (a) Assume the increase in housing prices has been linear. Find an equation for the line representing price, P, in terms of t. Use this equation to complete column (a) of Table 1.35. Work with the price in units of $1000.

(b) If instead the housing prices have been rising exponentially, determine an equation of the form $P = P_0 a^t$ which would represent the change in housing prices from 1970–1990, and complete column (b) of Table 1.35.

(c) On the same set of axes, sketch the functions represented in column (a) and column (b) of Table 1.35.

TABLE 1.35

t	(a) Linear growth price in $1000 units	(b) Exponential growth price in $1000 units
0	50	50
10		
20	100	100
30		
40		

19. Owing to improved seed types and new agricultural techniques, the grain production of a region has been increasing. Over a 20-year period, annual production (in millions of tons) was as follows:

1970	1975	1980	1985	1990
5.35	5.90	6.49	7.05	7.64

At the same time the population (in millions) was:

1970	1975	1980	1985	1990
53.2	56.9	60.9	65.2	69.7

(a) Find a linear or exponential function which approximately fits each set of data. (Pick whichever type of function fits better.)

(b) If this region was self-supporting in this grain in 1970, was it self-supporting between 1970 and 1990? (Being self-supporting means that each person has enough of the grain.) How does the amount of grain each person has in later years compare?

(c) What are your predictions for the future if the trends continue?

20. The article "Job Scene: What's Hot, What's Not in the Next 10 Years", appearing in the *Chicago Tribune* in July 1996, contained the following quote about predicted change in jobs between 1995 and 2005:

• Employment is projected to increase to 144.7 million from 127 million – only **14** percent. The growth rate was **24** percent from 1983 to 1994.

• Jobs in services and retail trades are expected to increase by **16.2** million. Business services, health and education will account for **9.1** million new jobs.

- Manufacturing will lose **1.3** million jobs, a continuation of its decline.
- Jobs for those with master's degrees will grow the most, **28** percent.

(a) The first number in bold is 14. Is this an absolute change, a relative change, an absolute rate of change, or a relative rate of change? Use the information given to find each of the other three measures of change, corresponding to 14.

(b) Identify each of the numbers in bold as an absolute change, a relative change, an absolute rate of change, or a relative rate of change.

21. Suppose a town has a population of 1000. Fill in the values of the population in Table 1.36 if

(a) each year, the town has an absolute growth rate of 50 people per year.

(b) each year, the town has a relative growth rate of 5% per year.

TABLE 1.36

Year	0	1	2	3	4	...	10
Population	1000					...	

22. Figure 1.69 shows several possible graphs of a city's population against time. Match each description below to a graph and write a description to match each of the remaining graphs.

(a) The city's population increased at 5% per year.

(b) The city's population increased at 8% per year.

(c) The city's population increased by 5000 people per year.

(d) The city's population was stable.

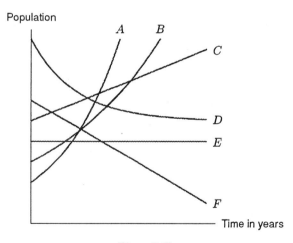

Figure 1.69

23. In 1994, the world's population was 5.6 billion and was growing at a rate of about 1.2% per year.

(a) Write a formula for the world's population as a function of time, t, in years since 1994.

(b) Predict the average rate of change in the world's population between 1994 and 2000. Give units with your answer.

(c) Predict the average rate of change in the world's population between 2010 and 2020. Give units with your answer.

Countries with very high inflation rates often publish monthly rather than yearly inflation figures, because monthly figures are less alarming. Problems 24–25 involve such high rates, which are called *hyperinflation.*

24. In 1989, US inflation was 4.6% a year. In 1989 Argentina had an inflation rate of about 33% a month.

    (a) What is the yearly equivalent of Argentina's 33% monthly rate?
    (b) What is the monthly equivalent of the US 4.6% yearly rate?

25. Between December 1988 and December 1989, Brazil's inflation rate was 1290% a year. (This means that between 1988 and 1989, prices increased by a factor of $1 + 12.90 = 13.90$.)

    (a) What would an article which cost 1000 cruzados (the Brazilian currency unit) in 1988 cost in 1989?
    (b) What was Brazil's monthly inflation rate during this period?

1.6 POWER FUNCTIONS

We have seen functions which are increasing and functions which are decreasing. We have also seen functions whose graphs are concave up and functions whose graphs are concave down. Often, however, a quantity might be increasing for some values and decreasing for others, or its graph may be concave up for some values and concave down for others. One of the ways to model relationships such as these is to use a power function. The functions $y = x^2$ and $y = x^3$ are examples of power functions. The graph of $y = x^2$ is shown in Figure 1.70. Notice that it is decreasing for negative x and increasing for positive x. The graph of $y = x^3$ is shown in Figure 1.71. Notice that it is concave down for negative x and concave up for positive x.

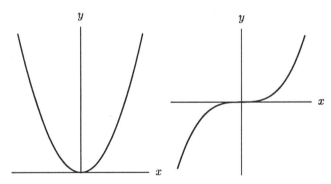

Figure 1.70: Graph of $y = x^2$ **Figure 1.71:** Graph of $y = x^3$

> In general, a **power function** has the form
> $$y = f(x) = kx^p$$
> where k and p are any constants.

In this section we will compare various power functions with one another and with the exponential functions. Since $y = f(x) = mx$ (with m a constant) is also a power function (because $x = x^1$, the first power), linear functions are included in the comparison too.

Positive Integral Powers: $y = x$, $y = x^2$, $y = x^3$, . . .

First, we'll look at functions of the form $f(x) = x^n$, with n a positive integer. Figures 1.72 and 1.73 show that the graphs of these functions fall into two groups: the odd powers and the even powers. All the odd powers (x, x^3, x^5, and so on) are increasing everywhere and their graphs are symmetric about the origin. All odd powers above $n = 1$ have a bend, or "seat," at the origin. The even powers, on the other hand, are first decreasing and then increasing, making them ∪-shaped with symmetry about the y-axis. The even powers are concave up everywhere, whereas the odd ones (greater than 1) are concave down for negative x and concave up for positive x. All odd and even powers, however, go through the points $(0,0)$ and $(1,1)$.

Figure 1.74 shows that the higher the power of x, the faster the function climbs. For large values of x (in fact, for all $x > 1$), $y = x^5$ is above $y = x^4$, which is above $y = x^3$, and so on. Not only are the higher powers larger, but they are *much* larger. This is because if $x = 100$, for example, 100^5 is one hundred times as big as 100^4 which is one hundred times as big as 100^3. As x gets larger (written as $x \to \infty$), any positive power of x completely swamps all lower powers of x. We say that, as $x \to \infty$, higher powers of x *dominate* lower powers.

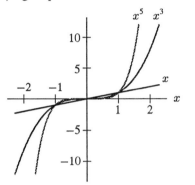

Figure 1.72: Odd powers of x *Figure 1.73:* Even powers of x

The close-up view near the origin in Figure 1.75 shows an entirely different story. For x between 0 and 1, the order is reversed: x^3 is bigger than x^4, which is bigger than x^5. (Try $x = 0.1$ to confirm this.) The fact that higher powers of x climb faster is true for large values of x but not for small ones. For big values of x, the highest powers are largest; for values of x near zero, smaller powers dominate.

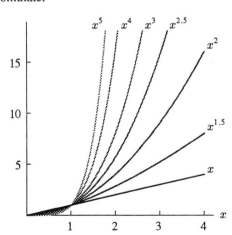

Figure 1.74: Powers of x: Which is largest for large values of x?

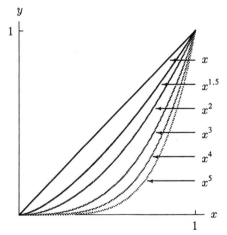

Figure 1.75: Between 0 and 1: Small powers of x dominate

Zero and Negative Integral Powers: $y = x^0$, $y = x^{-1}$, $y = x^{-2}$, . . .

The function $y = x^0 = 1$ has a graph that is a horizontal line. Rewriting

$$y = x^{-1} = \frac{1}{x} \quad \text{and} \quad y = x^{-2} = \frac{1}{x^2}$$

makes it easier to see that as x increases, the denominators increase and the functions decrease. The graphs of $y = x^{-1}$ and $y = x^{-2}$ have both the x- and y-axes as asymptotes. (See Figure 1.76.) For $x > 1$, the graph of $y = x^{-2}$ is below that of $y = x^{-1}$, and both must stay below $y = x^0 = 1$.

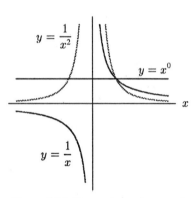

Figure 1.76: Comparison of zero and negative powers of x

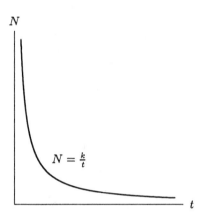

Figure 1.77: Number of nonsense syllables recalled, N, against time, t

---

Example 1 Psychologists test memory by presenting a person with a list of nonsense syllables (such as kab, nuv, mip) to memorize[11]. After the list has been removed, the psychologist will wait a certain amount of time and then ask the subject how many of the syllables he or she remembers. These experiments have shown that the number of nonsense syllables, N, that the subject can remember is a decreasing function of the time, t, since the list was removed. We can model this by saying that N is inversely proportional to t. Write a formula for N as a function of t and plot a graph of N against t.

Solution As more time elapses between the memorization and the recall, we expect subjects to remember fewer and fewer of the nonsense syllables, so it makes sense that N is a decreasing function of t. When we say N is inversely proportional to t, we are saying that N is proportional to the reciprocal of t. In other words,

$$N = k\left(\frac{1}{t}\right)$$

with k positive. Notice that, as we expect, when t is small, N is large, and when t is large, N is small. For any (positive) value of k, the graph has the shape shown in Figure 1.77. Both axes are asymptotes, showing that as the time tends to infinity, the number remembered tends to zero. The power function $N = k(1/t) = kt^{-1}$ differs from exponential decay in that it is undefined for $t = 0$, so this graph does not cross the vertical axis. In addition, it approaches the horizontal axis more slowly than the exponential decay function.

---

[11] A. Searleman and D. Herrmann, *Memory from a Broader Perspective*, (New York: McGraw-Hill, 1994)

Positive Fractional Powers: $y = x^{1/2}$, $y = x^{1/3}$, $y = x^{3/2}$, ...

The function giving the side of a square, s, in terms of its area, A, involves a root, or fractional power:

$$s = \sqrt{A} = A^{1/2}.$$

Similarly, the equation relating the average number of species found on an island and the size of the island involves a fractional power. If N is the number of species and A is the area of the island, observations have shown[12] that approximately

$$N = k\sqrt[3]{A} = kA^{1/3}$$

where k is a constant depending on the region of the world in which the island is found.

We will now look at functions of the form $y = x^{m/n} = \sqrt[n]{x^m}$. Since some fractional powers such as $x^{1/2}$ involve roots and are defined only for positive x and 0, we frequently restrict the domain of positive fractional powers of x to $x \geq 0$. Many calculators will not allow you to raise a negative number to a fractional power.

Figure 1.78 shows that for large x (in fact, all $x > 1$), the graph of $y = x^{1/2}$ is below the graph of $y = x$, and $y = x^{1/3}$ is below $y = x^{1/2}$. This is reasonable since, for example, $10^{1/2} = \sqrt{10} \approx 3.16$ and $10^{1/3} = \sqrt[3]{10} \approx 2.15$, so $10^{1/3} < 10^{1/2} < 10$. Between $x = 0$ and $x = 1$, the situation is reversed, and $y = x^{1/3}$ is on top. (Why?) Not surprisingly, $y = x^{3/2}$ is between $y = x$ and $y = x^2$ for all x.

The other important feature to notice about the graphs of $y = x^{1/2}$ and $y = x^{1/3}$ is that they bend in a direction opposite to that of the graphs of $y = x^2$ and x^3. For example, the graph of $y = x^2$ is climbing faster and faster as x increases; it is concave up. On the other hand, the graphs of $y = x^{1/2}$ and $y = x^{1/3}$ are climbing slower and slower; they are concave down. Despite this, all these functions do become infinitely large as x increases.

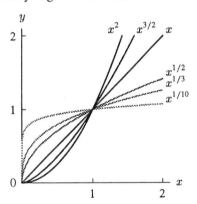

Figure 1.78: Comparison of some
fractional powers of x

What Effect Do Coefficients Have?

We know that $x^2 < x^3$ for all $x > 1$. But which is larger, $50x^2$ or x^3? Eventually, $50x^2 < x^3$ too. In fact, $50x^2 < x^3$ for all $x > 50$. (See Figure 1.79.) The graphs of $y = x^2$ and $y = x^3$ cross at $x = 1$, whereas graphs of $y = 50x^2$ and $y = x^3$ cross at $x = 50$. Thus, the effect of the factor of 50 is to change the point at which the graphs cross. However, x^3 ends up on top in both cases: provided the coefficients are positive, as $x \to \infty$, the higher power is always larger eventually.

---

[12]*Scientific American* (September 1989) 112.

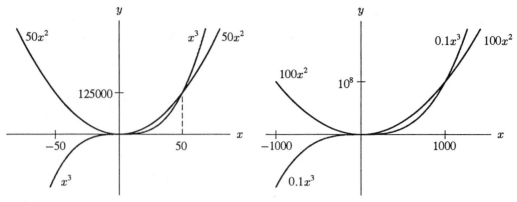

Figure 1.79: Graph of $y = x^3$ lies above graph of $y = 50x^2$ for large positive x

Figure 1.80: For large positive x, $y = 0.1x^3$ dominates $y = 100x^2$

What if the coefficient is negative? A negative coefficient reflects the graph about the x-axis, as we see in Figure 1.81.

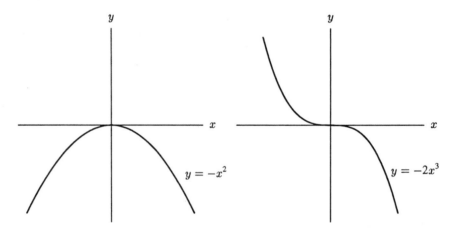

Figure 1.81: The effect of a negative coefficient

Example 2   Which of $y = 100x^2$ and $y = 0.1x^3$ is larger as $x \to \infty$?

Solution   Since $x \to \infty$, we are looking at large positive values of x, where the higher power will eventually be larger, or dominate. Thus, $y = 0.1x^3$ will be larger. (See Figure 1.80.)

Example 3   Sketch a global view of $f(x) = -x^3$, $g(x) = 40x^4$, and $h(x) = -0.1x^5$. Which function has the largest positive values as $x \to \infty$? Which function has the largest positive values as $x \to -\infty$ (i.e., as x gets more and more negative)?

Solution   As $x \to \infty$, $g(x) = 40x^4$ is the only function which is positive. As $x \to -\infty$, the graph of $h(x) = -0.1x^5$ is eventually (for $x < -400$) above the graph of the other functions, so $h(x)$ has the largest positive values as $x \to -\infty$ (See Figure 1.82.) Notice that, for large x, the values of $f(x) = -x^3$ are so much smaller in magnitude than the values of the other functions that the graph of $f(x)$ cannot be seen in the far-away view.

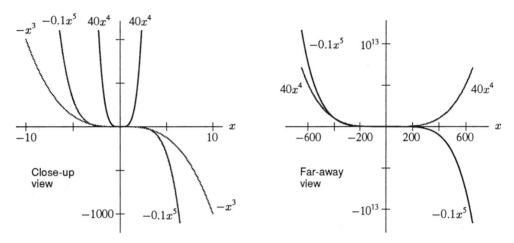

Figure 1.82: As $x \to \infty$, $g(x) = 40x^4$ dominates; as $x \to -\infty$, $h(x) = -0.1x^5$ dominates

Exponentials and Power Functions: Which Dominate?

In everyday language, *exponential* is often used to imply very fast growth. But do exponential functions always grow faster than power functions?

Let's consider $y = 2^x$ and $y = x^3$. The close-up, or local, view in Figure 1.83(a) shows that between $x = 2$ and $x = 5$, the graph of $y = 2^x$ lies below the graph of $y = x^3$. But the more global, or far-away, view in Figure 1.83(b) shows that the exponential function $y = 2^x$ eventually overtakes $y = x^3$. And Figure 1.83(c), which gives a very far-away, or global, view, shows that, for large x, x^3 is insignificant compared to 2^x. Indeed, 2^x is growing so much faster than x^3 that its graph appears almost vertical—in comparison to the more leisurely climb of x^3.

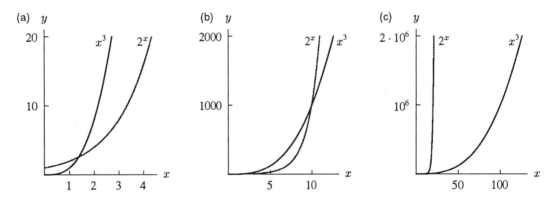

Figure 1.83: Comparison of $y = 2^x$ and $y = x^3$: $y = 2^x$ eventually dominates $y = x^3$

In fact, *every* exponential growth function eventually dominates *every* power function. Although an exponential function may be below a power function for some values of x, if you look at large enough x-values, a^x (with $a > 1$) will eventually dominate x^n, no matter what n is. Two more examples are presented in Figure 1.84 and Table 1.37.

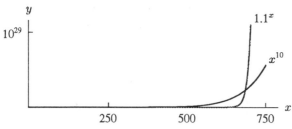

Figure 1.84: Exponential function eventually dominates power function

TABLE 1.37   *Comparison of x^{100} and 1.01^x*

x	x^{100}	1.01^x
10^4	10^{400}	$1.6 \cdot 10^{43}$
10^5	10^{500}	$1.4 \cdot 10^{432}$
10^6	10^{600}	$2.4 \cdot 10^{4321}$

Definition and Properties of Exponents

Below we list the definitions and rules that are used to manipulate exponents.

Definition of Zero, Negative, and Fractional Exponents

$$a^0 = 1, \quad a^{-1} = \frac{1}{a}, \quad \text{and, in general, } a^{-x} = \frac{1}{a^x}$$

$$a^{1/2} = \sqrt{a}, \quad a^{1/3} = \sqrt[3]{a}, \quad \text{and, in general, } a^{1/n} = \sqrt[n]{a}.$$

Rules for Computing Using Exponents

1. $a^x \cdot a^t = a^{x+t}$ For example, $2^4 \cdot 2^3 = (2 \cdot 2 \cdot 2 \cdot 2) \cdot (2 \cdot 2 \cdot 2) = 2^7$.

2. $\dfrac{a^x}{a^t} = a^{x-t}$ For example, $\dfrac{2^4}{2^3} = \dfrac{2 \cdot 2 \cdot 2 \cdot 2}{2 \cdot 2 \cdot 2} = 2^1$.

3. $(a^x)^t = a^{xt}$ For example, $(2^3)^2 = 2^3 \cdot 2^3 = 2^6$.

Problems for Section 1.6

1. Sketch graphs of $y = x^{1/2}$ and $y = x^{2/3}$ on the same axes. Which function has larger values as $x \to \infty$?

2. What happens to the value of $y = x^4$ as $x \to \infty$? As $x \to -\infty$?

3. What happens to the value of $y = -x^7$ as $x \to \infty$? As $x \to -\infty$?

4. Sketch a graph of $y = x^{-4}$.

5. On a graphing calculator or computer, plot graphs of the following functions, first for $-5 \leq x \leq 5$, $-100 \leq y \leq 100$, and then for $-1.2 \leq x \leq 1.2$, $-2 \leq y \leq 2$.
   (a) $y = x, y = x^3, y = x^6, y = x^9$
   (b) $y = x, y = x^4, y = x^7, y = x^{10}$
   Observe the general shape of these functions: Do the odd powers have the same general shape? What about the even powers? Which function is largest in magnitude for big x? For x near 0? Is this what you expected?

6. Global pictures of the three functions $y = x^5, y = 100x^2$, and $y = 3^x$ are shown in Figure 1.85. Which function corresponds to which curve?

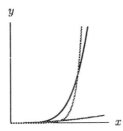

Figure 1.85

7. Table 1.38 shows the average fuel efficiency (number of miles per gallon) of all U.S. automobiles in the years shown.[13] If we modeled this data with a smooth curve, discuss the shape of the curve. Which of the following six functions does the shape most closely resemble: a linear function, an exponential function, or one of the following four power functions, x^2, $-x^2$, x^3, or $-x^3$?

   TABLE 1.38

Year	1940	1950	1960	1970	1980	1986
MPG	14.8	13.9	13.4	13.5	15.5	18.3

8. Recall that the cost curve for a company gives the cost, C, to produce a certain quantity, q. A cost curve will be increasing (since it costs more to produce more items) but most cost curves are not perfectly linear. Usually costs rise steeply for small quantities and then tend to level off at higher quantities. However, at very large quantities, the cost will again rise steeply. Sketch a graph of a cost curve with these properties. Which of the following six functions does the shape of your graph most closely resemble: a linear function, an exponential function, or one of the following four power functions, x^2, $-x^2$, x^3 or $-x^3$?

9. Find the average rate of change between $x = 0$ and $x = 10$ of each of the following functions: $y = x, y = x^2, y = x^3$, and $y = x^4$. Which has the largest average rate of change? Sketch graphs of the four functions, and draw lines whose slopes represent these average rates of change.

10. Simplify each of the following: (a) $8^{2/3}$ (b) $9^{-3/2}$

11. Do some calculations using specific values of x to verify that $y = x^{1/3}$ is above $y = x^{1/2}$ and that $y = x^{1/2}$ is above $y = x$ for $0 < x < 1$.

12. (a) Use a graphing calculator (or a computer) to plot the graphs of x^3, x^4, and x^5 on the interval $-0.1 \leq x \leq 0.1$. Determine an appropriate range for y so that all powers will be distinguishable in the viewing rectangle.
    (b) Plot the same graphs for $-100 \leq x \leq 100$, and determine an appropriate range for y.

[13]C. Schaufele and N. Zumoff, *Earth Algebra, Preliminary Version*, (New York: Harper Collins, 1993) 91

13. By hand, sketch global graphs of $f(x) = x^3$ and $g(x) = 20x^2$ on the same axes. Which function has larger values as $x \to \infty$?

14. By hand, sketch graphs of $f(x) = x^5$, $g(x) = -x^3$, and $h(x) = 5x^2$ on the same axes. Which has the largest positive values as $x \to \infty$? As $x \to -\infty$?

15. By trial and error, use a calculator to find to two decimal places the point near $x = 10$ at which $y = 2^x$ and $y = x^3$ cross.

16. Use a graphing calculator to find the point(s) of intersection of the graphs of $y = (1.06)^x$ and $y = 1 + x$.

17. For what values of x is $4^x > x^4$?

18. For what values of x is $3^x > x^3$? (Note: You will need to think about how to deal with the fact that the graphs of 3^x and x^3 are relatively close together for values of x near 3.)

19. According to the April 1991 issue of *Car and Driver*, an Alfa Romeo going at 70 mph requires 177 feet to stop. Assuming that the stopping distance is proportional to the square of velocity, find the stopping distances required by an Alfa Romeo going at 35 mph and at 140 mph (its top speed).

20. Suppose that the demand equation for a product shows that the quantity demanded is inversely proportional to the price of the product.

    (a) Sketch a graph of this demand function.
    (b) There should be no vertical intercept on your graph. What does this tell you about the relationship between price and quantity?
    (c) There should also be no horizontal intercept on your graph. What does this tell you about the relationship between price and quantity?

21. Values of three functions are given in Table 1.39. One function is of the form $y = ab^t$, one is of the form $y = at^2$, and one is of the form $y = bt^3$. Which function is which?

    TABLE 1.39

t	$f(t)$	t	$g(t)$	t	$h(t)$
2.0	4.40	1.0	3.00	0.0	2.04
2.2	5.32	1.2	5.18	1.0	3.06
2.4	6.34	1.4	8.23	2.0	4.59
2.6	7.44	1.6	12.29	3.0	6.89
2.8	8.62	1.8	17.50	4.0	10.33
3.0	9.90	2.0	24.00	5.0	15.49

22. Use a graphing calculator (or a computer) to graph $y = x^4$ and $y = 3^x$. Determine the appropriate domains and ranges that will give each of the graphs in Figure 1.86.

    (a) 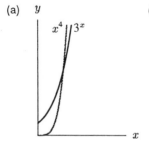   (b) 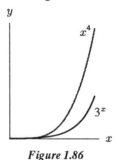   (c)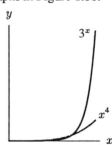

    Figure 1.86

1.7   FITTING FORMULAS TO DATA

We have seen many formulas in this chapter, and you may wonder where these formulas come from. For example, how does a company determine the demand curve for a product? Some of the formulas we have seen are exact. If you deposit $1000 in the bank and it earns 5% interest each year, then the amount in the bank at time t is exactly given by $P(t) = 1000(1.05)^t$. Most of the time, however, the formulas we use are approximations, often constructed from tables of data. Finding such a formula is often the first step in constructing a mathematical model.

A company wants to understand the relationship between the amount spent on advertising, a, in thousands of dollars, and the total sales, S, in thousands of dollars. The company has been experimenting and has collected data that shows, for example, that when they spend $3000 on advertising, total sales are $100,000. All of this data is shown in Table 1.40.

TABLE 1.40   *Find a formula to fit this data*

a(advertising in $1000s)	3	4	5	6
S(sales in $1000s)	100	120	140	160

You can see that the data in Table 1.40 is linear and so we got lucky this time: it is easy to find a formula to fit the data. The slope of the line is 20 and we can determine that the vertical intercept is 40, so the line is

$$S = 40 + 20a.$$

Now suppose that the data collected by the company is shown in Table 1.41. We are not so lucky in this case: the data is not linear. In general, it is a very difficult problem to find a formula to fit the data exactly. We must be satisfied with a formula that is close to the data.

TABLE 1.41   *Find a formula to approximate this data*

a(advertising in $1000s)	3	4	5	6
S(sales in $1000s)	105	117	141	152

Fitting a Linear Function To Data

The data in Table 1.41 has been plotted in Figure 1.87. The relationship is not linear, since not all the data points fall on one line, but we see that it is nearly linear. It seems reasonable to approximate this relationship with a line. One line that fits the data reasonably well is

$$S = 40 + 20a.$$

Figure 1.88 shows this line and the data.

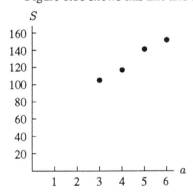

Figure 1.87: The sales data from Table 1.41

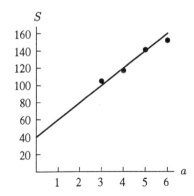

Figure 1.88: How well does this line fit the data?

The Regression Line

Is there a line that fits the data better than the one in Figure 1.88? If so, how do we find it? The process of fitting a line to a set of data is called *linear regression*, and the line of best fit is called the *regression line*. Fortunately, many calculators and computer programs will give you the regression line directly if you enter the data points.

For the data in Table 1.41, the regression line is

$$S = 54.5 + 16.5a.$$

This line fits the data better than any other line. See Figure 1.89. If you have technology available that will find a regression line, you should learn how to use it.

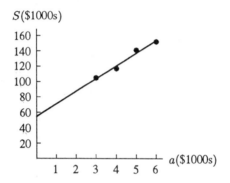

Figure 1.89: The regression line

Using the Regression Line to Make Predictions

Now that we have a formula for sales, we can use it to make predictions. Suppose that we want to predict total sales if we spend $3500 on advertising. We substitute $a = 3.5$ into the regression line:

$$S = 54.5 + 16.5(3.5) = 112.25.$$

The regression line predicts sales of $112,250. To see that this is a reasonable estimate, we can compare it to the entries in Table 1.41. When $a = 3$, we have $S = 105$, and when $a = 4$, we have $S = 117$. Our predicted sales of $S = 112.25$ when $a = 3.5$ makes sense because it falls between 105 and 117. Figure 1.90 shows where our predicted point is in relation to the data points. Of course, if we spent $3500 on advertising, sales would probably not be exactly $112,250. The regression equation allows us to make predictions, but does not provide exact results.

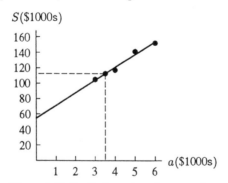

Figure 1.90: Predicting sales when spending $3500 on advertising

Example 1   Predict total sales given advertising expenditures of $4800 and $10,000.

Solution   When $4800 is spent on advertising, $a = 4.8$, and so we have

$$S = 54.5 + 16.5(4.8) = 133.7.$$

Sales are predicted to be $133,700. When $10,000 is spent on advertising, $a = 10$, and so we have

$$S = 54.5 + 16.5(10) = 219.5.$$

Sales are predicted to be $219,500.

Consider the two predictions made in Example 1 at $a = 4.8$ and $a = 10$. We should have much more confidence in the accuracy of the prediction when $a = 4.8$, because we are making a prediction on an interval we already know something about. The prediction for $a = 10$ is very questionable, since we are using the regression line beyond the limits of our knowledge from the data values in Table 1.41.

When we make a prediction between two data points, our estimate is called an *interpolation*. If instead we estimate the value of S for a value of a beyond those given in the data, our estimate is called an *extrapolation*. In general interpolation is safer than extrapolation.

Interpreting the Slope

The slope of a linear function is the change in the dependent variable divided by the change in the independent variable. In the regression line above, the slope is 16.5, and this tells us the change in sales divided by the change in advertising expenditures. Since 16.5 is the same as $\frac{16.5}{1}$, the slope tells us that S will go up by about 16.5 whenever a goes up by 1. If we increase advertising expenses by $1000, we expect sales to increase by about $16,500. In general, the slope tells you the expected change in the dependent variable given a unit change in the independent variable.

Example 2   A company has collected some data on the cost of producing different quantities of its product. The data is given in Table 1.42. Use a calculator or computer to find the regression line. Plot the points and the line to check your answer. Interpret the slope of the line. (If you do not have technology which can find a regression line, plot the points and visually estimate the regression line.)

TABLE 1.42   *Cost to produce different quantities*

q(quantity in units)	25	50	75	100	125
C(cost in dollars)	500	625	689	742	893

Solution   Using technology, we see that the regression line is

$$C = 418.9 + 3.612q.$$

The regression line is shown with the data in Figure 1.91, and it appears to fit the data well. The slope of the line is 3.612. We expect cost to increase about 3.612 dollars for every additional unit produced.

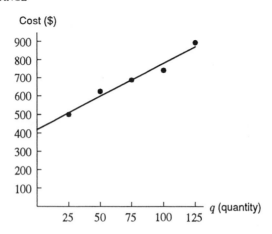

Figure 1.91: The regression line fits the data well

Regression When the Relationship Is Not Linear

Table 1.43 shows the number of cars imported into the United States from Japan between 1964 and 1971.[14] These points are plotted in Figure 1.92.

TABLE 1.43 *New passenger cars imported into the US from Japan*

Year since 1964	0	1	2	3	4	5	6	7
Cars	16,023	23,538	56,050	70,304	169,849	260,005	381,338	703,672

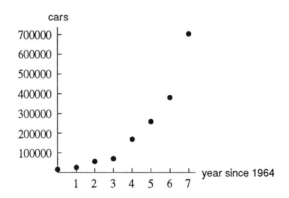

Figure 1.92: Is this data linear?

Does the data in Figure 1.92 look linear? Not really. In fact, it may remind you of an exponential function. With this data, it makes more sense to fit an exponential function than a linear function. Finding the exponential function of best fit is called *exponential regression*. Many calculators and

---

[14]*The World Almanac 1995*

computer software packages will give you the exponential regression function when you enter the data.

Using a graphing instrument, we learn that the best exponential function to fit the data in Table 1.43 is

$$C = 15{,}862.61 \cdot (1.7268)^t$$

where C is the number of imported Japanese cars and t is years since 1964. The graph of the data and this exponential function is given in Figure 1.93. It appears to fit the data very well.

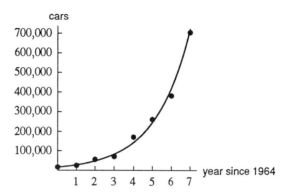

Figure 1.93: The exponential regression function

Since the base of this exponential function is 1.7268, we see that during this time period, sales of Japanese cars in the US were increasing at the rate of about 73% per year.

Calculators and computers will do linear regression, exponential regression, logarithmic regression, quadratic regression, and more. When you are trying to find the best formula for a set of data, the first step is to look at a graph of the data and try to identify the appropriate family of functions to use.

Problems for Section 1.7

1. Table 1.44 shows the Gross National Product of the United States for several different years, in constant 1982 dollars.

   TABLE 1.44

Year	1960	1970	1980	1989
GNP (in billions)	1665	2416	3187	4118

   (a) Plot these data, with GNP on the vertical axis. Does a line seem to fit the data reasonably well?
   (b) Find the regression line for this data, and graph it with the data.
   (c) Use the regression line to estimate the GNP in 1985. In 2000. Which estimate should you have more confidence in, and why?

2. The acidity of a solution is measured by its pH, with lower pH values indicating that the solution is more acidic. A study of acid rain was undertaken in Colorado between 1975 and 1978, in which the acidity of precipitation was measured for 150 consecutive weeks. The data followed a generally linear pattern, and the regression line was determined to be

    $$P = 5.43 - 0.0053t,$$

    where P is the pH of the rain and t is the number of weeks into the study.[15]

    (a) Is the pH level increasing or decreasing over the period of the study? What does this tell you about the level of acidity in the precipitation?
    (b) According to the line, what was the pH at the beginning of the study? At the end of the study ($t = 150$)?
    (c) What is the slope of the regression line? Explain what this slope is telling you about the change in pH over time.

3. In a study of 21 of the best American female runners, researchers measured the average stride rate, S, (number of steps per second) at different speeds, v, measured in feet per second. The data are given in Table 1.45.[16]

    TABLE 1.45

Speed (ft/sec)	15.86	16.88	17.50	18.62	19.97	21.06	22.11
Stride rate	3.05	3.12	3.17	3.25	3.36	3.46	3.55

    (a) Use a calculator or computer to find the regression line for these data, using stride rate as the dependent variable.
    (b) Plot the regression line and the data on the same axes. Does the line appear to fit the data well?
    (c) Use the regression line to predict the stride rate when the speed is 18 ft/sec. Use the regression line to predict the stride rate when the speed is 10 ft/sec. Which prediction should you have more confidence in? Why?

4. The level of carbon dioxide in the atmosphere has been increasing, due largely to increased energy consumption and to deforestation of the earth. The CO_2 concentration in parts per million (ppm) is given for four different years in Table 1.46.[17] The measurements were made at the Mauna Loa Observatory in Hawaii.

    TABLE 1.46

Year	1965	1970	1980	1988
CO_2 concentration (in ppm)	319.9	325.3	338.5	351.3

    (a) Plot these data. Do they look approximately linear?

---

[15] William M. Lewis and Michael C. Grant, "Acid Precipitation in the Western United States," *Science* 207 (1980) 176-177.

[16] R.C. Nelson, C.M. Brooks, and N.L. Pike, "Biomechanical Comparison of Male and Female Distance Runners." *The Marathon: Physiological, Medical, Epidemiological, and Psychological Studies*, ed. P. Milvy, (New York: New York Academy of Sciences, 1977) 793-807.

[17] Christopher Schaufele and Nancy Zumoff, *Earth Algebra*, (New York: Harper-Collins, 1993).

    (b)   Use a calculator or computer to find the regression line for these data. (If you do not have technology available, use your graph of the data to visually estimate the best line.)

    (c)   What CO_2 concentration does your line predict for the four years 1965, 1970, 1980, and 1988? Compare these predictions to the actual values shown in Table 1.46. Is the line relatively accurate for these four years?

    (d)   What carbon dioxide concentration does the line predict for the year 1995?

5.   In Problem 4, we modeled the increase in carbon dioxide concentration as a linear function of time. However, if we include data for carbon dioxide concentration from as far back as 1900, it appears that the data are more exponential than linear. (They looked linear in Problem 4 because we were only looking at a small piece of the graph.) If we use an exponential regression function to model CO_2 concentration since 1900, we obtain the function

$$C = 272.27(1.0026)^t,$$

where C is the CO_2 concentration in ppm, and t is in years since 1900.

    (a)   What is the annual growth rate during this period? Interpret this rate in terms of the increase in the CO_2 concentration.

    (b)   What CO_2 concentration is given by the model for 1900? For 1980? Compare the estimate for 1980 to the actual value given in Table 1.46.

6.   Climbing health care costs continue to be a concern. Table 1.47 shows the average yearly per-capita (i.e. per person) health care expenditures for various years. Does a linear or an exponential model appear to fit these data best? Find the linear or exponential regression function (whichever you decide is best) for these data. Graph the function with the data and assess how well it fits the data.

TABLE 1.47 *Health care costs*

Year	Per capita expenditure ($)
1970	349
1975	591
1980	1055
1985	1596
1987	1987[20]

7.   The number, N, of passenger cars in the United States (in millions) is shown in Table 1.48[21], where t is in years since 1940.

TABLE 1.48

t (years since 1940)	0	10	20	30	40	46
N (millions of cars)	27.5	40.3	61.7	89.3	121.6	135.4

    (a)   Plot the data, with number of passenger cars as the dependent variable.

---

[20]Yes, the last number is correct. The expenditures in 1987 were $1987 per person.
[21]*Statistical Abstracts of the United States*

(b)   Does a linear or exponential model appear to fit the data better?

(c)   We use a linear model first: Find the regression line for these data. Graph it with the data. Use the regression line to predict the number of passenger cars in the year 2000 ($t = 60$).

(d)   Interpret the slope of the regression line found in part (c) in terms of the number of passenger cars.

(e)   Now we try an exponential model: Find the exponential regression function for these data. Graph it with the data. Use the exponential function to predict the number of passenger cars in the year 2000 ($t = 60$). Compare your prediction with the prediction obtained from the linear model.

(f)   What annual rate of growth in number of US passenger cars does your exponential model show?

8.   Table 1.49 gives the population of the world in billions, in years since 1950.

TABLE 1.49

Year (since 1950)	0	10	20	30	40	44
World Population (in billions)	2.6	3.1	3.7	4.5	5.4	5.6

(a)   Plot these data. Does a linear or exponential model seem to fit the data best?

(b)   Use a calculator or computer to find the exponential regression function.

(c)   What annual growth rate does the exponential function show?

(d)   Use the exponential regression function to predict the population of the world in the year 2000. In the year 2050. Comment on the relative confidence you should have in these two estimates.

9.   All field goal attempts and successes were analyzed in the National Football League and American Football League in 1969, and the percentage of successes is shown in Table 1.50.

TABLE 1.50

X = yards from the goal line	14.5	24.5	34.5	44.5	52.0
Y = percentage of tries that were successful	0.90	0.75	0.54	0.29	0.15

(a)   Graph the data, treating the field goal success rate as the dependent variable, and discuss whether a linear or an exponential model provides the best fit.

(b)   Find the best linear regression function, and graph it with the data. Interpret the slope of the regression line in terms of success in kicking field goals and distance from the goal line.

(c)   Find the best exponential regression function, and graph it with the data. What success rate does this function predict from a distance of 50 yards?

(d)   Now that you have looked at the graphs, which model seems to fit the data best?

10.   In this chapter, we have developed "a library of functions". Any of these functions can be used to model data. We have discussed linear and exponential regression in this chapter, but it is also possible to fit a power function such as x^2 to data. Most graphing calculators will do regression using any of these families of functions. The job of the person analyzing the data, therefore, is to determine which family of functions is best for a given data set. Graphs of several different data sets are shown in Figure 1.94. In each case, indicate whether the best function for the data appears to be a *linear function*, an *exponential function*, or a *power function*.

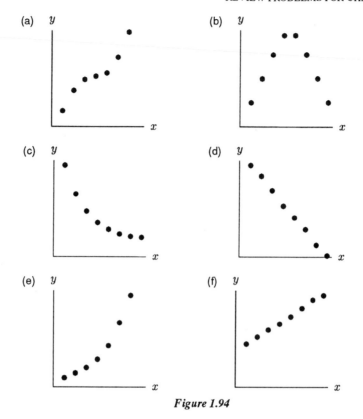

Figure 1.94

REVIEW PROBLEMS FOR CHAPTER ONE

1. A car starts out slowly and then goes faster and faster until a tire blows out. Sketch a possible graph of the distance the car has traveled as a function of time.

2. Having left home in a hurry, I'd only gone a short distance when I realized I hadn't turned off the washing machine, and so I went back to do so. I then set out again immediately. Sketch my distance from home as a function of time.

3. The graph in Figure 1.95 shows how the usage of household gas, e.g., for cooking, varies with the time of day in Ankara, the capital city of Turkey. Give a possible explanation for the shape of the graph.

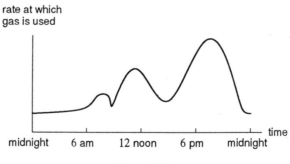

Figure 1.95

4. Sketch a possible graph for a function that is decreasing everywhere, concave up for negative x and concave down for positive x.

5. Consider the graph in Figure 1.96.

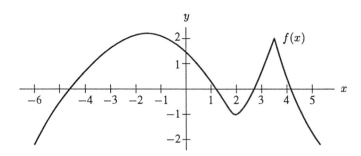

Figure 1.96

(a) How many zeros does this function have? Approximately where are they?
(b) Give approximate values for $f(2)$ and $f(4)$.
(c) Is the function increasing or decreasing near $x = -1$? How about near $x = 3$?
(d) Is the graph concave up or concave down near $x = 2$? How about near $x = -4$?
(e) List all intervals (approximately) on which the function is increasing.

6. Table 1.51 gives the average temperature in Wallingford, Connecticut, for the first 10 days in March 1990.

(a) Over which intervals was the average temperature increasing? Decreasing?
(b) Find a pair of consecutive intervals over which the average temperature was increasing at a decreasing rate. Find another pair of consecutive intervals over which the average temperature was increasing at an increasing rate.

TABLE 1.51

Date in March	1	2	3	4	5	6	7	8	9	10
Average temperature (°F)	42°	42°	34°	25°	22°	34°	38°	40°	49°	49°

7. For tax purposes, you may have to report the value of your assets, such as cars or refrigerators. The value you report depreciates, or drops, with time. The idea is that a car you originally paid $10,000 for may be worth only $5000 a few years later. The simplest way to calculate the value of your asset is using "straight-line depreciation," which assumes that the value is a linear function of time. If a $950 refrigerator depreciates completely in seven years, find a formula for its value as a function of time.

8. An airplane uses a fixed amount of fuel for takeoff, a (different) fixed amount for landing, and a fixed amount per mile when it is in the air. How does the total quantity of fuel required depend on the length of the trip? Write a formula for the function involved. Explain the meaning of the constants in your formula.

9. Sketch reasonable graphs for the following. Pay particular attention to the concavity of the graphs, and explain your reasoning.

(a) The total revenue generated by a car rental business, plotted against the amount spent on advertising.
(b) The temperature of a cup of hot coffee standing in a room, plotted as a function of time.

10. The six graphs in Figure 1.97 show frequently observed patterns of age-specific cancer incidence rates, in number of cases per 1000, as a function of age.[22]

    (a) For each of the six graphs, write a sentence explaining the effect of age on the cancer rate.

    (b) Which graph shows a relatively high incidence rate for children? Suggest a type of cancer that behaves this way.

    (c) Which graph shows a brief decrease in the incidence rate at around age 50? Suggest a type of cancer that might behave this way.

    (d) Which graph or graphs might represent a cancer that is caused by toxins which build up in the body over time (for example, lung cancer)? Explain.

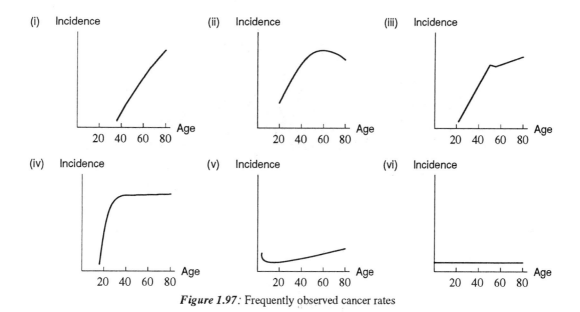

Figure 1.97: Frequently observed cancer rates

11. Table 1.52 shows the atmospheric concentration of carbon dioxide (in parts per million) between 1960 (the first year in which it was systematically measured) and 1990.[23]

    (a) Find the average rate of change of the concentration of carbon dioxide between 1960 and 1990. Give units with your answer and interpret it in terms of carbon dioxide.

    (b) Plot the data, and use technology to find the regression line for carbon dioxide concentration against year. (The slope should be approximately the same as the average rate of change found in part (a).) Use the regression line to predict the concentration of carbon dioxide in the atmosphere in the year 2000.

TABLE 1.52   *Carbon dioxide in the atmosphere*

Year	1960	1965	1970	1975	1980	1985	1990
Carbon Dioxide (ppm)	316.8	319.9	325.3	331.0	338.5	345.7	354.0

---

[22]Abraham M. Lilienfeld, *Foundations of Epidemiology*, (New York: Oxford University Press, 1976) 155.

[23]Lester R. Brown, et. al., *Vital Signs 1994*, (New York: W. W. Norton and Company, 1994) 67.

12. Table 1.53 shows the cumulative number of cases of AIDS worldwide (in thousands) between 1980 and 1993.[24]

    (a) Find the absolute increase in the number of AIDS cases between 1988 and 1989. Find the absolute increase between 1992 and 1993.

    (b) Find the relative increase in the number of AIDS cases between 1988 and 1989. Find the relative increase between 1992 and 1993.

    TABLE 1.53 *Cumulative cases of AIDS worldwide*

Year	AIDS cases (in thousands)	Year	AIDS Cases (in thousands)
1980	0	1987	493
1981	1	1988	813
1982	7	1989	1275
1983	21	1990	1892
1984	65	1991	2701
1985	143	1992	3657
1986	279	1993	4820

13. Sketch the graph of a function defined for $x \geq 0$ with all of the following properties. (There are lots of possible answers.)

    (a) $f(0) = 2$.
    (b) $f(x)$ is increasing for $0 \leq x < 1$.
    (c) $f(x)$ is decreasing for $1 < x < 3$.
    (d) $f(x)$ is increasing for $x > 3$.
    (e) $f(x) \to 5$ as $x \to \infty$.

14. When a new product is advertised, more and more people try it. However, the rate at which new people try it slows as time goes on.

    (a) Sketch a graph of the total number of people who have tried such a product against time.
    (b) What do you know about the concavity of the graph?

15. An amusement park operator has fixed costs of $5000 per day and variable costs averaging $2 per customer. The amusement park fee is $7.

    (a) How many customers does the park need in order to make a profit?
    (b) Find the cost and revenue functions and graph them on the same axes. Mark the break-even point on the graph.

16. Values of the functions $F(t)$, $G(t)$, and $H(t)$ are listed in the following table. Determine which one is concave up, which one is concave down, and which one is linear.

t	$F(t)$	$G(t)$	$H(t)$
10	15	15	15
20	22	18	17
30	28	21	20
40	33	24	24
50	37	27	29
60	40	30	35

---

[24] Lester R. Brown, et. al., *Vital Signs 1994*, (New York: W. W. Norton and Company, 1994) 103.

17. Since the opening up of the West, the US population has moved westward. To observe this, we look at the "population center" of the US, which is the point at which the country would balance if it were a flat plate with no weight, and every person had equal weight. In 1790 the population center was east of Baltimore, Maryland. It has been moving westward ever since, and in 1990 it crossed the Mississippi river to Steelville, Missouri (southwest of St. Louis). During the second half of this century, the population center has moved about 50 miles west every 10 years.

    (a) Express the approximate position of the population center as a function of time, measured in years from 1990. Measure position westward from Steelville, along the line running through Baltimore.

    (b) The distance from Baltimore to Steelville is a bit over 700 miles. Could the population center have been moving at roughly the same rate for the last two centuries?

    (c) Could the function in part (a) continue to apply for the next three centuries? Why or why not? [Hint: You may want to look at a map. Note that distances are in air miles and are not driving distances.]

18. Figure 1.98 shows the age-adjusted death rates (in deaths per 100,000) from different types of cancer among males in the United States, between 1930 and 1967.[25]

    (a) Discuss how the death rate has changed for the different types of cancers.

    (b) For which type of cancer has the average rate of change between 1930 and 1967 been the largest? Estimate the average rate of change for this cancer type. Interpret your answer.

    (c) For which type of cancer has the average rate of change between 1930 and 1967 been the most negative? Estimate the average rate of change for this cancer type. Interpret your answer.

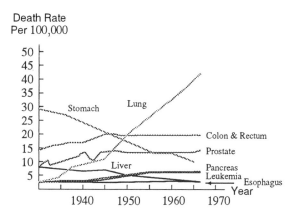

Figure 1.98: Cancer death rates

19. Each of the functions described by the data in Table 1.54 is increasing over its domain, but each increases in a different way. Which of the graphs in Figure 1.99 best fits each function?

[25]Abraham M. Lilienfeld, *Foundations of Epidemiology*, (New York; Oxford University Press, 1976) 67.

TABLE 1.54

x	f(x)		x	g(x)		x	h(x)
1	1		3.0	1		10	1
2	2		3.2	2		20	2
4	3		3.4	3		28	3
7	4		3.6	4		34	4
11	5		3.8	5		39	5
16	6		4.0	6		43	6
22	7		4.2	7		46.5	7
29	8		4.4	8		49	8
37	9		4.6	9		51	9
47	10		4.8	10		52	10

(a)   (b)   (c)

Figure 1.99

CHAPTER TWO

RATE OF CHANGE: THE DERIVATIVE

In Chapter 1, we studied the average rate of change of a function on an interval. In this chapter, we investigate the *instantaneous* rate of change of a function at a single point. How can we define the rate of change of a quantity at a single instant when, by focusing on the instant, we stop the change? The search for an answer to this question will lead us to the key concept of the *derivative*, which forms the basis for the study of calculus.

The derivative can be interpreted geometrically as the slope of a curve and physically as the rate of change at a point. Derivatives have a wide range of applications; they can be used to represent everything from fluctuations in interest rates to the rate at which fish are dying to the rate of the growth of a tumor.

2.1 INSTANTANEOUS RATE OF CHANGE

In Chapter 1, we looked at the average rate of change of a function over an interval. In this chapter, we consider the rate of change of a function at a single point. The rate of change at a single point is sometimes called the *instantaneous rate of change*, or simply, the *rate of change*. The notion of rate of change at a given instant is surprisingly subtle and difficult to define precisely.

We saw in Chapter 1 that the average rate of change of distance with respect to time is the average velocity. Therefore the instantaneous rate of change of distance will be the velocity at a single instant in time. What do we mean by this? Consider the statement "At the instant it crossed the finish line, the horse was traveling at 42 mph." How can such a claim be substantiated? A photograph taken at that instant will show the horse motionless—it is no help at all. There is some paradox in trying to quantify the property of motion at a particular instant in time, since by focusing on a single instant you stop the motion!

A similar difficulty arises whenever we attempt to measure the rate of change of anything—for example, oil leaking out of a damaged tanker. The statement "One hour after the ship's hull ruptured, oil was leaking at a rate of 200 barrels per second" seems not to make sense. You could argue that at any given instant *no* oil is leaking.

Problems of motion were of central concern to Zeno and other philosophers as early as the fifth century B.C. The modern approach, made famous by Newton's calculus, is to stop looking for a simple notion of speed at an instant, and instead to look at speed over small intervals containing the instant. This method sidesteps the philosophical problems mentioned earlier but brings new ones of its own.

We will begin our investigation by taking a closer look at velocity.

Instantaneous Velocity

If you drive for 4 hours and cover 200 miles, then your average velocity is $200/4 = 50$ miles per hour. Of course, this does not mean that you are traveling at exactly 50 mph the entire trip. Your instantaneous velocity at a given instant during the trip will be shown on your speedometer, and it is this quantity that we investigate now.

TABLE 2.1 *Height of the grapefruit above the ground*

t (sec)	0	1	2	3	4	5	6
y (feet)	6	90	142	162	150	106	30

We threw a grapefruit in the air in Chapter 1, and we return to an investigation of that grapefruit. Table 2.1 gives the height of the grapefruit at time t. Suppose we would like to determine the velocity of the grapefruit at, say, $t = 1$. How fast is the grapefruit going exactly one second after we let go? We can use average velocities to help us estimate this quantity.

We saw in Chapter 1 that the average velocity on the interval $0 \leq t \leq 1$ is 84 ft/sec, and the average velocity on the interval $1 \leq t \leq 2$ is 52 ft/sec. Notice that the average velocity before $t = 1$ is more than the average velocity after $t = 1$ since the grapefruit is slowing down. We would expect the velocity *at* $t = 1$ to be between these two average velocities. How can we find a more accurate measure of the velocity at *exactly* $t = 1$? We'll have to look at what happens near $t = 1$ in more detail. Suppose that we could measure the height of the grapefruit at any instant. Then we could find the average velocities on either side of $t = 1$ over smaller and smaller intervals. This is done in Figure 2.1.

We would expect to define the instantaneous velocity at $t = 1$ to be between average velocities on either side of $t = 1$. We see in Figure 2.1 that, as the size of the interval shrinks, the values of the velocity before $t = 1$ and the velocity after $t = 1$ get closer together. By the smallest interval in Figure 2.1, both velocities are 68.0 ft/sec (to one decimal place), so we will define the velocity at $t = 1$ to be 68.0 ft/sec (to one decimal place).

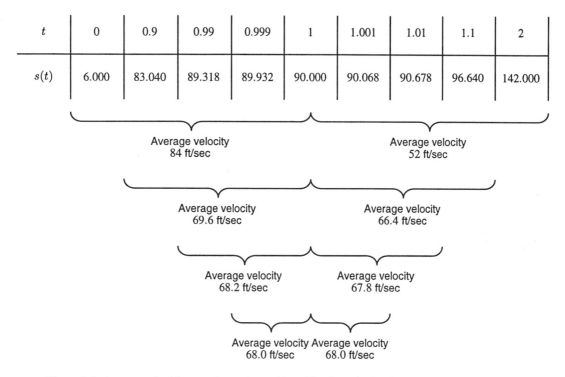

t	0	0.9	0.99	0.999	1	1.001	1.01	1.1	2
$s(t)$	6.000	83.040	89.318	89.932	90.000	90.068	90.678	96.640	142.000

Average velocity 84 ft/sec   Average velocity 52 ft/sec

Average velocity 69.6 ft/sec   Average velocity 66.4 ft/sec

Average velocity 68.2 ft/sec   Average velocity 67.8 ft/sec

Average velocity 68.0 ft/sec   Average velocity 68.0 ft/sec

Figure 2.1: Average velocities over intervals on either side of $t = 1$: showing successively smaller intervals

Of course, if we showed more decimal places, average velocities before and after $t = 1$ would no longer agree even in the smallest interval. To calculate the velocity at $t = 1$ to more decimal places, we would have to take smaller and smaller intervals on either side of $t = 1$ until the average velocities agree to the number of decimal places we want. The velocity at $t = 1$ is then defined to be this common average velocity.

Defining Instantaneous Velocity Using the Idea of a Limit

When you take smaller and smaller intervals near $t = 1$, it turns out that the average velocities for the grapefruit are always just above or just below 68 ft/sec. It seems natural, then, to define velocity at the instant $t = 1$ to be 68 ft/sec. This is called the *instantaneous velocity* at this point, and its definition depends on our being adequately convinced that smaller and smaller intervals will provide average speeds that come arbitrarily close to 68. Modern mathematics has a name for this process: it is called *taking the limit*.

> The **instantaneous velocity** of an object at time t is defined to be the limit of the average velocity of the object over shorter and shorter time intervals around t.

Make sure that you see how we have replaced the original difficulty of computing velocity at a point by a search for an argument to convince ourselves that the average velocities do approach a number as the time intervals shrink in size. In a sense, we have traded one hard question for another, since we don't yet have any idea how to be certain what number the average velocities are approaching. In our thought experiment, the number seems to be exactly 68, but what if it were 68.000001? How can we be sure that we have taken small enough intervals?

For most practical purposes, it is not likely to be important whether the velocity is exactly 68 or 68.000001. Showing that the limit is exactly 68 requires more precise knowledge of how the velocities were calculated and of the limiting process; we will see this in Chapter 5.

Instantaneous Rate of Change

We can define the instantaneous rate of change of any function $y = f(t)$ at a point $t = a$. We simply mimic what we did for velocity, namely, look at the average rate of change over smaller and smaller intervals.

> The **instantaneous rate of change** of f at a, also called the rate of change of f at a, is defined to be the limit of the average rates of change of f over shorter and shorter intervals around a.

Since the average rate of change equals a difference quotient, we see that instantaneous rate of change equals a limit of difference quotients. In practice, we often approximate the derivative by one of these difference quotients.

Example 1   The quantity (in mg) of a drug in the blood at time t (in minutes) is given by

$$Q(t) = 25(0.8)^t.$$

Estimate the rate of change of the quantity at $t = 3$ and interpret your answer.

Solution   We estimate the rate of change at $t = 3$ by computing the average rate of change over intervals near $t = 3$. If we use the intervals $2 \leq t \leq 3$ and $3 \leq t \leq 4$, we have

t	2	3	4
$Q(t)$	16.00	12.80	10.24

Average Rate of change
$$= \frac{12.80 - 16.0}{3 - 2}$$
$$= -3.20.$$

Average Rate of change
$$= \frac{10.24 - 12.80}{4 - 3}$$
$$= -2.56.$$

We estimate that the rate of change at $t = 3$ is between -3.20 and -2.56. We can make our estimate as accurate as we like by choosing our intervals small enough. Let's look at the average rate of change over the intervals $2.99 \leq t \leq 3$ and $3 \leq t \leq 3.01$:

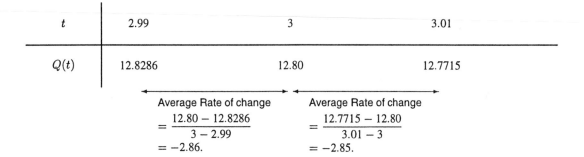

t	2.99	3	3.01
$Q(t)$	12.8286	12.80	12.7715

Average Rate of change
$$= \frac{12.80 - 12.8286}{3 - 2.99}$$
$$= -2.86.$$

Average Rate of change
$$= \frac{12.7715 - 12.80}{3.01 - 3}$$
$$= -2.85.$$

A reasonable estimate for the rate of change of the quantity at $t = 3$ is -2.855. The units are the units of Q over the units of t, or mg/minute. Since the rate of change is negative, the quantity of the drug is decreasing. At $t = 3$, the quantity of the drug in the body is changing at a rate of about -2.855 mg/minute.

Visualizing Rate of Change: Slope of Curve

We saw in the previous chapter that average rate of change of a function over an interval is represented graphically as the slope of the secant line to its graph over the interval. Figure 2.2 shows the average rate of change between a and b. The next question is how to visualize the rate of change at a point, say a. We took average rates of change across smaller and smaller intervals ending at the point a. As the length of the interval shrinks, the slope of the secant line gets closer to the slope of the graph at a. See Figure 2.3.

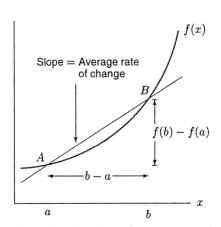

Figure 2.2: Visualizing the average rate of change of f between a and b

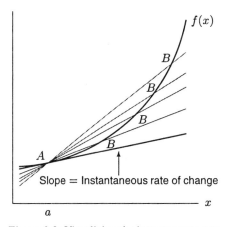

Figure 2.3: Visualizing the instantaneous rate of change of f at a

The cornerstone of the idea is the fact that, on a very small scale, the graphs of most functions look almost like straight lines. Imagine taking the graph of a function near a point and "zooming in" to get a close-up view. (See Figure 2.4.) The more you zoom in, the more the curve will appear to be a straight line. We call the slope of this line the *slope of the curve* at the point. Therefore, the slope of the magnified line is the instantaneous rate of change. Thus, we can say that

> The **instantaneous rate of change** of a function at a point is the slope of the graph at that point.

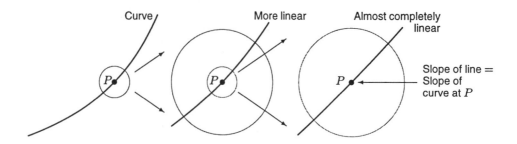

Figure 2.4: Finding the slope of a curve at a point by "zooming in"

Example 2   Let's throw the grapefruit in the air one more time. Figure 2.5 shows the height of the grapefruit plotted against time. (Note that this is not a picture of the grapefruit's path, which is straight up and down.) The points labeled A, B, C, D, E, F, G correspond to the values given in Table 2.1. Discuss when the velocity of the grapefruit is positive, when it is negative, and when it is zero.

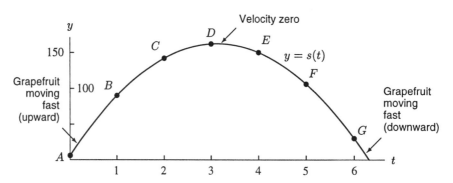

Figure 2.5: The height, y, of the grapefruit at time t

Solution   The velocity is given graphically by the slope of the curve in Figure 2.5. The slope is positive at points A, B, and C, and is negative at points E, F, and G. Since the curve is horizontal at point D, the slope is zero at this point.

The slope is positive for $t < 3$, so the velocity is positive for the first 3 seconds and the grapefruit is moving up. The slope is negative for $t > 3$, so the velocity is negative after the first three seconds and the grapefruit is moving down. The slope is zero at $t = 3$, and so the velocity is zero at $t = 3$. At the top of its path the grapefruit changes direction–from moving upward to moving downward–and at the moment it makes this change, the grapefruit's instantaneous velocity is zero.

Example 3   A car starts out slowly and then goes faster and faster. Sketch a graph of the distance the car has traveled as a function of time.

Solution   Since the car is moving, the distance it has traveled is an increasing function. The speed of the car is the slope of the distance graph. Therefore, the slope of the graph is small at first but increases, and so the graph gets steeper and steeper. One possible graph is shown in Figure 2.6.

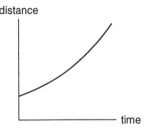

Figure 2.6: This car is speeding up

Estimating Instantaneous Rate of Change from a Table of Values

We have seen that the instantaneous rate of change of a function at a point is approximated by looking at the average rate of change over smaller and smaller intervals. Sometimes it is not possible to look at "smaller and smaller intervals"—this is the case, for example, if the function is given as a table of values such as that in Table 2.2. What is the best way to estimate the rate of change of this function at $t = 4$? We look at the average rate of change around $t = 4$ on the smallest interval possible. We could use the interval just to the left of the point (in this case, $t = 2$ to $t = 4$), or we could use the interval just to the right of the point (in this case, $t = 4$ to $t = 6$), or we could compute the average rate of change over both of these intervals and average the two answers. Each of these methods of approximating the rate of change at a point gives a reasonable answer. For simplicity, in this text, we will compute the average rate of change for the interval just to the right of the point in which we are interested.

TABLE 2.2   *Estimate the rate of change at t = 4*

t	0	2	4	6	8
y	54	75	90	102	110

For the function in Table 2.2, we have

$$\text{Instantaneous rate of change at } t = 4 \approx \frac{\Delta y}{\Delta t} = \frac{102 - 90}{6 - 4} = \frac{12}{2} = 6.$$

Example 4   Worldwide, the amount of land per person that is used for farming has been decreasing over time. Table 2.3 shows crop land (in acres per person) as a function of year[1].

TABLE 2.3   *Amount of land on earth, per person, used for farming*

Year	1800	1900	1950	1970
Crop land (acres/person)	3.51	2.03	1.24	0.84

(a)   What is the average annual rate of change in world crop land between 1800 and 1970?

(b)   Estimate the rate, in acres per person per year, at which crop land was decreasing in the year 1950.

[1]From *An Introduction to Population, Environment, Society*, by Lawrence Schaefer, (Hamden, CT, E-P Education Services, 1972, p.30.)

Solution   (a)   Between 1800 and 1970,

$$\text{Average rate of change} = \frac{0.84 - 3.51}{1970 - 1800} = \frac{-2.67}{170} = -0.0157.$$

The number of acres per person used for farming between 1800 and 1970 has been decreasing at an average rate of 0.0157 acres per person per year.

(b)   We want to estimate the instantaneous rate of change at the year 1950. We will use the interval from 1950 to 1970 to estimate the instantaneous rate of change at 1950:

$$\text{Rate of change in 1950} \approx \frac{0.84 - 1.24}{1970 - 1950} = \frac{-0.4}{20} = -0.02.$$

In 1950, the amount of farmland per person worldwide was decreasing at a rate of approximately 0.02 acres per person per year.

Problems for Section 2.1

1.   A car is driven at a constant speed. Sketch a graph of the distance the car has traveled as a function of time.

2.   A car is driven at an increasing speed. Sketch a graph of the distance the car has traveled as a function of time.

3.   A car starts at a high speed, and its speed then decreases slowly. Sketch a graph of the distance the car has traveled as a function of time.

4.   A bicyclist pedals at a fairly constant rate, with evenly spaced intervals of coasting. Sketch a graph of the distance she has traveled as a function of time.

5.   Suppose the distance (in feet) of an object from a point is given by $s(t) = t^2$, where t is measured in seconds.

   (a)   What is the average velocity of the object between $t = 3$ and $t = 5$?
   (b)   By using smaller and smaller intervals around 3, estimate the instantaneous velocity at time $t = 3$.

6.   The size, S, of a malignant tumor (in square millimeters) is given by $S = 2^t$, where t is the number of months since the tumor was discovered.

   (a)   What is the total change in the size of the tumor during the first six months? Give units with your answer.
   (b)   What is the average rate of change in the size of the tumor during the first six months? Give units with your answer.
   (c)   Estimate the rate at which the tumor is growing at $t = 6$. (Use smaller and smaller intervals.) Give units with your answer.

7.   Match the points labeled on the curve in Figure 2.7 with the given slopes.

Slope	Point
-3	
-1	
0	
1/2	
1	
2	

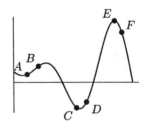

Figure 2.7

8. For the function shown in Figure 2.8, at what labeled points is the slope of the graph positive? Negative? At which labeled point does the graph have the greatest (i.e., most positive) slope? The least slope (i.e., negative and with the largest magnitude)?

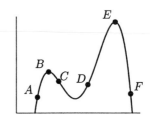

Figure 2.8

9. Find the average velocity over the interval $0 \leq t \leq 0.8$, and estimate the velocity at $t = 0.2$ of a car whose position, s, is given by the following table.

t (sec)	0	0.2	0.4	0.6	0.8	1.0
s (ft)	0	0.5	1.8	3.8	6.5	9.6

10. Suppose $y = f(x)$ graphed in Figure 2.9 represents the cost (in thousands of dollars) of manufacturing x kilograms of a chemical.

    (a) Is the average rate of change of the cost of producing x kilograms greater between $x = 0$ and $x = 3$, or between $x = 3$ and $x = 5$? Illustrate your answer graphically.

    (b) Is the instantaneous rate of change of the cost of producing x kilograms greater at $x = 1$ or at $x = 4$? Illustrate your answer graphically.

    (c) What are the units of this rate of change?

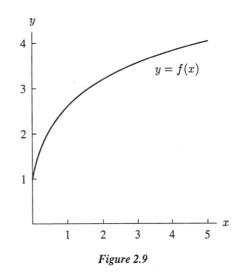

Figure 2.9

11. The population, P, of China, in billions, can be approximated by the function

$$P = 1.15(1.014)^t,$$

where t is the number of years since the start of 1993. According to this model, how fast is the population growing at the start of 1993 and at the start of 1995? Give your answers in millions of people per year.

12. Table 2.4 gives the percent of the US population living in urban areas as a function of year[2].

TABLE 2.4   *Percent of US population living in urban areas*

Year	1800	1830	1860	1890	1920	1950	1960	1970	1980
Percent urban	6	9	20	35	51	64	69.9	73.5	73.7

(a) Find the average rate of change in the percent of the US population living in urban areas between 1860 and 1960.

(b) Estimate the rate at which this percent is increasing at the year 1960.

(c) Estimate the rate of change of this function for the year 1830 and explain what it is telling you.

(d) Estimate the rate of change of this function for the year 1970 and explain what it is telling you.

(e) Is the rate of change of this function getting larger or smaller over time? What does this mean about how the population is changing?

(f) Is this function increasing or decreasing? Is its graph concave up or concave down?

13. A ball is tossed into the air from a bridge, and its height, y (in feet), above the ground t seconds after it is thrown is given by

$$y = f(t) = -16t^2 + 50t + 36.$$

(a) How high above the ground is the bridge?

(b) What is the average velocity of the ball for the first second?

(c) Approximate the velocity of the ball at $t = 1$ second.

(d) Graph the function f, and determine the maximum height the ball will reach. What should the velocity be at the time the ball is at the peak?

(e) Use the graph to decide at what time, t, the ball reaches its maximum height.

14. For the graph $y = f(x)$ shown in Figure 2.10, arrange the following numbers in ascending (i.e., smallest to largest) order:

- The slope of the graph at A.
- The slope of the graph at B.
- The slope of the graph at C.
- The slope of the line AB.
- The number 0.
- The number 1.

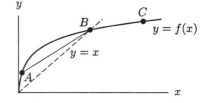

Figure 2.10

15. Suppose a particle is moving at varying velocity along a straight line and that $s = f(t)$ represents the distance of the particle from a point as a function of time, t. Sketch a possible graph for f if the average velocity of the particle between $t = 2$ and $t = 6$ is the same as the instantaneous velocity at $t = 5$.

2.2 THE DERIVATIVE

The instantaneous rate of change of a function f at a point a is so important that it is given its own name, the *derivative of f at a*, denoted $f'(a)$. If we want to emphasize that $f'(a)$ is the rate

[2] *Statistical Abstracts of the US*, 1985, US Department of Commerce, Bureau of the Census, p.22

of change of $f(x)$ as the variable x increases, we call $f'(a)$ the derivative of f *with respect to x at $x = a$*. Notice that the derivative is just a new name for a familiar concept: the rate of change of a function at a point.

We define $f'(a)$, the **derivative of f at a**, to be the instantaneous rate of change of f at a.

Example 1   Estimate $f'(2)$ if $f(x) = x^3$.

Solution   We know that $f'(2)$ is the derivative of $f(x) = x^3$ at 2. This is the same as the rate of change of x^3 at 2. We estimate this by looking at the average rate of change over intervals near 2. If we use the intervals $1.999 \leq x \leq 2$ and $2 \leq x \leq 2.001$, we see that

$$\frac{\text{Average rate of change}}{\text{on } 1.999 \leq x \leq 2} = \frac{2^3 - (1.999)^3}{2 - 1.999} = \frac{8 - 7.988}{0.001} = 12.0$$

$$\frac{\text{Average rate of change}}{\text{on } 2 \leq x \leq 2.001} = \frac{(2.001)^3 - 2^3}{2.001 - 2} = \frac{8.012 - 8}{0.001} = 12.0$$

It appears that the rate of change of $f(x)$ at $x = 2$ is 12, so we estimate $f'(2) = 12$.

Visualizing the Derivative: Slope of the Graph and Slope of the Tangent Line

We have already seen that the rate of change (the derivative) of a function at a point is equal to the slope of the graph at that point. Since the derivative is found by taking the average rate of change over smaller and smaller intervals, we can also think of the derivative as the slope of the tangent line to the graph at that point.

The derivative of a function at the point A is equal to
- The slope of the graph of the function at A.
- The slope of the tangent line to the graph at A.

The slope interpretation is often useful in gaining rough information about the derivative, as the following examples show.

Example 2   Let $f(x) = x^2$. Use a graph of this function to determine whether each of the following derivatives is positive, negative, or zero.
(a) $f'(1)$ (b) $f'(-1)$ (c) $f'(2)$ (d) $f'(0)$

Solution

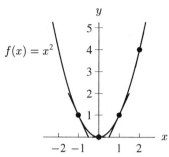

Figure 2.11: Tangent lines to
$f(x) = x^2$

The graph of $f(x) = x^2$ is shown in Figure 2.11, and tangent line segments have been drawn at the four points where $x = 1$, $x = -1$, $x = 2$, and $x = 0$. (Note: the tangent lines at $x = 0$ and $x = 2$ are hard to see because they are so close to the graph of x^2.) Since the derivative is equal to the slope of the tangent line at that point, we see that

(a) $f'(1)$ is positive,

(b) $f'(-1)$ is negative,

(c) $f'(2)$ is positive (and larger than $f'(1)$), and,

(d) $f'(0) = 0$ since the graph has a horizontal tangent at $x = 0$.

Example 3   Estimate the derivative of $f(x) = 2^x$ at $x = 0$ graphically and numerically.

Solution   Graphically: If you draw a tangent line at $x = 0$ to the exponential curve in Figure 2.12, you will see that it has a positive slope. Since the slope of the line BA is $(2^0 - 2^{-1})/(0 - (-1)) = 1/2$ and the slope of the line AC is $(2^1 - 2^0)/(1 - 0) = 1$, we know that the derivative is between $1/2$ and 1.

Numerically: To find the derivative at $x = 0$, we compute the average rate of change on intervals near 0.

$$\text{Average rate of change} \atop \text{on } -0.0001 \leq x \leq 0 = \frac{2^0 - 2^{-0.0001}}{0 - (-0.0001)} = \frac{1 - 0.999930688}{0.0001} = 0.69312$$

$$\text{Average rate of change} \atop \text{on } 0 \leq x \leq 0.0001 = \frac{2^{0.0001} - 2^0}{0.0001 - 0} = \frac{1.000069317 - 1}{0.0001} = 0.69317$$

It appears that the derivative is between 0.69312 and 0.69317. To three decimal places, $f'(0) = 0.693$.

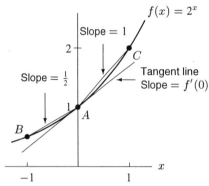

Figure 2.12: Graph of $y = 2^x$ showing the derivative at $x = 0$

Example 4   Find an approximate equation for the tangent line to the graph of $f(x) = 2^x$ at $x = 0$.

Solution   From the previous example, we know the slope of the tangent line is about 0.693. Since the line passes through the point $(0, 2^0) = (0, 1)$, we know the line has y intercept 1. Its equation is approximately

$$y = 0.693x + 1.$$

Example 5   The graph of a function $y = f(x)$ is shown in Figure 2.13. Indicate whether each of the following is positive or negative, and illustrate your answers graphically.

(a) $f'(1)$ (b) $\dfrac{f(3) - f(1)}{3 - 1}$ (c) $f(4) - f(2)$

(d)   The average rate of change of $f(x)$ between $x = 3$ and $x = 7$.

(e)   The instantaneous rate of change of $f(x)$ at $x = 3$.

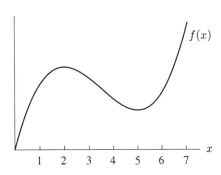

Figure 2.13: A graph of a function

Solution   (a)   Since $f'(1)$ is the slope of the graph at $x = 1$, we see in Figure 2.14 that $f'(1)$ is positive.

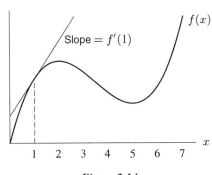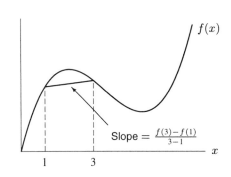

Figure 2.14   *Figure 2.15*

(b)   This difference quotient is the slope of the secant line between $x = 1$ and $x = 3$ and we see in Figure 2.15 that this slope is positive.

(c)   Since $f(4)$ is the value of the function at $x = 4$ and $f(2)$ is the value of the function at $x = 2$, the expression $f(4) - f(2)$ is the change in the function between $x = 2$ and $x = 4$. Since $f(4)$ lies below $f(2)$, this change is negative. See Figure 2.16.

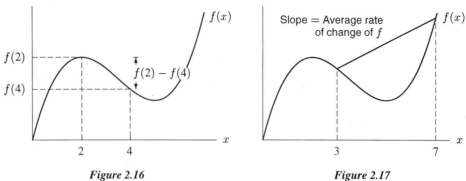

Figure 2.16   Figure 2.17

(d) The average rate of change is the slope of the secant line in Figure 2.17, which shows that this slope is positive.

(e) The instantaneous rate of change is the slope of the graph at $x = 3$, which we see from Figure 2.18 is negative.

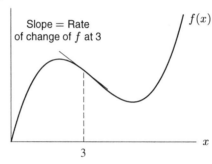

Figure 2.18

Finding the Derivative of a Function Given Numerically

If we are given a table of function values instead of a graph or a formula for a function, we can still estimate values of its derivative. To do this, we have to assume that the points in the table are close enough together that the function does not change wildly between them. We make this assumption in the following example.

Example 6   Table 2.5 gives values of a function $P = f(t)$.

(a) Does $f'(5)$ appear to positive or negative?

(b) Does $f'(20)$ appear to positive or negative?

(c) Estimate $f'(10)$.

TABLE 2.5

t	0	5	10	15	20	25
P	50	42	38	35	46	64

Solution   (a) Since the values of P go down as t goes from 0 to 5 to 10, we see that $f'(5)$ appears to be negative.

(b) Since the values of P go up as t goes from 15 to 20 to 25, we see that $f'(20)$ appears to be positive.

(c) We estimate $f'(10)$ using the difference quotient for the interval to the right of $t = 10$, as follows:

$$f'(10) = \frac{\Delta P}{\Delta t} = \frac{35 - 38}{15 - 10} = -0.6.$$

Finding Derivatives by Zooming In

We have seen that when we zoom in on a curve near a point, the curve will look like a line. We can estimate the derivative at a point by zooming in and estimating the slope of this line.

Example 7   By zooming in on the point $(0,0)$ on the graph of $f(x) = x^3 - x$, estimate the derivative of this function at $x = 0$.

Solution   Figure 2.19 shows successive graphs of $f(x) = x^3 - x$, with smaller and smaller scales. On the interval $-0.1 \le x \le 0.1$, the graph looks like a straight line of slope -1. Thus, the derivative of $x^3 - x$ at $x = 0$ is about -1.

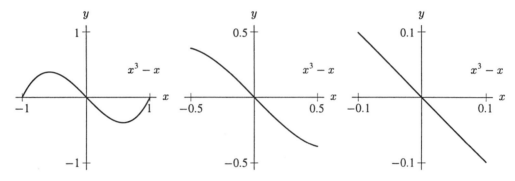

Figure 2.19: Zooming in on the graph of $x^3 - x$ near $x = 0$

Consider the function $f(x) = x^2$. What is $f'(1)$? We have seen several ways to answer this question. If we look at a graph of $f(x) = x^2$, and examine the slope at $x = 1$, we see that $f'(1)$ is positive and larger than 1. We can obtain a more accurate answer by zooming in on this point. Or we can use the formula $f(x) = x^2$ and a small interval around $x = 1$ (such as $1 \le x \le 1.001$) to estimate $f'(1)$ with a difference quotient:

$$f'(1) = \frac{f(1.001) - f(1)}{1.001 - 1} = \frac{(1.001)^2 - (1)^2}{0.001} = \frac{1.002001 - 1}{0.001} \approx 2.$$

What does it mean that the derivative of $f(x) = x^2$ at the point $x = 1$ is 2? Since the derivative is the rate of change, it means that for small changes in x, near $x = 1$, the change in $f(x) = x^2$ is about twice as big as the change in x. As an example, if x changes from 1 to 1.1, a net change of 0.1, $f(x)$ should change by about 0.2. Figure 2.20 shows this geometrically.

Table 2.6 shows the derivative of $f(x) = x^2$ numerically. Notice that near $x = 1$, every time the value of x increases by 0.001, the value of x^2 increases by approximately 0.002. Thus near $x = 1$ the graph is approximately linear with slope $0.002/0.001 = 2$.

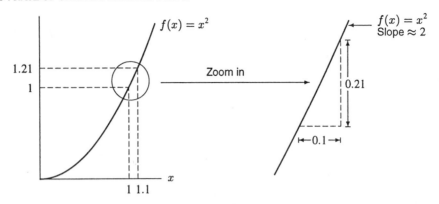

Figure 2.20: Graph of $f(x) = x^2$ near $x = 1$ has slope ≈ 2

TABLE 2.6 *Values of $f(x) = x^2$ near $x = 1$*

x	x^2	Difference in successive x^2 values
0.998	0.996004	
		0.001997
0.999	0.998001	
		0.001999
1.000	1.000000	
		0.002001
1.001	1.002001	
		0.002003
1.002	1.004004	
↑		↑
x increments		all approximately
0.001		0.002

Problems for Section 2.2

1. (a) Sketch a graph of $f(x) = 2 - x^3$, and use the graph to decide whether the derivative of $f(x)$ at $x = 1$ is positive or negative. Give reasons for your decision.
   (b) Use a small interval to estimate $f'(1)$.

2. Estimate $P'(0)$ if $P(t) = 200(1.05)^t$.

3. The graph of a function is given in Figure 2.21, and several points are labeled.
   (a) State whether the derivative of the function is positive, negative, or zero at each of the labeled points.
   (b) At which point is the derivative greatest (i.e. most positive)? At which point is the derivative smallest (i.e. most negative)?

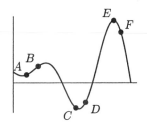

Figure 2.21

4. Let $f(x) = x + 2^x + 2^{-x}$.

   (a) Use the graph of $f(x)$ to decide if the derivative is positive, negative, or zero at the points $x = -1$, $x = 0$, and $x = 1$.

   (b) By zooming in on the graph of $f(x)$, estimate the value of $f'(0)$.

5. Estimate $f'(2)$ if $f(x) = x^3 - 2x$. Use a small interval.

6. Table 2.7 gives the percent, P, of households in the United States with a microwave oven as a function of time, t, in years since 1978.[3] We have $P = f(t)$.

   (a) Is $f'(6)$ positive or negative? What does this tell you about the percent of households with microwave ovens?

   (b) Estimate $f'(2)$. Estimate $f'(9)$. Explain what each is telling you, in terms of microwave ovens. Which is larger?

   TABLE 2.7   *Percent of U.S. households with microwave ovens*

t (years since 1978)	0	2	4	6	9	12
P (% with microwave)	8	14	21	34	61	79

7. (a) Let $g(t) = (0.8)^t$. Use a graph to determine whether $g'(2)$ is positive, negative, or zero.

   (b) Use a small interval to estimate $g'(2)$.

8. Figure 2.22 shows the number of farms in the United States, in millions, between 1930 and 1992. We can think of the number of farms, N, as a function of year, t, and we write $N = f(t)$.

   (a) Is $f'(1950)$ positive or negative? What does this tell you about the number of farms?

   (b) Which is more negative: $f'(1960)$ or $f'(1980)$? Explain.

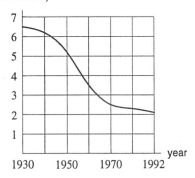

Figure 2.22: Number of farms

---

[3] *The World Almanac*, (New Jersey: Funk & Wagnalls, 1994), p.152.

9. Show how you can represent the following on a sketch similar to that in Figure 2.23.

    (a) $f(4)$ (b) $f(4) - f(2)$ (c) $\dfrac{f(5) - f(2)}{5 - 2}$ (d) $f'(3)$

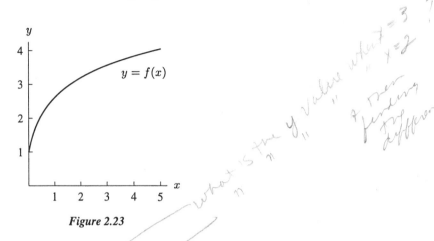

Figure 2.23

10. Consider the function $y = f(x)$ shown in Figure 2.23. For each of the following pairs of numbers, decide which is larger. Explain your answer.

    (a) $f(3)$ or $f(4)$?   (b) $f(3) - f(2)$ or $f(2) - f(1)$?

    (c) $\dfrac{f(2) - f(1)}{2 - 1}$ or $\dfrac{f(3) - f(1)}{3 - 1}$?   (d) $f'(1)$ or $f'(4)$?

11. With the function f given by Figure 2.23, arrange the following quantities in ascending order:
    $$0, \quad 1, \quad f'(2), \quad f'(3), \quad f(3) - f(2)$$

12. (a) Sketch graphs of the functions $f(x) = x^2$ and $g(x) = x^2 + 3$ on the same set of axes. What can you say about the slopes of the tangent lines to the two graphs at the point $x = 0$? $x = 1$? $x = 2$? $x = a$, where a is any value?

    (b) Explain why adding a constant to any function will not change the value of the derivative at any point.

13. If $f(x) = x^3 + 4x$, estimate $f'(3)$ using a table similar to Table 2.6 on page 100.

14. For $g(x) = x^5$, use tables similar to Table 2.6 on page 100 to estimate $g'(2)$ and $g'(-2)$. What relationship do you notice between $g'(2)$ and $g'(-2)$? Explain geometrically why this must occur.

2.3 THE DERIVATIVE FUNCTION

In the last section we looked at the derivative of a function at a fixed point. Now we'll consider what happens at a variety of points and see that, in general, the derivative takes on different values at different points and is itself a function.

First, remember that the derivative of a function at a point tells you the rate at which the value of the function is changing at that point. Geometrically, if you "zoom in" on a point in the graph until the graph looks like a line, the slope of that line is the derivative at that point. Equivalently, you can think of the derivative as the slope of the tangent line to the graph at the point, because as you "zoom in," the graph and the tangent line become indistinguishable.

Example 1   Estimate the derivative of the function given by the graph in Figure 2.24 at $x = -2, -1, 0, 1, 2, 3, 4, 5$.

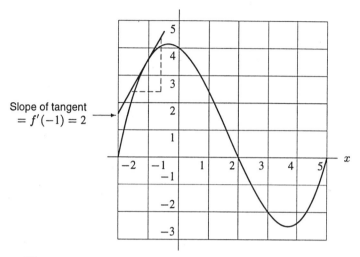

Figure 2.24: Estimating the derivative graphically as the slope of a tangent line

Solution   From the graph you can estimate the derivative at any point by placing a straight edge so that it forms the tangent line at that point, and then using the grid squares to estimate the slope of the straight edge. (For example, the tangent at $x = -1$ is drawn in Figure 2.24, and has a slope of about 2, so $f'(-1) \approx 2$.) Notice that the slope at $x = -2$ is positive and fairly large; the slope at $x = -1$ is positive but smaller. At $x = 0$, the slope is negative, by $x = 1$ it has become more negative, and so on. Some estimates of the derivative are listed in Table 2.8. You should check these values yourself. Are they reasonable? Is the derivative positive where you expect? Negative?

TABLE 2.8   *Estimated values of derivative of function in Figure 2.24*

x	-2	-1	0	1	2	3	4	5
Derivative at x	6	2	-1	-2	-2	-1	1	4

The important point to notice is that for every x-value, there's a corresponding value of the derivative. The derivative, therefore, is itself a function of x.

For any function f, we define the **derivative function**, f', by

$$f'(x) = \text{Rate of change of } f \text{ at } x.$$

Finding the Derivative of a Function Given Graphically

Example 2   Sketch the graph of the derivative of the function given by the graph in Figure 2.24.

Solution   Table 2.8 gives some values of this derivative which we can plot. However, it is a good idea to first identify some of the key features of the derivative graph from the graph of the original function. For example, when is the derivative zero? Since the slope is zero when the graph has a horizontal tangent line, we see that the derivative is zero at approximately $x = -0.5$ and $x = 3.7$. These are the x-intercepts of the derivative graph.

We can see from Figure 2.24 (repeated in Figure 2.25) that the function is increasing from $x = -2$ to about $x = -0.5$. Thus, the derivative is positive in this interval, and so we must draw the graph of f' above the x-axis from $x = -2$ to about $x = -0.4$. Between $x = -0.4$ and about $x = 3.7$ the function is decreasing, so the derivative is negative and its graph must be below the x-axis. Beyond $x = 3.7$ the function is increasing, so the derivative is positive again and the graph of f' is above the axis. Somewhere in the region where the derivative is negative, it is going to reach its lowest point; this will be at the point where the graph of the original function is decreasing most steeply. From Figure 2.24 we see that this occurs a little before $x = 2$, where the slope is slightly steeper than -2. Thus, our derivative graph should have a minimum value of slightly below -2 occurring slightly to the left of $x = 2$. With this in mind, and using the data in Table 2.8, we obtain Figure 2.25, which shows a graph of the derivative.

You should check for yourself that this graph of f' makes sense. Notice that at the points where f has large upward slope, such as $x = -2$, the graph of the derivative is far above the x-axis, as it should be, since the value of the derivative should be large there. On the other hand, at points where the slope is gentle, the graph of f' is close to the x-axis, since the derivative is small.

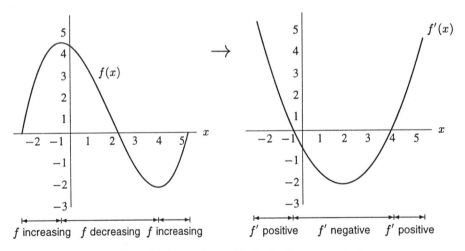

Figure 2.25: Function and derivative from Example 2

Finding the Derivative of a Function Given Numerically

If we are given a table of function values instead of a graph of the function, we can estimate values of the derivative.

Example 3   Table 2.9 gives values of $c(t)$, the concentration (mg/cc) of a drug in the bloodstream at time t (min). Construct a table of estimated values for $c'(t)$, the rate of change of $c(t)$ with respect to t.

TABLE 2.9   *Concentration as a function of time*

t (min)	0	0.1	0.2	0.3	0.4	0.5	0.6	0.7	0.8	0.9	1.0
$c(t)$ (mg/cc)	0.84	0.89	0.94	0.98	1.00	1.00	0.97	0.90	0.79	0.63	0.41

Solution   We want to estimate the derivative of c using the values in the table. To do this, we have to assume that the data points are close enough together that the concentration doesn't change wildly between them. From the table, we can see that the concentration is increasing between $t = 0$ and $t = 0.4$, so we'd expect a positive derivative there. However, the increase is quite slow, so we would expect the derivative to be small. The concentration doesn't change between 0.4 and 0.5, so we expect the derivative to be 0 there. From $t = 0.5$ to $t = 1.0$, the concentration starts to decrease, and the rate of decrease gets larger and larger, so we would expect the derivative to be negative and of greater and greater magnitude.

Using the data in the table, we can estimate the derivative for each value of t with a difference quotient. For example,

$$c'(0) \approx \frac{c(0.1) - c(0)}{0.1 - 0} = \frac{0.89 - 0.84}{0.1} = 0.5 \text{ mg/cc/min}.$$

Thus, we estimate that

$$c'(0) \approx 0.5.$$

Similarly, we get the estimates

$$c'(0.1) \approx \frac{c(0.2) - c(0.1)}{0.2 - 0.1} = \frac{0.94 - 0.89}{0.1} = 0.5$$

$$c'(0.2) \approx \frac{c(0.3) - c(0.2)}{0.3 - 0.2} = \frac{0.98 - 0.94}{0.1} = 0.4$$

$$c'(0.3) \approx \frac{c(0.4) - c(0.3)}{0.4 - 0.3} = \frac{1.00 - 0.98}{0.1} = 0.2$$

$$c'(0.4) \approx \frac{c(0.5) - c(0.4)}{0.5 - 0.4} = \frac{1.00 - 1.00}{0.1} = 0.0$$

and so on. These values are tabulated in Table 2.10. Notice that the derivative has small positive values up until $t = 0.4$, and then it gets more and more negative, as we expected. The slopes are shown on the graph of $c(t)$ in Figure 2.26.

TABLE 2.10
Derivative of concentration

t	$c'(t)$
0	0.5
0.1	0.5
0.2	0.4
0.3	0.2
0.4	0.0
0.5	−0.3
0.6	−0.7
0.7	−1.1
0.8	−1.6
0.9	−2.2

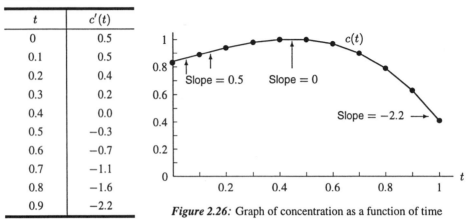

Figure 2.26: Graph of concentration as a function of time

Other Ways We Could Calculate the Derivative Numerically

In the previous example, our estimate for the derivative of $c(t)$ at $t = 0.2$ used the point to the right. We found the average rate of change between $t = 0.2$ and $t = 0.3$. However, we could equally well have gone to the left and used the rate of change between $t = 0.1$ and $t = 0.2$ to approximate the derivative at 0.2. For a more accurate result, we could average these slopes, getting the approximation

$$c'(0.2) \approx \frac{1}{2} \left(\begin{array}{c} \text{slope to left} \\ \text{of } 0.2 \end{array} + \begin{array}{c} \text{slope to right} \\ \text{of } 0.2 \end{array} \right) = \frac{0.5 + 0.4}{2} = 0.45.$$

Each of these methods of approximating the derivative gives a reasonable answer. For convenience, unless there is a reason to do otherwise, we will estimate the derivative by going to the right.

Finding the Derivative of a Function Given by a Formula

If we are given a formula for a function f, can we come up with a formula for f'? Using the definition of the derivative, we often can. Indeed, much of the power of calculus depends on our ability to find formulas for the derivatives of all the familiar functions. This is explained in detail in Chapter 5. In the next example, we see how to guess at a formula for the derivative.

Example 4   Find a formula for the derivative of $f(x) = x^2$.

Solution   We look for a pattern in the values of $f'(x)$. Table 2.11 contains values of $f(x) = x^2$ (rounded to three decimals), which we can use to estimate the values of $f'(1)$, $f'(2)$, and $f'(3)$.

TABLE 2.11   *Values of $f(x) = x^2$ near $x = 1$, $x = 2$, $x = 3$ (rounded to three decimals)*

x	x^2 (approx)	x	x^2 (approx)	x	x^2 (approx)
0.999	0.998	1.999	3.996	2.999	8.994
1.000	1.000	2.000	4.000	3.000	9.000
1.001	1.002	2.001	4.004	3.001	9.006
1.002	1.004	2.002	4.008	3.002	9.012

Near $x = 1$, x^2 increases by about 0.002 each time x increases by 0.001, so

$$f'(1) \approx \frac{0.002}{0.001} = 2.$$

Similarly,

$$f'(2) \approx \frac{0.004}{0.001} = 4$$

$$f'(3) \approx \frac{0.006}{0.001} = 6.$$

Knowing the value of f' at specific points can never tell us the formula for f', but it certainly can be suggestive: knowing $f'(1) \approx 2$, $f'(2) \approx 4$, $f'(3) \approx 6$ certainly suggests that $f'(x) = 2x$. In Chapter 5, we will prove that this is indeed the case.

What Does the Derivative Tell Us Graphically?

When the derivative, f', of a function is positive, the tangent to the graph of f is sloping up; when f' is negative, the tangent is sloping down. If $f' = 0$ everywhere, then the tangent is horizontal everywhere and so f is constant. Thus, the sign of the derivative f' tells us whether the function f is increasing or decreasing.

> If $f' > 0$ on an interval, then f is *increasing* over that interval.
> If $f' < 0$ on an interval, then f is *decreasing* over that interval.
> If $f' = 0$ on an interval, then f is *constant* over that interval.

Moreover, the magnitude of the derivative gives us the magnitude of the rate of change; so if f' is large (positive or negative), then the graph of f will be steep (up or down), whereas if f' is small the graph of f will slope gently. With this in mind, you can deduce a lot about the behavior of a function from the behavior of its derivative.

Example 5   Consider the step function $f(t)$ graphed in Figure 2.27. Sketch a graph of the derivative $f'(t)$.

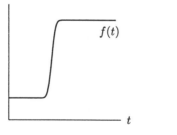

Figure 2.27: Step function

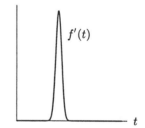

Figure 2.28: Spike derivative function

Solution   On the intervals where the graph of $f(t)$ is horizontal, the slope is zero, and so $f'(t) = 0$. This means the graph of $f'(t)$ lies on the horizontal axis on these intervals. On the small interval where the graph of $f(t)$ is not horizontal, it is increasing, and so $f'(t)$ is positive on this interval. (This means that the graph of $f'(t)$ lies above the horizontal axis.) The function $f(t)$ is increasing quite rapidly on this small interval, and so the derivative $f'(t)$ gets quite large here. A possible graph of $f'(t)$ is shown in Figure 2.28.

Problems for Section 2.3

For Problems 1–6, sketch a graph of the derivative function of each of the given functions.

1.
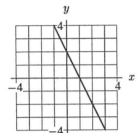

2.
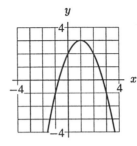

3.
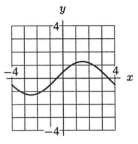

4.
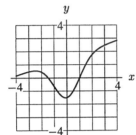

5.
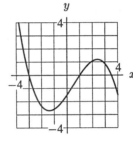

6.
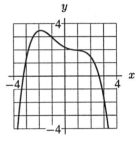

7. Sketch the graph of the derivative of the downward step function in Figure 2.29.

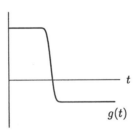

Figure 2.29: Downward
step function

8. (a) Sketch a smooth curve whose slope is everywhere positive and increasing gradually.
   (b) Sketch a smooth curve whose slope is everywhere positive and decreasing gradually.
   (c) Sketch a smooth curve whose slope is everywhere negative and increasing gradually (i.e., becoming less and less negative).
   (d) Sketch a smooth curve whose slope is everywhere negative and decreasing gradually (i.e., becoming more and more negative).

9. Given the numerical values shown, find approximate values for the derivative of $f(x)$ at each of the x-values given. Where is the rate of change of $f(x)$ positive? Where is it negative? Where does the rate of change of $f(x)$ seem to be greatest?

x	0	1	2	3	4	5	6	7	8
$f(x)$	18	13	10	9	9	11	15	21	30

10. Suppose $f(x) = \frac{1}{3}x^3$. Estimate $f'(2)$, $f'(3)$, and $f'(4)$. What do you notice? Can you guess a formula for $f'(x)$?

11. If $g(t) = t^2 + t$, estimate $g'(1)$, $g'(2)$, and $g'(3)$. Use these to guess a formula for $g'(t)$.

12. Draw a possible graph of $y = f(x)$ given the following information about its derivative.
    - $f'(x) > 0$ on $1 < x < 3$
    - $f'(x) < 0$ for $x < 1$ and $x > 3$
    - $f'(x) = 0$ at $x = 1$ and $x = 3$

13. Draw a possible graph of $y = f(x)$ given the following information about its derivative.
    - $f'(x) > 0$ for $x < -1$
    - $f'(x) < 0$ for $x > -1$
    - $f'(x) = 0$ at $x = -1$

14. In the graph of f in Figure 2.30, at which of the labeled x-values is

    (a) $f(x)$ greatest?   (b) $f(x)$ least?
    (c) $f'(x)$ greatest?   (d) $f'(x)$ least?

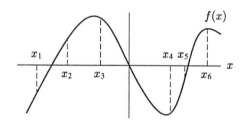

Figure 2.30

15. The solid waste generated each year by our cities has been increasing. Values for yearly solid waste (measured in millions of tons) as a function of year are given in Table 2.12[4].

    (a) Estimate the derivative of this function for the years 1960, 1965, 1970, 1975, and 1980.
    (b) Interpret these derivatives in terms of the amount of solid waste being generated.
    (c) Are the derivatives getting larger or smaller over time? What does this mean in terms of solid waste?

    TABLE 2.12   *Municipal solid waste in the US*

Year	1960	1965	1970	1975	1980	1984
Waste (millions of tons)	82.3	98.3	118.3	122.7	139.1	148.1

16. Let $f(x) = 5x + 2$.

    (a) Find the average rate of change between $x = 2$ and $x = 5$; between $x = 2$ and $x = 3$; and between $x = 2$ and $x = 2.1$.
    (b) What is $f'(2)$? Justify your answer in two different ways: first, by using your answers to part (a), and second, by using a graph of $f(x)$.
    (c) Let $f(x) = mx + b$ be any line. What is $f'(a)$ for any a?

    ---
    [4]*Statistical Abstracts of the US*, 1988 p193, Table 333

For Problems 17–22, sketch the graph of $f(x)$, and use this graph to sketch the graph of $f'(x)$.

17. $f(x) = x(x - 1)$ 18. $f(x) = 5x$ 19. $f(x) = x^3$

20. $f(x) = 4 + 2x - x^2$ 21. $f(x) = x^2 + 3$ 22. $f(x) = 2 - x^4$

For Problems 23–28, sketch the graph of $y = f'(x)$ for the function given.

23.
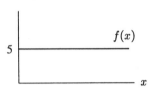

24.
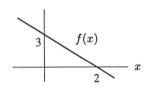

25.
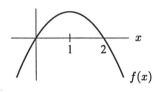

26.
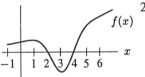

27.
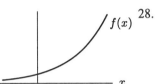

28.
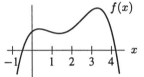

29. The amount of a certain bacteria in the blood (measured in number of bacteria per cubic centimeter) can be modeled by

$$f(t) = 100t(0.8)^t,$$

where t is measured in days since the person first became ill.

(a) How does the amount of bacteria vary with time? Sketch a graph of $f(t)$ from the time the person first becomes ill until three weeks later.

(b) Use the graph to decide when the population is a maximum. What is that maximum? Is there a minimum after the person first becomes ill? If so, when?

(c) Use the graph to decide when the population of bacteria is growing fastest. When is it decreasing fastest?

(d) Estimate roughly how fast the population is changing one week after the onset of the illness.

30. To give a patient an antibiotic slowly, the drug is injected into the muscle. (For example, penicillin for venereal disease is administered this way.) The quantity of the drug in the bloodstream starts out at zero, increases to a maximum, and then decreases to zero again.

(a) Sketch a possible graph of the quantity of the drug in the bloodstream as a function of time. Mark the time at which the drug is at a maximum by t_0.

(b) Describe in words how the rate at which the drug is entering or leaving the blood changes over time. Sketch a graph of this rate against time, marking t_0 on the time axis.

2.4 INTERPRETATIONS OF THE DERIVATIVE

We have already seen how the derivative can be interpreted as a slope and as a rate of change. In this section, you will see examples of other interpretations. The point of these examples is not to make a catalog of interpretations but to illustrate the process of obtaining them. Surprisingly, a new notation for the derivative can be quite helpful in discovering the correct interpretations.

An Alternative Notation for the Derivative

So far we have used the notation f' to stand for the derivative of the function f. An alternative notation for derivatives was introduced by the German mathematician Gottfried Wilhelm Leibniz (1646–1716) when calculus was first being developed in the seventeenth century. We know that $f'(x)$ is approximated by the average rate of change over smaller and smaller intervals. If the variable y depends on the variable x, that is, $y = f(x)$, then the average rate of change is given by $\frac{\Delta y}{\Delta x}$. We have

$$f'(x) \approx \frac{\Delta y}{\Delta x},$$

over a small interval. The new notation for the derivative is meant to remind us of this. If

$$y = f(x),$$

then we write

$$f'(x) = \frac{dy}{dx}.$$

Leibniz's notation is quite suggestive, especially if you think of the letter d in dy/dx as standing for "small difference in . . . ." The notation dy/dx reminds us that the derivative is a limit of ratios of the form

$$\frac{\text{Difference in } y\text{-values}}{\text{Difference in } x\text{-values}}.$$

It is always a good idea to have a mathematical symbol that lets us know where it came from and what it means: dy/dx does this and $f'(x)$ does not. The notation dy/dx is useful for determining the units for the derivative: the units for dy/dx are the units for y divided by (or "per") the units for x. The d/dx notation can also be very convenient. For example, it is a lot easier to say

"$\dfrac{d}{dx}(x^2 + 3x) = 2x + 3$" than it is to say "if $f(x) = x^2 + 3x$, then $f'(x) = 2x + 3$."

The separate entities dy and dx officially have no independent meaning: they are part of one notation. In fact, a good formal way to view the notation dy/dx is to think of d/dx as a single symbol meaning "the derivative with respect to x of . . .". Thus dy/dx could be viewed as

$$\frac{d}{dx}(y), \quad \text{meaning "the derivative with respect to } x \text{ of } y.\text{"}$$

On the other hand, many scientists and mathematicians really do think of dy and dx as separate entities representing "infinitesimally" small differences in y and x, even though it is difficult to say exactly how small "infinitesimal" is. It may not be formally correct, but it is very helpful to the intuition to think of dy/dx as a very small change in y divided by a very small change in x.

For example, recall that if $s = f(t)$ is the position of a moving object at time t, then $v = f'(t)$ is the velocity of the object at time t. Writing

$$v = \frac{ds}{dt}$$

directly reminds you of this fact, since it suggests a distance, ds, over a time, dt, and we know that distance over time is velocity. Similarly, we recognize

$$\frac{dy}{dx} = f'(x)$$

as the slope of the graph of $y = f(x)$ by remembering that slope is vertical rise, dy, over horizontal run, dx.

The disadvantage of the Leibniz notation is that it is rather awkward if you want to specify the x-value at which you are evaluating the derivative. To specify $f'(2)$, for example, we have to write

$$\frac{dy}{dx}\bigg|_{x=2}$$

Using Units to Interpret the Derivative

The following examples illustrate the fact that if you want to interpret a derivative in practical terms, it often helps to think about units of measurement.

For example, suppose $s = f(t)$ gives the position in meters of a body from a fixed point as a function of time, t, in seconds. Then knowing that

$$\frac{ds}{dt} = f'(2) = 10 \text{ meters/sec}$$

tells us that when $t = 2$ sec, the body is moving at a velocity of 10 meters/sec. This is an instantaneous velocity, meaning that if the body continued to move at this speed for a whole second, it would cover 10 meters. In practice, however, the velocity of the body is probably changing and so doesn't remain 10 meters/sec for long. Notice that the units of instantaneous velocity and of average velocity are the same.

> The units of the derivative of a function are the units of the dependent variable over the units of the independent variable.

We can interpret the derivative as the expected change in the dependent variable if the independent variable increases by 1 unit. Since the derivative usually is not constant over this one unit, this interpretation is only an approximation and will not be exact. It does provide us, however, with a helpful way of understanding the derivative, as we will see in the following examples.

> If the derivative of a function is not changing rapidly near a point, then it is approximately equal to the change in the function when the independent variable increases by 1 unit.

Example 1   Suppose that the length of time, L, (in hours) that a drug stays in a person's system is a function of the dose or quantity administered, q, in mg. We have $L = f(q)$.
   (a)   Interpret the statement $f(10) = 6$. Give units for 10 and for 6.
   (b)   Express the derivative of the function $L = f(q)$ in the Leibniz notation. If $f'(10) = 0.5$, what are the units of 0.5?
   (c)   Clearly interpret the statement $f'(10) = 0.5$ in terms of dose and duration.

Solution   (a)   We know that $f(q) = L$. Thus, in the statement $f(10) = 6$, we have $q = 10$ and $L = 6$, so the units are 10 mg and 6 hours. The statement $f(10) = 6$ tells us that a dose of 10 mg will last for 6 hours.

(b)   Since $L = f(q)$, we see that L depends on q and is the dependent variable. The derivative of this function is dL/dq. The units of the derivative are the units of L over the units of q, or hours per mg. In the statement $f'(10) = 0.5$, the 0.5 is a derivative and so the units are 0.5 hour per mg.

(c)   The statement $f'(10) = 0.5$ tells us that, at a dose of 10 mg, the rate of change of duration is 0.5 hours per mg. In other words, if we increase the dose by 1 mg, the drug would stay in the body approximately 30 minutes longer.

Example 2   The cost C (in dollars) of building a house A square feet in area is given by the function $C = f(A)$. What is the practical interpretation of the function $f'(A)$?

Solution   In the Leibniz notation,

$$f'(A) = \frac{dC}{dA}.$$

This is a cost divided by an area, so it is measured in dollars per square foot. You can think of dC as the extra cost of building an extra dA square feet of house. Thus, dC/dA is the additional cost per square foot. So if you are planning to build a house roughly A square feet in area, $f'(A)$ is the cost per square foot of the *extra* area involved in building a slightly larger house, and is called the *marginal cost*. The marginal cost is not necessarily the same thing as the average cost per square foot for the entire house, since once you are already set up to build a large house, the cost of adding a few square feet could be comparatively small.

Example 3   The cost of extracting T tons of ore from a copper mine is $C = f(T)$ dollars. What does it mean to say that $f'(2000) = 100$?

Solution   In the Leibniz notation,

$$f'(2000) = \left.\frac{dC}{dT}\right|_{T=2000}$$

Since C is measured in dollars and T is measured in tons, dC/dT must be measured in dollars per ton. So the statement

$$\left.\frac{dC}{dT}\right|_{T=2000} = 100$$

says that when 2000 tons of ore have already been extracted from the mine, the cost of extracting the next ton is approximately \$100. Another way of saying this is that it costs about \$100 to extract ton number 2000 or 2001. Note that this may well be different from the cost of extracting the tenth ton, which is likely to be more accessible.

Example 4   If $q = f(p)$ gives the number of pounds of sugar produced when the price per pound is p dollars, then what are the units and the meaning of

$$\left.\frac{dq}{dp}\right|_{p=3} = 50?$$

Solution   The units of dq/dp are the units of q over the units of p, or pounds/dollar. The statement

$$\frac{dq}{dp}\bigg|_{p=3} = f'(3) = 50 \text{ pounds/dollar}$$

tells us that the rate of change of q with respect to p is 50 when $p = 3$. This means that when the price is \$3, the quantity produced is increasing at 50 pounds/dollar. This is an instantaneous rate of change, meaning that if the rate were to remain 50 pounds/dollar and if the price were to increase by a whole dollar, the quantity produced would increase by 50 pounds. In fact, the rate probably doesn't remain constant and so the quantity produced would probably not be exactly 50 pounds. Notice, however, that the units of the derivative and of the average rate of change are again the same. This is because the units of the instantaneous and the average rate of change are *always* the same.

Example 5   You are told that water is flowing through a pipe at a rate of 10 cubic feet per second. Interpret this rate as the derivative of some function.

Solution   You might think at first that the statement has something to do with the velocity of the water, but in fact a flow rate of 10 cubic feet per second could be achieved either with very slowly moving water through a large pipe, or with very rapidly moving water through a narrow pipe. If we look at the units—cubic feet per second—we realize that we are being given the rate of change of a quantity measured in cubic feet. But a cubic foot is a measure of volume, so we are being told the rate of change of a volume. If you imagine all the water that is flowing through ending up in a tank somewhere and let $V(t)$ be the volume of the tank at time t, then we are being told that the rate of change of $V(t)$ is 10, or that

$$V'(t) = \frac{dV}{dt} = 10.$$

Example 6   Climbing health care costs have been a source of concern for some time. The data[5] in Table 2.13 show the average yearly health care expenditures per person for various years since 1970. Use this data to estimate average per-capita expenditures in 1988 and 1995.

TABLE 2.13   *Health care costs*

Year	Per capita expenditure (\$)
1970	349
1975	591
1980	1055
1985	1596
1987	1987[8]

Solution   Health care costs have clearly been increasing throughout the last 20 years. Between 1985 and 1987 they increased $(1987 - 1596)/2 = \$195.50$ per year. To make estimates beyond 1987 we will assume that costs continue to climb at the same rate. Therefore,

---

[5]Data from the Department of Health and Human Services, reported by John Wright, ed., *Universal Almanac 1990* (Kansas City, MO: Andrews & McMeel, 1990)

[8]Yes, the last number is correct. The expenditures in 1987 were \$1987 per person.

$$\text{Costs in 1988} = \text{Costs in 1987} + \text{Change in Costs}$$
$$\approx \$1987 + \$195.50 = \$2182.50.$$

Since 1995 is 8 years beyond 1987,

$$\text{Costs in 1995} \approx \$1987 + \$195.50(8) = \$3551.$$

You should realize that the estimate for 1988 in the preceding example is much more likely to be close to the true value than the estimate for 1995. The further we extrapolate from the given data, the less accurate we are likely to be. It is unlikely that the rate of change of health care costs will stay at \$195.50/year all the way until 1995.

Graphically, what we have done is to extend the line joining the points for 1985 and 1987 to make projections for the future. (See Figure 2.31.) You might be concerned that we used only the last two pieces of data to make the estimates. Isn't there valuable information to be gained from the rest of the data? Yes, indeed—though there's no fixed way of taking this information into account. You might look at the rate of change for the years before 1985 and take an average. Alternatively, you might use linear or exponential regression to construct a mathematical model, and use the model to estimate the future costs.

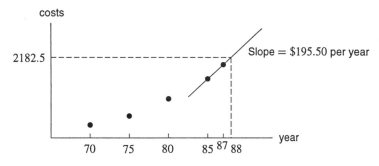

Figure 2.31: Graph of health care costs

Problems for Section 2.4

1. The weight, W, in lbs, of a child is a function of the age, a, of the child, in years. We have $W = f(a)$.
   (a) Do you expect $f'(a)$ to be positive or negative? Why?
   (b) What does $f(8) = 45$ tell you? Give units for 8 and for 45.
   (c) What are the units of $f'(a)$? Clearly explain what $f'(a)$ tells you in terms of age and weight.
   (d) What does $f'(8) = 4$ tell you about age and weight?
   (e) As a increases, do you expect $f'(a)$ to increase or decrease? Explain.

2. In Chapter 1, we saw that the thickness of the eggshell of pelican eggs, P, in mm, depends on the concentration, c, of PCBs in the eggshell, measured in ppm (parts per million). We have $P = f(c)$.
   (a) The derivative $f'(c)$ is negative. What does this tell you?
   (b) Interpret the statements $f(200) = 0.28$ and $f'(200) = -0.0005$ in terms of PCBs and pelican eggs. Be sure to use units appropriately.

3. Let $f(x)$ be the elevation in feet of the Mississippi river x miles from its source. What are the units of $f'(x)$? What can you say about the sign of $f'(x)$? (Assume that $0 \leq x \leq$ length of the river.)

4. Let $g(t)$ be the height, in inches, of Amelia Earhart (one of the first woman airplane pilots) t years after her birth. What are the units of $g'(t)$? What can you say about the signs of $g'(10)$ and $g'(30)$? (Assume that $0 \leq t < 39$, the age at which Amelia Earhart's plane disappeared.)

5. Suppose $C(r)$ is the total cost of paying off a car loan borrowed at an annual interest rate of $r\%$. What are the units of $C'(r)$? What is the practical meaning of $C'(r)$? What is its sign?

6. Suppose $P(t)$ is the monthly payment on a mortgage which will take t years to pay off. What are the units of $P'(t)$? What is the practical meaning of $P'(t)$? What is its sign?

7. After investing \$1000 at an annual interest rate of 7% compounded continuously for t years, your balance is \$$B$, where $B = f(t)$. What are the units of dB/dt? What is the financial interpretation of dB/dt?

8. An economist is interested in how the price of a certain commodity affects its sales. Suppose that at a price of \$$p$, a quantity q of the commodity is sold. If $q = f(p)$, explain in economic terms the meaning of the statements $f(10) = 240,000$ and $f'(10) = -29,000$.

9. The temperature, T, in degrees Fahrenheit, of a cold yam placed in a hot oven is given by $T = f(t)$, where t is the time in minutes since the yam was put in the oven.

   (a) What is the sign of $f'(t)$? Why?
   (b) What are the units of $f'(20)$? What is the practical meaning of the statement $f'(20) = 2$?

10. Investing \$1000 at an annual interest rate of $r\%$, compounded continuously, for 10 years gives you a balance of \$$B$, where $B = g(r)$. What is a financial interpretation of the statements

    (a) $g(5) \approx 1649$?
    (b) $g'(5) \approx 165$? What are the units of $g'(5)$?

11. Table 2.14 shows the public debt of the United States between 1980 and 1993, as a function of year.[9]

    (a) Estimate the derivative of this function in 1993. Give units with your answer and interpret it.
    (b) Use your answer to part (a) to estimate the public debt in 1994, and in the year 2000. Which answer should you have more confidence in and why?

TABLE 2.14 *Public Debt of the United States*

Year	Debt (billions)	Year	Debt (billions)
1980	907.7	1987	2350.3
1981	997.9	1988	2602.3
1982	1142.0	1989	2857.4
1983	1377.2	1990	3233.3
1984	1572.3	1991	3665.3
1985	1823.1	1992	4064.6
1986	2125.3	1993	4351.2

[9]*The World Almanac*, 1995

12. Suppose that $f(t)$ is a function, that $f(25) = 3.6$ and that $f'(25) = -0.2$. Use this information to estimate $f(26)$ and $f(30)$.

13. If $g(v)$ is the fuel efficiency of a car going at v miles per hour (i.e., $g(v) = $ the number of miles per gallon at v mph), what are the units of $g'(55)$? What is the practical meaning of the statement $g'(55) = -0.54$?

14. Let P be the total petroleum reservoir on earth in the year t. (In other words, P represents the total quantity of petroleum, including what's not yet discovered, on earth at time t.) Assume that no new petroleum is being made and that P is measured in barrels. What are the units of dP/dt? What is the meaning of dP/dt? What is its sign? How would you set about estimating this derivative in practice? What would you need to know to make such an estimate?

15. (a) If you jump out of an airplane without a parachute, you will fall faster and faster until wind resistance causes you to approach a steady velocity, called a *terminal* velocity. Sketch a graph of your velocity against time.
    (b) Explain the concavity of your graph.
    (c) Assuming wind resistance to be negligible at $t = 0$, what natural phenomenon is represented by the slope of the graph at $t = 0$?

16. A company's revenue from car sales, R (measured in thousands of dollars), is a function of advertising expenditure, a, also measured in thousands of dollars. Suppose $R = f(a)$.
    (a) What does the company hope is true about the sign of f'?
    (b) What does the statement $f'(100) = 2$ mean in practical terms? How about $f'(100) = 0.5$?
    (c) Suppose the company plans to spend about \$100,000 on advertising. If $f'(100) = 2$, should the company spend slightly more or slightly less than \$100,000 on advertising? What if $f'(100) = 0.5$?

17. Let $P(x) = $ the number of people in the US of height $\leq x$ inches. What is the meaning of $P'(66)$? What are its units? Estimate $P'(66)$ (using common sense). Is $P'(x)$ ever negative? [Hint: You may want to approximate $P'(66)$ by a difference quotient, using $h = 1$. Also, you may use the fact that the US population is about 250 million, and that 66 inches = 5 feet 6 inches.]

TABLE 2.15 *US population (in millions), 1790–1990*

Year	Population	Year	Population	Year	Population	Year	Population
1790	3.9	1850	23.1	1910	92.0	1970	205.0
1800	5.3	1860	31.4	1920	105.7	1980	226.5
1810	7.2	1870	38.6	1930	122.8	1990	248.7
1820	9.6	1880	50.2	1940	131.7		
1830	12.9	1890	62.9	1950	150.7		
1840	17.1	1900	76.0	1960	179.0		

18. Table 2.15 gives the US population figures between 1790 and 1990.
    (a) Estimate the rate of change of the population for the years 1900, 1945, and 1990.
    (b) When, approximately, was the rate of change of the population greatest?
    (c) Estimate the US population in 1956.
    (d) Based on the data from the table, what would you predict for the census in the year 2000?

19. (a) In Problem 18, we thought of the US population as a smooth function of time. To what extent is this justified? What happens if we zoom in at a point of the graph? What about events such as the Louisiana Purchase? Or the moment of your birth?

    (b) What do we in fact mean by the rate of change of the population for a particular time t?

    (c) Give another example of a real-world function which is not smooth but is usually treated as such.

2.5 THE SECOND DERIVATIVE

Since the derivative is itself a function, we can often calculate its derivative. For a function f, the derivative of its derivative is called the *second derivative*, and written f'' (read "f double-prime"). If $y = f(x)$, the second derivative can also be written as $\dfrac{d^2y}{dx^2}$, which means $\dfrac{d}{dx}\left(\dfrac{dy}{dx}\right)$, the derivative of $\dfrac{dy}{dx}$.

What Does the Second Derivative Tell Us?

Recall that the derivative of a function tells you whether the function is increasing or decreasing:

> If $f' > 0$ on an interval, then f is *increasing* over that interval.
> If $f' < 0$ on an interval, then f is *decreasing* over that interval.

Since f'' is the derivative of f',

> If $f'' > 0$ on an interval, then f' is *increasing* over that interval.
> If $f'' < 0$ on an interval, then f' is *decreasing* over that interval.

So the question becomes: What does it mean for f' to be increasing or decreasing? The case in which f' is increasing is shown in Figure 2.32, where the graph of f is bending upward, or is *concave up*. In the case when f' is decreasing, shown in Figure 2.33, the graph is bending downward, or is *concave down*.

> $f'' > 0$ means f' is increasing, so the graph of f is concave up on that interval,
> $f'' < 0$ means f' is decreasing, so the graph of f is concave down on that interval.

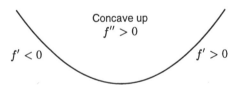

Figure 2.32: Meaning of f'': The slope increases from left to right, so f'' is positive and f is concave up

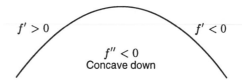

Figure 2.33: Meaning of f'': The slope decreases from left to right, so f'' is negative and f is concave down

---

Example 1   For the functions whose graphs are given in Figure 2.34, decide where their second derivatives are positive and where they are negative.

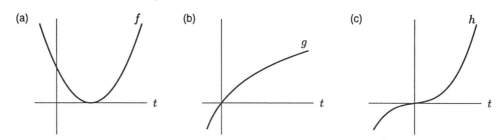

Figure 2.34: What signs do the second derivatives have?

Solution   From the graphs it appears that
(a) $f'' > 0$ everywhere, because the graph of f is concave up everywhere.
(b) $g'' < 0$ everywhere, because the graph is concave down everywhere.
(c) $h'' > 0$ for $t > 0$, because the graph of h is concave up there; $h'' < 0$ for $t < 0$, because the graph of h is concave down there.

Interpretation of the Second Derivative as a Rate of Change

If we think of the derivative as a rate of change, then the second derivative is a rate of change of a rate of change. If the second derivative is positive, the rate of change is increasing; if the second derivative is negative, the rate of change is decreasing.

The second derivative is often a matter of practical concern. In 1985 a newspaper headline reported the Secretary of Defense as saying that Congress and the Senate had cut the defense budget. As his opponents pointed out, however, Congress had merely cut the rate at which the defense budget was increasing.[10] In other words, the derivative of the defense budget was still positive (the budget was increasing), but the second derivative was negative (the budget's rate of increase had slowed).

Example 2   A population, P, growing in a confined environment often follows a *logistic* growth curve, like the graph shown in Figure 2.35. Describe how the rate at which the population is increasing changes over time. What is the sign of the second derivative d^2P/dt^2? What is the practical interpretation of t_0 and L?

---

[10]In the *Boston Globe*, March 13, 1985, Representative William Gray (D–Pa.) was reported as saying: "It's confusing to the American people to imply that Congress threatens national security with reductions when you're really talking about a reduction in the increase."

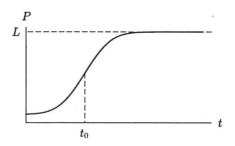

Figure 2.35: Logistic growth curve

Solution   Initially, the population is increasing, and at an increasing rate. Thus, initially dP/dt is increasing and so $d^2P/dt^2 > 0$. At t_0, the rate at which the population is increasing is a maximum. Thus at time t_0 the population is growing fastest. Beyond t_0, the rate at which the population is growing is decreasing, and so $d^2P/dt^2 < 0$. At t_0, the concavity of the graph changes from positive to negative, and $d^2P/dt^2 = 0$.

The quantity L represents the limiting value of the population which is approached as t tends to infinity. L is called the *carrying capacity* of the environment and represents the maximum population that the environment can support.

Example 3   Table 2.16 shows the number of abortions per year, A, performed in the US in the year t (as reported to the Center for Disease Control and Prevention). Suppose these data points lie on a smooth curve $A = f(t)$.

TABLE 2.16   *Abortions reported in the US (1972–1985)*

Year, t	1972	1976	1980	1985
Number of abortions reported, A	586,760	988,267	1,297,606	1,328,570

(a)   Estimate dA/dt for the time intervals shown between 1972 and 1985.
(b)   What can you say about the sign of d^2A/dt^2 during the period 1972–1985?

Solution   (a)   For each time interval we can calculate the average rate of change of the number of abortions per year over this interval. For example, between 1972 and 1976

$$\frac{dA}{dt} \approx \begin{array}{c} \text{Average rate} \\ \text{of change} \end{array} = \frac{\Delta A}{\Delta t} = \frac{988{,}267 - 586{,}760}{1976 - 1972} \approx 100{,}377.$$

Approximate values of dA/dt are listed in Table 2.17:

TABLE 2.17   *Rate of change of number of abortions reported*

Time	1972–1976	1976–1980	1980–1985
Average rate of change, $\Delta A/\Delta t$	100,377	77,335	6,193

(b) Since the approximate values $\Delta A/\Delta t$ of dA/dt are decreasing so dramatically for 1976–1985, we can be pretty certain that dA/dt also decreases, and so d^2A/dt^2 is negative for this period. For 1972–1976, the sign of d^2A/dt^2 is less clear; abortion data from 1968 would help. The graph of A against t in Figure 2.36 confirms this; the graph appears to be concave down for 1976–1985, but straight for 1972–1976. The fact that dA/dt is positive tells us that the number of abortions reported has increased during the period 1972–1985. The fact that d^2A/dt^2 is negative for 1976–1985 tells us that the rate of increase has slowed over this period.

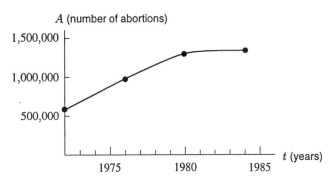

Figure 2.36: How the number of reported abortions in the US is changing with time

Problems for Section 2.5

1. (a) If f'' is positive on an interval, then f' is _____ on that interval, and f is _____ on that interval.
    (b) If f'' is negative on an interval, then f' is _____ on that interval, and f is _____ on that interval.

2. (a) Sketch the graph of a function whose first and second derivatives are everywhere positive.
    (b) Sketch the graph of a function whose second derivative is everywhere negative but whose first derivative is everywhere positive.
    (c) Sketch the graph of a function whose second derivative is everywhere positive but whose first derivative is everywhere negative.
    (d) Sketch the graph of a function whose first and second derivatives are everywhere negative.

3. (a) Sketch a smooth curve whose slope is both positive and increasing at first but later is positive and decreasing.
    (b) Sketch the graph of the first derivative of the function whose graph is the curve in part (a).
    (c) Sketch the graph of the second derivative of the function whose graph is the curve in part (a).

4. Graphs of two functions are given in Figure 2.37. For each function,
    (a) Estimate the intervals over which the derivative is positive and the intervals over which the derivative is negative.
    (b) Estimate the intervals over which the second derivative is positive and the intervals over which the second derivative is negative.

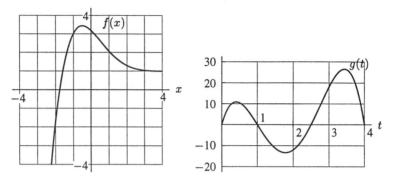

Figure 2.37

5. Table 2.18 shows tables of values for two functions, $s(t)$ and $w(t)$. For each function:
   (a) Is the derivative positive or negative?
   (b) Is the second derivative positive or negative?

TABLE 2.18

t	0	1	2	3	4	5
$s(t)$	12	14	17	20	31	55

t	100	110	120	130	140	150
$w(t)$	10.7	6.3	4.2	3.5	3.3	3.2

6. "Winning the war on poverty" has been described cynically as slowing the rate at which people are slipping below the poverty line. Assuming that this is happening:
   (a) Sketch a graph of the total number of people in poverty against time.
   (b) If N is the number of people below the poverty line at time t, what are the signs of dN/dt and d^2N/dt^2?

7. In economics, *total utility* refers to the total satisfaction from consuming some commodity. According to the economist Samuelson[11]:

   > As you consume more of the same good, the total (psychological) utility increases. However, ... with successive new units of the good, your total utility will grow at a slower and slower rate because of a fundamental tendency for your psychological ability to appreciate more of the good to become less keen.

   (a) Sketch the total utility as a function of the number of units consumed.
   (b) In terms of derivatives, what is Samuelson telling us?

8. Let $P(t)$ represent the price of a share of stock of a corporation at time t. What does each of the following statements tell us about the signs of the first and second derivatives of $P(t)$?
   (a) "The price of the stock is rising faster and faster."
   (b) "The price of the stock is close to bottoming out."

---

[11]From Paul A. Samuelson, *Economics*, 11th edition (New York: McGraw-Hill, 1981).

9. IBM-Peru uses second derivatives to assess the relative success of various advertising campaigns. They assume that all campaigns produce some increase in sales. If a graph of sales against time shows a positive second derivative during a new advertising campaign, what does this suggest to IBM management? Why? What does a negative second derivative during a new campaign suggest?

10. Given the following data:

x	0	0.2	0.4	0.6	0.8	1.0
$f(x)$	3.7	3.5	3.5	3.9	4.0	3.9

(a) Estimate $f'(0.6)$ and $f'(0.5)$.   (b) Estimate $f''(0.6)$.

(c) Where do you think the maximum and minimum values of f occur in the interval $0 \le x \le 1$?

11. An industry is being charged by the Environmental Protection Agency (EPA) with dumping unacceptable levels of toxic pollutants in a lake. Over a period of several months, an engineering firm makes daily measurements of the rate at which pollutants are being discharged into the lake.

Suppose the engineers produce a graph similar to either Figure 2.38(a) or Figure 2.38(b). For each case, give an idea of what argument the EPA might make in court against the industry and of the industry's defense.

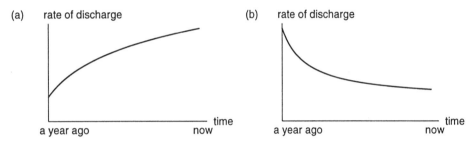

Figure 2.38: Discharge of pollutants

12. A high school principal is concerned about the drop in the percentage of high school students who graduate from her school, shown in the table below.

Year entering high school, t	Percent graduating, P
1977	62.4
1980	54.1
1983	48.0
1986	43.5
1989	41.8

(a) Estimate dP/dt for each of the three-year intervals between 1977 and 1989.

(b) Does d^2P/dt^2 appear to be positive or negative between 1977 and 1989?

(c) Explain why the values of P and dP/dt are troublesome to the principal.

(d) Explain why the sign of d^2P/dt^2 and the magnitude of dP/dt in the year 1986 may give the principal some cause for optimism.

13. The graph of f' (not f) is given in Figure 2.39. At which of the marked values of x is
    (a) $f(x)$ greatest?   (b) $f(x)$ least?   (c) $f'(x)$ greatest?   (d) $f'(x)$ least?
    (e) $f''(x)$ greatest?   (f) $f''(x)$ least?

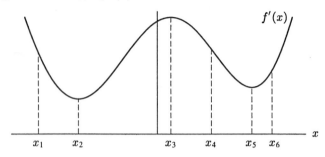

Figure 2.39: Note that this is a graph of f' against x

14. Which of the points labeled by letters in the graph of f in Figure 2.40 have
    (a) f' and f'' nonzero and of the same sign?
    (b) At least two of f, f', and f'' equal to zero?

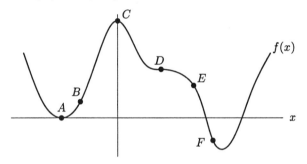

Figure 2.40

2.6 MARGINAL COST AND REVENUE

Management decisions within a particular firm or industry usually depend on the costs and revenues involved. In this section we will look at applications of the derivative to the cost and revenue functions.

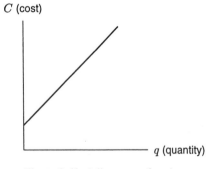

Figure 2.41: A linear cost function

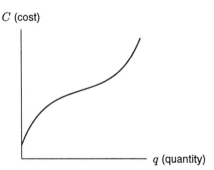

Figure 2.42: A nonlinear cost function

In Chapter 1, we considered primarily linear cost functions (See Figure 2.41.) In practice, however, cost functions usually have graphs of the general shape shown in Figure 2.42. Recall that the intercept on the C-axis represents the fixed costs, which are incurred even if nothing is produced. (This includes, for instance, the cost of the machinery needed to begin production.) The cost function increases quickly at first and then more slowly because producing larger quantities of a good is usually more efficient than producing smaller quantities—this is called economy of scale. At still higher production levels, the cost function starts to increase faster again as resources become scarce, and sharp increases may occur when new factories have to be built. Thus, the graph of the cost function $C(q)$ starts out concave down and becomes concave up later on.

Recall that if the price, p, is a constant, then the revenue function is $R = pq$ and the graph of R against q is a straight line through the origin with slope equal to the price. See Figure 2.43. In practice, for large values of q, the market may become glutted, causing the price to drop and giving R the shape in Figure 2.44.

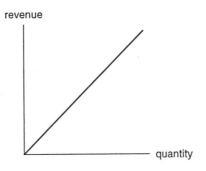

Figure 2.43: Revenue: Constant price

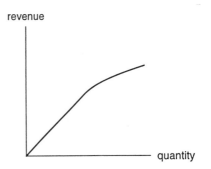

Figure 2.44: Revenue: Decreasing price

Example 1   If cost, C, and revenue, R, are given by the graph in Figure 2.45, for what production quantities does the firm make a profit?

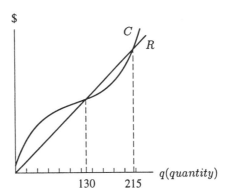

Figure 2.45: Costs and revenues for
Example 1

Solution   The firm makes a profit whenever revenues are greater than costs, that is, when $R > C$. The graph of R is above the graph of C approximately when $130 < q < 215$. Production between 130 units and 215 units will generate a profit.

Marginal Analysis

Many economic decisions are based on an analysis of the costs and revenues "at the margin." Let's look at this idea through an example.

Suppose you are running an airline and you are trying to decide whether to offer an additional flight. How should you decide? We'll assume that the decision is to be made purely on financial grounds: if the flight will make money for the company, it should be added. Obviously you need to consider the costs and revenues involved. Since the choice is between adding this flight and leaving things the way they are, the crucial question is whether the *additional costs* incurred are greater or smaller than the *additional revenues* generated by the flight. These additional costs and revenues are called the *marginal costs* and *marginal revenues*.

Suppose $C(q)$ is the function giving the cost of running q flights. If the airline had originally planned to run 100 flights, its costs would be $C(100)$. With the additional flight, its costs would be $C(101)$. Therefore,

$$\text{Marginal cost} = C(101) - C(100).$$

Now

$$C(101) - C(100) = \frac{C(101) - C(100)}{101 - 100},$$

and this quantity is the average rate of change of cost between 100 and 101 flights. In Figure 2.46 the average rate of change is the slope of the secant line. If the graph of the cost function is not curving too fast near the point, the slope of the secant line will be close to the slope of the tangent line there. Therefore, the average rate of change is close to the instantaneous rate of change. Since these rates of change are not very different, many economists choose to define marginal cost, MC, as the instantaneous rate of change of cost with respect to quantity:

$$\boxed{\text{Marginal cost} = MC = C'(q).}$$

Similarly, if the revenue generated by q flights is $R(q)$, then the additional revenue generated by increasing the number of flights from 100 to 101 is

$$\text{Marginal revenue} = R(101) - R(100).$$

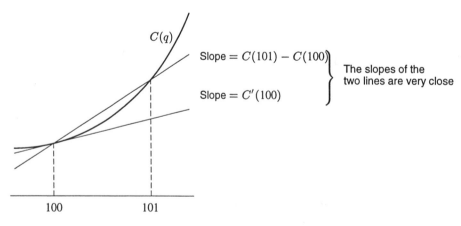

Figure 2.46: Marginal cost: Slope of one of these lines

Now $R(101) - R(100)$ is the average rate of change of revenue between 100 and 101 flights. As before, the average rate of change is usually almost equal to the instantaneous rate of change, so economists often define

$$\boxed{\text{Marginal revenue} = MR = R'(q).}$$

Example 2   The graph of a cost function is given in Figure 2.47. Does it cost more to produce the 500th item or the 2000th? Does it cost more to produce the 3000th item or the 4000th? At approximately what production level is marginal cost smallest? What is the total cost at this production level?

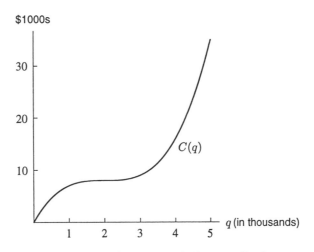

Figure 2.47: Where is marginal cost smallest?

Solution   The cost to produce an additional item is the marginal cost. Since marginal cost is the derivative of the cost function, it is given graphically by the slope of the graph of the cost function. Since the slope of the cost function in Figure 2.47 is greater at $q = 0.5$ (when the quantity produced is 0.5 thousand, or 500) than at $q = 2$, we see that it costs more to produce the 500th item than the 2000th item. Since the slope is greater at $q = 4$ than $q = 3$, it costs more to produce the 4000th item than the 3000th item.

The slope of the cost function is approximately zero around $q = 2$, and is positive everywhere else, so the slope is smallest at $q = 2$. The marginal cost is smallest at a production level of 2000 units. Since $C(2) \approx 8$, the total cost at a production level of 2000 units is about $8000.

Example 3   If $C(q)$ and $R(q)$ for the airline are given in Figure 2.48, should the company add the 101st flight?

Solution   The marginal revenue is the slope of the revenue line. The marginal cost is the slope of the graph of C at the point 100. From Figure 2.48, you can see that the slope at the point A is smaller than the slope at B, so $MC < MR$ for $q = 100$. This means that the airline will make more in extra revenue than it will spend in extra costs if it runs another flight, so it should go ahead and run the 101st flight.

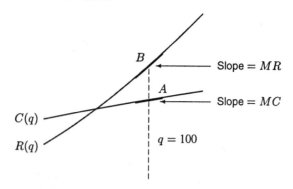

Figure 2.48: Cost and revenue for Example 3

Since MC and MR are derivative functions, they can be estimated from the graphs of total cost and total revenue.

Example 4   If R and C are given by the graphs in Figure 2.49, sketch graphs of $MR = R'(q)$ and $MC = C'(q)$.

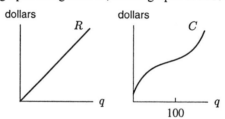

Figure 2.49: Total revenue and total cost for
Example 4

Solution   The revenue graph is a line through the origin, with equation

$$R = pq$$

where p represents the constant price, so the slope is p and

$$MR = R'(q) = p.$$

The total cost is increasing, so the marginal cost is always positive. For small q values, the graph of the cost function is concave down, so the marginal cost is decreasing. For larger q, say $q > 100$, the graph of the cost function is concave up and the marginal cost is increasing. Thus the marginal cost has a minimum at about $q = 100$. (See Figure 2.50.)

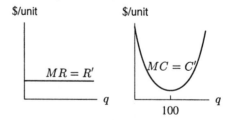

Figure 2.50: Marginal revenue and costs for Example 4

Maximizing Profit

Now let's look at how to maximize total profit, given functions for revenue and cost. This is clearly a fundamental issue for any producer of goods. The next example looks at this problem and arrives at a result which is generally valid.

Example 5   Estimate the maximum profit if the revenue and cost are given by the curves R and C, respectively, in Figure 2.51.

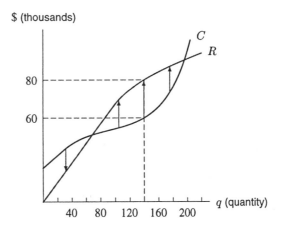

Figure 2.51: Maximum profit at $q = 140$

Solution   The profit is represented by the vertical difference between the curves and is marked by the vertical arrows on the graph. When revenue is below cost, the company is taking a loss; when revenue is above cost, the company is making a profit. The maximum profit must occur between about $q = 70$ and $q = 200$, which is the range where the company is making a profit. Profit is maximized when the vertical distance between the curves is as large as possible (and revenue is above cost). On the graph, we see that this occurs at approximately $q = 140$.

The maximum profit occurs at a production level of about 140 units. Since profit is revenue minus cost, the actual profit at this production level is the vertical distance between the two curves at $q = 140$. We see that the maximum profit = $80,000 - $60,000 = $20,000.

Maximum Profit occurs where $MR = MC$

For the cost and revenue functions given in Figure 2.51, we saw that profit is maximized at a production level of $q = 140$. Let's examine this more closely. If we zoom in on the cost and revenue functions around $q = 140$, we have the graph given in Figure 2.52.

At any production level q_1 to the left of 140, we see in Figure 2.52 that marginal cost is less than marginal revenue. The company will make money by producing more units, so it makes sense to increase production (toward a production level of 140). At any production level q_2 to the right of 140, we see that the marginal cost is greater than marginal revenue. The company loses money by producing more units (or, equivalently, makes money by producing fewer units). Production should be adjusted down, again pushing production toward $q = 140$.

What about the marginal revenue and marginal cost at $q = 140$? Since MC < MR to the left of 140, and MC > MR to the right of 140, we must have MC = MR at $q = 140$. You should verify

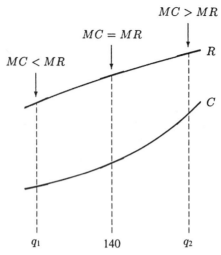

Figure 2.52: Maximum profit occurs where
$MC = MR$

this by noting that the slopes of the tangent lines to the cost and revenue graphs in Figure 2.52 are equal at $q = 140$. Profit is maximized at the point where the slopes of the cost and revenue graphs are equal.

> The Maximum (or minimum) profit can occur when
>
> Marginal cost = Marginal revenue.

Problems for Section 2.6

1. The cost and revenue functions for a charter bus company are shown in Figure 2.53. Should the company add a 50th bus? How about a 100th? Explain your answers using marginal revenue and cost.

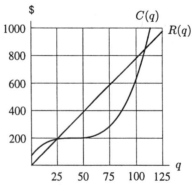

Figure 2.53: Cost and revenue for a charter bus company

2. In Figure 2.54, is marginal cost greater at $q = 5$ or $q = 30$? At $q = 20$ or $q = 40$? Justify your answers graphically.

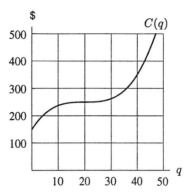

Figure 2.54: A cost function

3. A manufacturer reports the cost and revenue functions shown in Figure 2.55. Sketch graphs, as a function of quantity, of
    (a) Total profit,   (b) Marginal cost,   (c) Marginal revenue.
    Label the points q_1 and q_2 on your graphs.

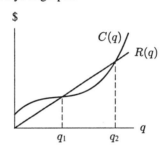

Figure 2.55: Cost and revenue for a good

4. Let $C(q)$ be the total cost of producing a quantity q of a certain good. (See Figure 2.56.)
    (a) What is the meaning of $C(0)$?
    (b) Describe in words how the marginal cost changes as the quantity produced increases.
    (c) Explain the concavity of the graph (in terms of economics).
    (d) Explain the economic significance (in terms of marginal cost) of the point at which the concavity changes.
    (e) Do you think the graph of $C(q)$ looks like this for all types of goods?

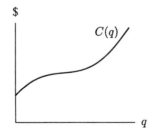

Figure 2.56: Cost of producing a good

5. Table 2.19 shows cost and revenue values for different production levels, q. At approximately what production level is profit maximized?

    TABLE 2.19

q	0	100	200	300	400	500
$R(q)$	0	500	1000	1500	2000	2500
$C(q)$	700	900	1000	1100	1300	1900

6. The cost function of a paper recycling plant is given in the following table. Estimate the marginal cost at $q = 2000$, and give units with your answer. Interpret your answer in terms of cost. At approximately what production level does marginal cost appear smallest?

q (tons of recycled paper)	1000	1500	2000	2500	3000	3500
$C(q)$ (dollars)	2500	3200	3640	3825	3900	4400

7. The graphs of the cost and revenue functions for a certain company are given in Figure 2.57. Approximately what quantity should be produced if the company wants to maximize profits? Explain your answer graphically.

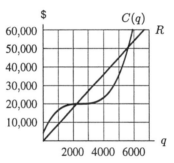

    Figure 2.57: What quantity
    maximizes profits?

8. Cost and revenue functions are given in Figure 2.57.
    (a) At a production level of $q = 3000$, is marginal cost or marginal revenue greater? Explain clearly what this tells you about whether the company should increase or decrease production.
    (b) Answer the same questions for a production level of $q = 5000$.

9. Marginal revenue and marginal cost at different production levels are given in Table 2.20. Estimate the production level that will maximize profit. Justify your answer.

    TABLE 2.20

q	1000	2000	3000	4000	5000	6000
MR	78	76	74	72	70	68
MC	100	80	70	65	75	90

10. At a production level of 1000 items, the total cost is $5000 and the marginal cost is $25. Use this information to estimate the costs of producing 1001 items, of producing 999 items, and of producing 1100 items.

11. Let $C(q)$ represent the total cost of producing q items. Then $C'(q)$ gives the marginal cost in dollars per item. Suppose a company determines that $C(15) = 2300$ and $C'(15) = 108$.

   (a) Estimate the total cost of producing 16 items.
   (b) Estimate the total cost of producing 14 items.

12. Let $C(q)$ represent the cost, $R(q)$ the revenue and $\pi(q)$ the total profit, where q is the number of items produced. Assume that $C(q)$, $R(q)$, and $\pi(q)$ are all measured in dollars.

   (a) If $C'(50) = 75$ and $R'(50) = 84$, approximately how much profit will be earned by the 51st item?
   (b) If $C'(90) = 71$ and $R'(90) = 68$, approximately how much profit will be earned by the 91st item?
   (c) If $\pi(q)$ reaches its maximum value when $q = 78$, how do you think $C'(78)$ and $R'(78)$ will compare? Explain.

13. Figure 2.58 shows graphs of marginal cost and marginal revenue. Estimate the production level that will maximize profit. Explain your reasoning.

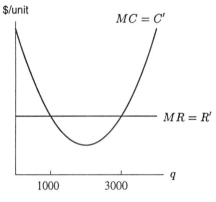

Figure 2.58

14. In Figure 2.59, find the marginal cost when the level of production is 10,000 units and interpret it.

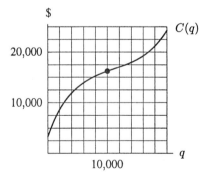

Figure 2.59

15.   In Figure 2.60, find the marginal revenue when the level of production is 600 units and interpret it.

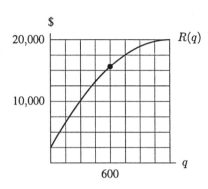

Figure 2.60

REVIEW PROBLEMS FOR CHAPTER TWO

1.   For $-3 \leq x \leq 7$, use a graphing calculator or computer to graph

$$f(x) = (x^3 - 6x^2 + 8x)(2 - 3^x).$$

(a)   How many zeros does f have in this interval?
(b)   Is f increasing or decreasing at $x = 0$? At $x = 2$? At $x = 4$?
(c)   On which interval is the average rate of change of f greater: $-1 \leq x \leq 0$ or $2 \leq x \leq 3$?
(d)   Is the instantaneous rate of change of f greater at $x = 0$ or $x = 2$?

2.   For the function $f(x) = 3^x$, estimate $f'(1)$. From the graph of $f(x)$, would you expect your estimate to be greater than or less than $f'(1)$?

3.   Let $f(x) = x^3$. Estimate the derivatives $f'(1)$, $f'(3)$, and $f'(5)$. Then guess a general formula for $f'(x)$.

For Problems 4–5, sketch a graph of the derivative function of the given functions.

4.

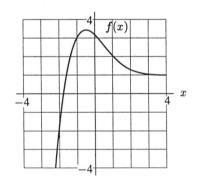

5.

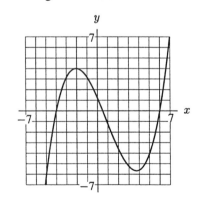

Sketch the graphs of the derivatives of the functions shown in Problems 6–10. Be sure your sketches are consistent with the important features of the original functions.

6.
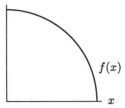

7.
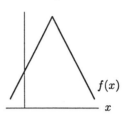

8.
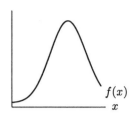

9.
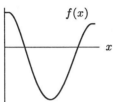

10.
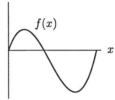

11. Draw a possible graph of $y = f(x)$ given the following information about its derivative:
    - For $x < -2$, $f'(x) > 0$ and the derivative is increasing.
    - For $-2 < x < 1$, $f'(x) > 0$ and the derivative is decreasing.
    - At $x = 1$, $f'(x) = 0$.
    - For $x > 1$, $f'(x) < 0$ and the derivative is decreasing (getting more and more negative).

12. A function $f(x)$ is shown in Figure 2.61.
    (a) At which of the labeled points is $f'(x)$ positive? Negative? Zero?
    (b) At which labeled point is f' largest? At which labeled point is f' most negative?

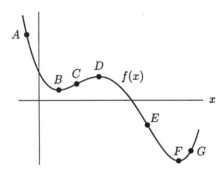

Figure 2.61

13. For a function $f(x)$, we know that $f(20) = 68$ and $f'(20) = -3$. Estimate $f(21), f(19)$ and $f(25)$. Explain your reasoning.

14. For one package mailing company, the cost, C, in dollars to send a package is a function of the weight, w, in pounds, of the package. We have $C = f(w)$.
    (a) In the statement $f(12) = 5$, what are the units of 12? What are the units of 5? Explain what this is telling us in terms of the cost to mail packages.
    (b) Do you expect the derivative f' to be positive or negative? Why?
    (c) In the statement $f'(12) = 0.4$, what are the units of 12? What are the units of 0.4? Explain what this is telling us in terms of the cost to mail packages.

15.   Given the following table of values:

x	0	0.2	0.4	0.6	0.8	1.0
$f(x)$	3.7	3.5	3.5	3.9	4.0	3.9

(a)   Estimate $f'(0.6)$.
(b)   Estimate an equation of the tangent line to $y = f(x)$ at $x = 0.6$.
(c)   Using this equation, estimate $f(0.7)$, $f(1.2)$, and $f(1.4)$. Which of these estimates do you feel most confident about? Why?

16.   Given the following data about a function, f:

x	6.5	7.0	7.5	8.0	8.5	9.0
$f(x)$	10.3	8.2	6.5	5.2	4.1	3.2

(a)   Estimate $f'(7.0)$, $f'(8.5)$ and $f'(6.75)$.
(b)   Estimate the rate of change of f' at $x = 7$.
(c)   Find, approximately, an equation of the tangent line to the graph of f at $x = 7$.
(d)   Estimate $f(6.8)$.

17.   Suppose you put a yam in a hot oven, maintained at a constant temperature of 200°C. As the yam picks up heat from the oven, its temperature rises.[12]

(a)   Draw a possible graph of the temperature T of the yam against time t (minutes) since it is put into the oven. Explain any interesting features of the graph, and in particular explain its concavity.
(b)   Suppose that, at $t = 30$, the temperature T of the yam is 120° and increasing at the (instantaneous) rate of 2°/min. Using this information, plus what you know about the shape of the T graph, estimate the temperature at time $t = 40$.
(c)   Suppose in addition you are told that at $t = 60$, the temperature of the yam is 165°. Can you improve your estimate of the temperature at $t = 40$?
(d)   Assuming all the data given so far, estimate the time at which the temperature of the yam is 150°.

18.   The graphs of the revenue and cost functions for a certain company are given in Figure 2.62.

(a)   Estimate the marginal cost at $q = 400$.
(b)   Should the company produce the 500th item? Explain.
(c)   Estimate the quantity which maximizes profit.

---

[12]From Peter D. Taylor, *Calculus: The Analysis of Functions* (Toronto: Wall & Emerson, Inc., 1992)

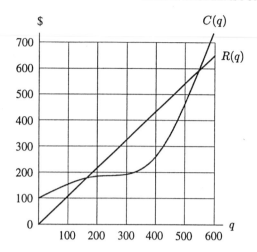

Figure 2.62: Cost and revenue functions

19. A software firm is trying to determine how much money to invest each year in research and development. An extensive study indicates that if they invest x million dollars each year in research and development then they can expect to make a profit, $\pi(x)$, given in thousands of dollars, by

$$\pi(x) = -0.01x^4 + 0.33x^3 + 2.69x^2 + 4.35x - 500.$$

   (a) Use a graphing calculator or computer to sketch the graph of $\pi(x)$.
   (b) For what values of x is the company going to make a profit?
   (c) For what values of x is $\pi(x)$ increasing and for what values of x is $\pi(x)$ decreasing?
   (d) For what values of x is $\pi'(x)$ increasing and for what values of x is $\pi'(x)$ decreasing?
   (e) Using the fact that $\pi'(x) = -0.04x^3 + 0.99x^2 + 5.38x + 4.35$, graph $\pi'(x)$ and check your answer to part (d).
   (f) Which value of x maximizes the rate of change of $\pi(x)$ with respect to x?

20. If the population, in millions, of Mexico during the 1980s was $P(t) = 68.4(1.026)^t$, with t measured in years after 1980, find the rate of growth, in people/year, in 1986.

21. To study traffic flow, the city installs a device on Main Street at 4:00 AM. The device counts the cars driving past, and records the total periodically. The resulting data is plotted on a graph with time (in hours) on the horizontal axis and the number of cars on the vertical axis. The graph is shown below; it is the graph of the function

$$C(t) = \text{the total number of cars that have passed by after } t \text{ hours.}$$

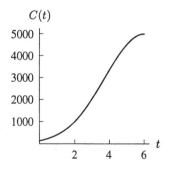

(a) When is the traffic flow the greatest?

(b) Estimate $C'(2)$.

(c) What does $C'(2)$ mean in practical terms?

22. Alone in your dim, unheated room, you light a single candle rather than curse the darkness. Depressed with the situation, you walk directly away from the candle, sighing. The temperature (in degrees Fahrenheit) and illumination (in % of one candle power) decrease as your distance (in feet) from the candle increases. In fact, you have tables showing this information.

TABLE 2.21

distance (feet)	temperature (° F)
0	55
1	54.5
2	53.5
3	52
4	50
5	47
6	43.5

TABLE 2.22

distance (feet)	illumination (%)
0	100
1	85
2	75
3	67
4	60
5	56
6	53

You are cold when the temperature is below 40°. You are in the dark when the illumination is at most 50% of one candle power.

(a) Two graphs are sketched below; one is temperature as a function of distance and one is illumination as a function of distance. Which is which? Explain.

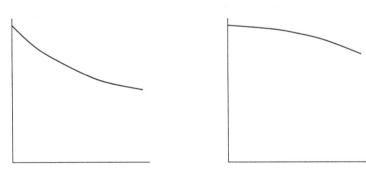

Figure 2.63   *Figure 2.64*

(b) What is the average rate at which the temperature is changing when the illumination drops from 75% to 56%?

(c) You can still read your watch when the illumination is about 65%. Can you still read your watch at 3.5 feet? Explain.

(d) Suppose you know that at 6 feet the instantaneous rate of change of the temperature is −4.5° F/ft and the instantaneous rate of change of illumination is −3% candle power/ft. Estimate the temperature and the illumination at 7 feet.

(e) Are you in the dark before you are cold, or vice-versa?

23. Students were asked to evaluate $f'(4)$ from the following table which shows the values of the function f:

x	1	2	3	4	5	6
$f(x)$	4.2	4.1	4.2	4.5	5.0	5.7

- Student A estimated the derivative as $f'(4) \approx \dfrac{f(5) - f(4)}{5 - 4} = 0.5$.
- Student B estimated the derivative as $f'(4) \approx \dfrac{f(4) - f(3)}{4 - 3} = 0.3$.
- Student C suggested that they should split the difference and estimate the average of these two results, that is, $f'(4) \approx \frac{1}{2}(0.5 + 0.3) = 0.4$.

 (a) Sketch the graph of f, and indicate how the three estimates are represented on the graph.
 (b) Explain which answer is likely to be best.

24. Let $g(x) = \sqrt{x}$ and $f(x) = kx^2$, where k is a constant.

 (a) Find the slope of the tangent line to the graph of g at the point $(4, 2)$.
 (b) Find the equation of this tangent line.
 (c) If the graph of f contains the point $(4, 2)$, find k.
 (d) Where, other than point $(4, 2)$, does the graph of f intersect the tangent line?

25. A circle with center at the origin and radius of length 5 has equation $x^2 + y^2 = 25$. Graph the circle.

 (a) Just from looking at the graph, what can you say about the slope of the line tangent to the circle at the point $(0, 5)$? What about the slope of the tangent at $(5, 0)$?
 (b) Estimate the slope of the tangent to the circle at the point $(3, -4)$ by graphing the tangent carefully at that point.
 (c) Use the result of part (b) and the symmetry of the circle to find slopes of the tangents drawn to the circle at $(-3, 4)$, $(-3, -4)$, and $(3, 4)$.

26. A person with a certain liver disease first exhibits larger and larger concentrations of certain enzymes (called SGOT and SGPT) in the blood. As the disease progresses, the concentration of these enzymes drops, first to the predisease level and eventually to zero (when almost all of the liver cells have died). Monitoring the levels of these enzymes allows doctors to track the progress of a patient with this disease. If $C = f(t)$ is the concentration of the enzymes in the blood as a function of time,

 (a) Sketch a possible graph of $C = f(t)$.
 (b) Mark on the graph the intervals where $f' > 0$ and where $f' < 0$.
 (c) What does $f'(t)$ represent, in practical terms?

27. A continuous function defined for all x has the following properties:
 - f is increasing.   • f is concave down.   • $f(5) = 2$.   • $f'(5) = \frac{1}{2}$.

 (a) Sketch a possible graph for f.
 (b) How many zeros does f have?
 (c) What can you say about the location of the zeros?
 (d) What is $\lim\limits_{x \to -\infty} f(x)$?
 (e) Is it possible that $f'(1) = 1$?
 (f) Is it possible that $f'(1) = \frac{1}{4}$?

CHAPTER THREE

ACCUMULATED CHANGE: THE DEFINITE INTEGRAL

In Chapter 2 we learned how to find the rate of change of a function, which led us to the derivative. Now we will consider the reverse process of how to find the total change from the rate of change. This will lead us to the second key concept, the *definite integral*, which computes the accumulated change in a function. We will discover that the definite integral can also be used to calculate the area under a curve and the average value of a function.

We end this chapter with the Fundamental Theorem of Calculus, which tells us that we can use a definite integral to get information about a function from its derivative. Calculating derivatives and calculating definite integrals are, in a sense, reverse processes.

3.1 ACCUMULATED CHANGE

In Chapter 2, we learned how to use the derivative to find the rate of change of a function from total change. Here we see how to go in the other direction. If we know the rate of change, how can we determine the total change?

One of the easiest rates of change to understand is the rate of change of distance, which is velocity. We start by considering how to find distance traveled given the velocity function.

How Do We Measure Distance Traveled?

The rate of change of distance is velocity. If we are given the rate of change (the velocity), can we find the distance traveled? Suppose we know that the velocity was 50 miles per hour throughout a four hour trip. Since

$$\text{Distance} = \text{velocity} \times \text{time},$$

we have

$$\text{Distance traveled on the trip} = (50 \text{ miles/hour}) \times (4 \text{ hours})$$
$$= 200 \text{ miles}.$$

In the next example, the velocity is not constant.

Example 1   Suppose that we travel 30 miles/hour for 2 hours, then 40 miles/hour for 1/2 hour, then 20 miles/hour for 4 hours. How far do we travel?

Solution   We compute the distances traveled for each of the three legs of the trip, and add them together to find the total distance traveled:

$$\text{Distance} = \text{velocity} \times \text{time} + \text{velocity} \times \text{time} + \text{velocity} \times \text{time}$$
$$= (30 \text{ miles/hour})(2 \text{ hours}) + (40 \text{ miles/hour})(1/2 \text{ hour}) + (20 \text{ miles/hour})(4 \text{ hours})$$
$$= 60 \text{ miles} + 20 \text{ miles} + 80 \text{ miles}$$
$$= 160 \text{ miles}.$$

We travel 160 miles on this trip.

Visualizing Distance on the Velocity Graph

Example 2   Sketch a graph of the velocity function for the trip described in Example 1.

Solution   The velocity is 30 miles/hour for the first 2 hours, 40 miles/hour for the next 1/2 hour, and 20 miles/hour for the last 4 hours. The entire trip lasts $2 + 1/2 + 4 = 6.5$ hours, so we need a scale on our horizontal (time) axis running from 0 to 6.5. Between $t = 0$ and $t = 2$, the velocity is constant at 30 miles/hour, so the velocity graph is a horizontal line at 30. Likewise, between $t = 2$ and $t = 2.5$, the velocity graph is a horizontal line at 40, and between $t = 2.5$ and $t = 6.5$, the velocity graph is a horizontal line at 20. The graph is shown in Figure 3.1.

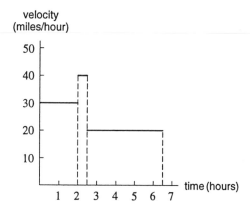

Figure 3.1: Velocity graph for Example 1

How can we visualize distance traveled on the velocity graph given in Figure 3.1? The velocity graph looks like the top edges of three rectangles. The distance traveled on the first leg of the journey is (30 miles/hour)(2 hours), which is the height times the width of the first rectangle in the velocity graph. The distance traveled on the first leg of the trip is equal to the area of the first rectangle. Likewise, the distances traveled during the second and third legs of the trip are equal to the areas of the second and third rectangles in the velocity graph. It appears that distance traveled is equal to the area under the velocity graph.

In Figure 3.2, the area under the velocity graph in Figure 3.1 is shaded in. Since this area is three rectangles, we have

$$
\begin{aligned}
\text{Area} &= \text{height} \times \text{width} + \text{height} \times \text{width} + \text{height} \times \text{width} \\
&= (30)(2) + (40)(1/2) + (20)(4) \\
&= 60 + 20 + 80 \\
&= 160.
\end{aligned}
$$

The area under the velocity graph is equal to distance traveled.

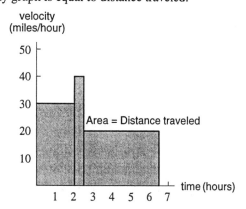

Figure 3.2: The area under the velocity graph

A Thought Experiment: How Far Did the Car Go?

In the above example, the velocity was constant over intervals. Of course, this is not always the case; we now look at an example where the velocity is continually changing.

One Second Velocity Data

Suppose a car is moving with increasing velocity. First, let's suppose we are given the velocity every second, as in Table 3.1:

TABLE 3.1 *Velocity of car every second*

Time (sec)	0	1	2	3	4	5
Velocity (ft/sec)	20	30	38	44	48	50

How far has the car moved? Since we don't know how fast the car is moving at every moment, we can't figure the distance out exactly, but we can make an estimate. Since the velocity is increasing, the car moves at a velocity of 20 ft/sec or more for the first second. The car moves at least (20 ft/sec)(1 sec) = 20 ft during the first second. Likewise, it goes at least 30 feet during the next second, at least 38 feet during the third second, at least 44 feet during the fourth second, and at least 48 feet during the last second.

During the five-second period it goes at least

$$20 + 30 + 38 + 44 + 48 = 180 \text{ feet.}$$

Thus, 180 feet is an underestimate of the total distance moved during the five seconds.

To get an overestimate, we can reason this way: In the first second the car moved at most (30 ft/sec)(1 sec) = 30 feet, in the next second it moved at most 38 feet, in the third second it moved at most 44 feet, and so on. Therefore, altogether it moved at most

$$30 + 38 + 44 + 48 + 50 = 210 \text{ feet.}$$

Thus, the total distance moved is between 180 and 210 feet:

$$180 \text{ feet} \leq \text{Total distance traveled} \leq 210 \text{ feet.}$$

There is a difference of 30 feet between our upper and lower estimates.

Half Second Velocity Data

What if we wanted a more accurate estimate? Then we should ask for more frequent velocity measurements, say every 0.5 seconds. See Table 3.2.

TABLE 3.2 *Velocity of car every half second*

Time (sec)	0	0.5	1.0	1.5	2.0	2.5	3.0	3.5	4.0	4.5	5.0
Velocity (ft/sec)	20	26	30	35	38	42	44	46	48	49	50

As before, we get a lower estimate for each half second by using the velocity at the beginning of that half second. During the first half second the velocity is at least 20 ft/sec, and so the car travels at least (20 ft/sec)(0.5 sec) = 10 feet. During the next half second the car moves at least (26)(0.5) = 13 feet, and so on. So now we can say

$$\begin{aligned} \text{Lower estimate} &= (20)(0.5) + (26)(0.5) + (30)(0.5) + (35)(0.5) + (38)(0.5) \\ &\quad + (42)(0.5) + (44)(0.5) + (46)(0.5) + (48)(0.5) + (49)(0.5) \\ &= 189 \text{ feet.} \end{aligned}$$

Notice that this is higher than our old lower estimate of 180 feet.

We get a new upper estimate by considering the velocity at the end of each half second. During the first half second the velocity is at most 26 ft/sec, and so the car moves at most $(26 \text{ ft/sec})(0.5 \text{ sec}) = 13$ feet; in the next half second it moves at most $(30)(0.5) = 15$ feet, and so on. Thus,

$$
\begin{aligned}
\text{Upper estimate} = {} & (26)(0.5) + (30)(0.5) + (35)(0.5) + (38)(0.5) + (42)(0.5) \\
& + (44)(0.5) + (46)(0.5) + (48)(0.5) + (49)(0.5) + (50)(0.5) \\
= {} & 204 \text{ feet.}
\end{aligned}
$$

This is lower than our old upper estimate of 210 feet. Now we know that

$$189 \text{ feet} \leq \text{Total distance traveled} \leq 204 \text{ feet.}$$

Notice that the difference between our new upper and lower estimates is now 15 feet, half of what it was before. By halving our interval of measurement, we have halved the difference between the upper and lower estimates.

Visualizing Distance on the Velocity Graph: One Second Data

We can represent both upper and lower estimates for distance traveled on a graph of the velocity. The velocity can be graphed by plotting the data in Table 3.1 and drawing a smooth curve through the points. (See Figure 3.3.)

The area of the first dark rectangle is $(20)(1) = 20$, the lower estimate of the distance moved in the first second. The area of the second dark rectangle is $(30)(1) = 30$, the lower estimate for the distance moved in the next second. Therefore, the total area of the dark rectangles represents the lower estimate for the total distance moved during the five seconds.

If the dark and light rectangles are considered together, the first area is $(30)(1) = 30$, the upper estimate for the distance moved in the first second. Continuing this calculation shows that the upper estimate for the total distance is represented by the area of the dark and light rectangles together.

Visualizing Distance on the Velocity Graph: Half Second Data

The data for the velocities measured each half second is in Figure 3.4.

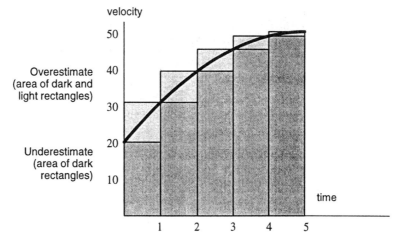

Figure 3.3: Velocity measured each second

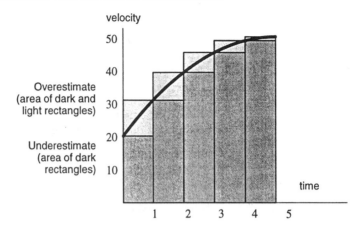

Figure 3.4: Velocity measured each half second

The area of the dark rectangles again represents the lower estimate, and the area of the dark and light rectangles together represents the upper estimate. We see that the underestimate and overestimate are closer together in Figure 3.4 than in Figure 3.3.

Finding Total Change from Rate of Change

We have seen how to use the rate of change of distance (the velocity) to calculate the total distance traveled. We can use these same methods to find total change from any rate of change.

Example 3   A city's population grows at the rate of 5000 people/year for 3 years, and then grows at the rate of 3000 people/year for the next 4 years. What is the total change in the population of the city during this 7 year period?

Solution   Notice that the units of *people/year* remind us that we are being given a rate of change of the population (people) over time (years). If the rate of change is constant, we know that

Total change in population = (rate of change per year) × (number of years).

Thus, the total change in this population is

$$\text{Total change} = (5000 \text{ people/year})(3 \text{ years}) + (3000 \text{ people/year})(4 \text{ years})$$
$$= 15{,}000 \text{ people} + 12{,}000 \text{ people}$$
$$= 27{,}000 \text{ people}.$$

The change in population was 27,000 people. Notice that this does not tell you the population of the city at the end of 7 years; it tells you the change. If the population of the city were initially 100,000 people, it would be 127,000 people at the end of the 7-year period. If it were initially 50,000, it would now be 77,000, and so on.

Example 4   A new video game has been put on the market, and the rate of sales (in games per week) is recorded over a 20-week period. The results are shown in Table 3.3, with *t* in weeks since the game went on the market. Assuming that the rate of sales increased throughout the 20-week period, estimate the total sales of the game during this period.

TABLE 3.3   *Weekly sales of a video game*

Time (weeks)	0	5	10	15	20
Rate of sales (games per week)	0	585	892	1875	2350

Solution   How many games were sold during the first five weeks? During this time, sales went from 0 (the moment the game went on the market) to 585 games per week during the fifth week. If we assume that 585 games were sold every week during the first five weeks, we will clearly have an overestimate for total sales of the video game during this period. Since

$$\text{Total sales} = \text{Rate of sales per week} \times \text{Number of weeks},$$

an overestimate for the first five weeks is (585 games/week)(5 weeks) = 2925 games. We can use similar overestimates for each of the five week periods to calculate an overestimate for the entire twenty-week period:

$$\text{Overestimate for total sales} = (585)(5) + (892)(5) + (1875)(5) + (2350)(5) = 28,510 \text{ games.}$$

We can find an underestimate for total sales by taking the lower value for rate of sales during each of the five week periods. The underestimate is given by

$$\text{Underestimate for total sales} = (0)(5) + (585)(5) + (892)(5) + (1875)(5) = 16,760 \text{ games.}$$

Thus, the total sales of the game during the twenty-week period is between 16,760 and 28,510 games. Our best single estimate of total sales is the average of these two numbers:

$$\text{total sales} = \frac{16,760 + 28,510}{2} = 22,635 \text{ games.}$$

Finding Total Change More Precisely: n and Δt

We use the notation Δt for the change in t, or the size of the t-intervals used to measure total change. In Example 4, we recorded the rate of change every 5 weeks, and so $\Delta t = 5$. We use n to represent the number of subintervals of length Δt. In Example 4, we divided the twenty-week period into subintervals each of width 5, and so there are 4 subintervals. In Example 4, we have $n = 4$.

In the thought experiment on page 143, we estimated the distance a car traveled during a five-second period. When we measured the velocity every second, we used $\Delta t = 1$ and $n = 5$. When we measured the velocity every half-second, we used $\Delta t = 0.5$ and $n = 10$. From the graphical representations of these estimates in Figure 3.3 and Figure 3.4, we see that the number of subintervals n also represents the number of rectangles in the graphical representation.

Example 5   A bacteria population increases at a rate of

$$f(t) = 3 + 0.1t^2$$

million bacteria per hour during a twenty-hour period, where t is measured in hours since the start of the period. Find an underestimate of the total change in the number of bacteria

(a)   using $\Delta t = 4$ hours,
(b)   using $\Delta t = 2$ hours,
(c)   using $\Delta t = 1$ hour.

In each case, include a graphical representation of the estimate.

Solution   (a)   The rate of change is $f(t) = 3 + 0.1t^2$. If we use $\Delta t = 4$, we measure the rate every 4 hours. These values are given in Table 3.4. An underestimate for the population change during the first 4 hours is (3.0 million/hour)(4 hours) = 12 million. An underestimate of total change is

$$\text{Total Change} \approx (3.0)(4) + (4.6)(4) + (9.4)(4) + (17.4)(4) + (28.6)(4)$$
$$= 252.0 \text{ million bacteria.}$$

An underestimate for the increase in the bacteria population is 252 million more bacteria during this 20 hour period. A graph of rate of change against time is given in Figure 3.5 (a), and the shaded rectangles in (a) represent the underestimate we just computed. We have $n = 5$, corresponding to the five rectangles in the graph.

TABLE 3.4 *Rate of change with $\Delta t = 4$*

t (hours)	0	4	8	12	16	20
$f(t)$ (million bacteria/hour)	3.0	4.6	9.4	17.4	28.6	43.0

(b) If we use $\Delta t = 2$, we need a table of values with measurements every 2 hours. See Table 3.5. We have

$$\text{Total change} \approx (3.0)(2) + (3.4)(2) + (4.6)(2) + \cdots + (35.4)(2)$$
$$= 288.0 \text{ million bacteria.}$$

We see that the bacteria population increased by at least 288 million during this period. We have $n = 10$, and we see in Figure 3.5 (b) that this estimate is more accurate than that found in part (a).

TABLE 3.5 *Rate of change with $\Delta t = 2$*

t (hours)	0	2	4	6	8	10	12	14	16	18	20
$f(t)$ (million bacteria/hour)	3.0	3.4	4.6	6.6	9.4	13.0	17.4	22.6	28.6	35.4	43.0

(c) If we use $\Delta t = 1$, then $n = 20$, and it can be shown that we have

$$\text{Total change} \approx 307.0 \text{ million bacteria.}$$

The shaded area in Figure 3.5 (c) represents this estimate, and we see that it is the most accurate of the three.

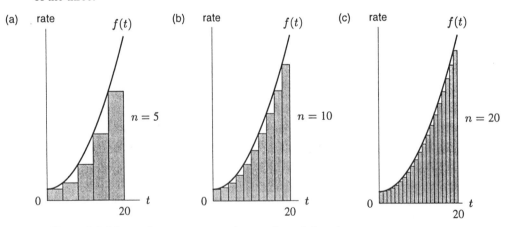

Figure 3.5: More and more accurate estimates of population change

Notice in Example 5 that as n gets larger, the estimate improves and the area of the shaded rectangles more closely matches the area under the curve. If we let n get arbitrarily large, we would find total change exactly, and the rectangles would exactly match the area under the curve.

Problems for Section 3.1

1. A car comes to a stop five seconds after the driver slams on the brakes. While the brakes are on, the following velocities are recorded:

Time since brakes applied (sec)	0	1	2	3	4	5
Velocity (ft/sec)	88	60	40	25	10	0

   (a) Give lower and upper estimates of the distance the car traveled after the brakes were applied.
   (b) On a sketch of velocity against time, show the lower and upper estimates of part (a).

2. Roger decides to run a marathon. Roger's friend Jeff rides behind him on a bicycle and clocks his pace every 15 minutes. Roger starts out strong, but after an hour and a half he is so exhausted that he has to stop. The data Jeff collected is summarized below:

Time spent running (min)	0	15	30	45	60	75	90
Speed (mph)	12	11	10	10	8	7	0

   (a) Assuming that Roger's speed is always decreasing, give upper and lower estimates for the distance Roger ran during the first half hour.
   (b) Give upper and lower estimates for the distance Roger ran in total during the entire hour and a half.

3. Coal gas is produced at a gasworks. Pollutants in the gas are removed by scrubbers, which become less and less efficient as time goes on. Measurements made at the start of each month showing the rate at which pollutants are escaping in the gas are as follows:

Time (months)	0	1	2	3	4	5	6
Rate pollutants are escaping (tons/month)	5	7	8	10	13	16	20

   (a) Make an overestimate and an underestimate of the total quantity of pollutants that escaped during the *first month*.
   (b) Make an overestimate and an underestimate of the total quantity of pollutants that escaped during the six months.

4. Draw a graph of the rate of sales against time for the video game data given in Example 4. Represent graphically the overestimate and the underestimate calculated in that example.

5. If the rate of change of a quantity is given by $f(t) = t^2 + 1$ between $t = 0$ and $t = 8$, find an underestimate and an overestimate of total change using

  (a) $\Delta t = 4$,
  (b) $\Delta t = 2$,
  (c) $\Delta t = 1$.

  Sketch a graph of $f(t)$ and shade rectangles to represent each of your six answers. What is n in each case?

6. Table 3.6 shows the annual rate of change in the population of the world (in millions), for selected years between 1950 and 1990.

  (a) Use this data to estimate the total change in the world's population between 1950 and 1990. Explain how you arrived at your answer.

  (b) The population of the world was 2555 million people in 1950 and 5295 million people in 1990. What was the total change in the population? How does this compare with your estimate in part (a)?

TABLE 3.6 *Rate of change of the world's population*

Year	1950	1960	1970	1980	1990
Rate of change (million people per year)	37	41	78	77	86

7. Figure 3.6 shows the graph of the velocity, v, of an object (in m/sec). Estimate the total distance the object traveled between $t = 0$ and $t = 6$.

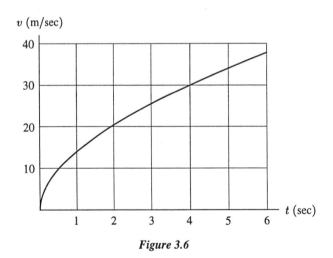

Figure 3.6

8. Your velocity is given by $v(t) = t^2 + 1$ in m/sec, with t in seconds. Estimate the distance traveled between $t = 0$ and $t = 5$. Explain how you arrived at your estimate.

9. As this country's relatively rich coal deposits are depleted, a larger fraction of the country's coal will come from strip-mining, and it will become necessary to strip-mine larger and larger areas for each ton of coal. The graph in Figure 3.7 shows an estimate of the number of acres

of land per million tons of coal that will be defaced during strip-mining as a function of the number of million tons removed, starting from the present day.

(a) Estimate the total number of acres defaced in extracting the next 4 million tons of coal (measured from the present day). Draw four rectangles under the curve, and compute their area.

(b) Reestimate the number of acres defaced using rectangles above the curve.

(c) Use your answers to parts (a) and (b) to get a better estimate of the actual number of acres defaced.

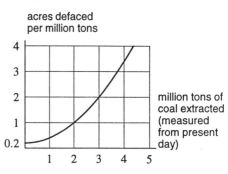

Figure 3.7

3.2 THE DEFINITE INTEGRAL

In the previous section, we learned how to estimate the total change in a function from its rate of change. In this section, we learn how to find the total change exactly, by looking at estimates that are more and more accurate. This is similar to what we did in Chapter 2, where we first learned how to estimate the rate of change, and then we saw how to define the rate of change at a point by taking more and more accurate estimates. We express the total change as a limit of estimates in much the same way we expressed rate of change as a limit of average rates of change.

Determining Total Change Precisely

Suppose we want to know the total change of a quantity over the time interval $a \leq t \leq b$. Let the rate of change at time t be given by the function $f(t)$. Suppose we take measurements of $f(t)$ at equally spaced times $t_0, t_1, t_2, \ldots, t_n$ with time $t_0 = a$ and time $t_n = b$. The time interval between any two consecutive measurements is

$$\Delta t = \frac{b - a}{n}$$

where Δt (read "delta t") means the change, or increment, in t.

During the first time interval, the rate of change can be approximated by $f(t_0)$, so the total change is approximately

$$f(t_0)\Delta t.$$

During the second time interval, the rate of change is about $f(t_1)$, so the total change is about

$$f(t_1)\Delta t.$$

Continuing in this way and adding up all the estimates, we get an estimate for the total change. In the last interval, the rate of change is approximately $f(t_{n-1})$, so the last term is $f(t_{n-1})\Delta t$.

$$\begin{array}{c} \text{Total change} \\ \text{between } a \text{ and } b \end{array} \approx f(t_0)\Delta t + f(t_1)\Delta t + f(t_2)\Delta t + \cdots + f(t_{n-1})\Delta t.$$

This is called a *left-hand sum* because we used the value of the rate of change from the left end of each time interval. It is represented by the sum of the areas of the rectangles in Figure 3.8.

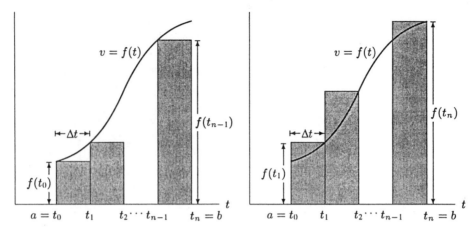

Figure 3.8: Left-hand sum   *Figure 3.9:* Right-hand sum

We can also calculate a *right-hand sum* by using the value of the rate of change at the right end of each time interval. In that case the estimate for the first interval is $f(t_1)\Delta t$, for the second interval it is $f(t_2)\Delta t$, and so on. The estimate for the last interval is now $f(t_n)\Delta t$, so

$$\begin{array}{c}\text{Total change}\\ \text{between } a \text{ and } b\end{array} \approx f(t_1)\Delta t + f(t_2)\Delta t + f(t_3)\Delta t + \cdots + f(t_n)\Delta t.$$

The right-hand sum is represented by the area of the rectangles in Figure 3.9.

If f is a decreasing function, the left- and right-hand sums are represented by areas of rectangles as in Figures 3.10 and 3.11.

To find exactly the total change between a and b, we take the limit of the sums, as n, the number of subdivisions of the interval, goes to infinity. The notation

$$\lim_{n \to \infty}$$

is read "the limit as n goes to infinity" and signifies that we want the number that is approximated by using larger and larger values of n, without bound. As n gets larger, the sum of the areas of the rectangles approaches the area under the curve between $t = a$ and $t = b$. So we conclude that

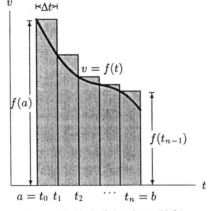

Figure 3.10: Left-hand sum if f is decreasing

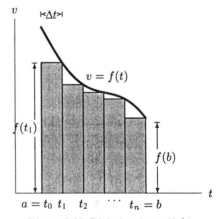

Figure 3.11: Right-hand sum if f is decreasing

$$\begin{aligned}
\text{Total change between } t = a \text{ and } t = b &= \lim_{n\to\infty} (\text{Left-hand sum}) \\
&= \lim_{n\to\infty} \left[f(t_0)\Delta t + f(t_1)\Delta t + \cdots + f(t_{n-1})\Delta t \right] \\
&= \text{Area under the graph of } f(t) \text{ from } t = a \text{ to } t = b,
\end{aligned}$$

and

$$\begin{aligned}
\text{Total change between } t = a \text{ and } t = b &= \lim_{n\to\infty} (\text{Right-hand sum}) \\
&= \lim_{n\to\infty} \left[f(t_1)\Delta t + f(t_2)\Delta t + \cdots + f(t_n)\Delta t \right] \\
&= \text{Area under the graph of } f(t) \text{ from } t = a \text{ to } t = b.
\end{aligned}$$

Writing Left and Right Sums Using Sigma Notation

Both the left-hand and right-hand sums can be written more compactly using *sigma*, or summation, notation. The symbol $\sum$ is a capital sigma, or Greek letter "S". We write

$$\text{Right-hand sum} = \sum_{i=1}^{n} f(t_i)\Delta t = f(t_1)\Delta t + f(t_2)\Delta t + \cdots + f(t_n)\Delta t,$$

where the $\sum$ tells us to add terms of the form $f(t_i)\Delta t$. The "$i = 1$" at the base of the sigma sign tells us to start at $i = 1$, and the "n" at the top tells us to stop at $i = n$.

In the left-hand sum we start at $i = 0$ and stop at $i = n - 1$, so we write

$$\text{Left-hand sum} = \sum_{i=0}^{n-1} f(t_i)\Delta t = f(t_0)\Delta t + f(t_1)\Delta t + \cdots + f(t_{n-1})\Delta t.$$

The Definition of the Definite Integral

For any function you are ever likely to meet, the limits of the left- and right-hand sums will exist and be equal. In such cases, we define the *definite integral* to be the limit of these sums. When $f(t) \geq 0$, the definite integral represents the area between the graph of $f(t)$ and the t-axis, from $t = a$ to $t = b$. The definite integral is well approximated by a left- or right-hand sum if n is large enough.

The **definite integral** of f from a to b, written

$$\int_a^b f(t)\, dt,$$

is the limit of the left-hand or right-hand sum with n subdivisions as n gets arbitrarily large. In other words,

$$\int_a^b f(t)\, dt = \lim_{n\to\infty} (\text{Left-hand sum}) = \lim_{n\to\infty} \left(\sum_{i=0}^{n-1} f(t_i)\Delta t \right)$$

and

$$\int_a^b f(t)\, dt = \lim_{n\to\infty} (\text{Right-hand sum}) = \lim_{n\to\infty} \left(\sum_{i=1}^{n} f(t_i)\Delta t \right).$$

Each of these sums is called a *Riemann sum*, f is called the *integrand*, and a and b are called the *limits of integration*.

The "$\int$" notation comes from an old-fashioned "S", which stands for "sum" in the same way that $\sum$ does. The "dt" in the integral comes from the factor Δt. Notice that the limits on the $\sum$ sign are 0 and $n - 1$ for the left-hand sum, or 1 and n for the right-hand sum, whereas the limits on the $\int$ sign are a and b.

Calculating Definite Integrals

Example 1   Find the left-hand and right-hand sums with $n = 2$ and $n = 10$ for $\int_1^2 \frac{1}{t}\, dt$. How do the values of these sums compare with the true value of the integral?

Solution   Here $a = 1$ and $b = 2$, so for $n = 2$, $\Delta t = (2 - 1)/2 = 0.5$. Therefore, $t_0 = 1$, $t_1 = 1.5$ and $t_2 = 2$. (See Figure 3.12.) Thus,

$$\text{Left-hand sum} = f(1)\Delta t + f(1.5)\Delta t$$
$$= 1(0.5) + \frac{1}{1.5}(0.5)$$
$$\approx 0.8333,$$

and

$$\text{Right-hand sum} = f(1.5)\Delta t + f(2)\Delta t$$
$$= \frac{1}{1.5}(0.5) + \frac{1}{2}(0.5)$$
$$\approx 0.5833.$$

From Figure 3.12 you can see that the left-hand sum is bigger than the area under the curve and the right-hand sum is smaller, so the area under the graph of $f(t) = 1/t$ from $t = 1$ to $t = 2$ is between 0.5833 and 0.8333. Thus,

$$0.5833 < \int_1^2 \frac{1}{t}\, dt < 0.8333.$$

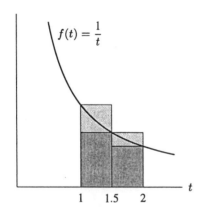

Figure 3.12: Approximating $\int_1^2 \frac{1}{t}\, dt$ with $n = 2$

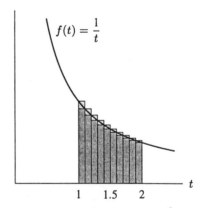

Figure 3.13: Approximation of $\int_1^2 \frac{1}{t}\, dt$ with $n = 10$

When $n = 10$, $\Delta t = 0.1$ (see Figure 3.13), so

$$\text{Left-hand sum} = f(1)\Delta t + f(1.1)\Delta t + \cdots + f(1.9)\Delta t$$
$$= \left(1 + \frac{1}{1.1} + \cdots + \frac{1}{1.9}\right) 0.1$$
$$\approx 0.7188,$$

and

$$\text{Right-hand sum} = f(1.1)\Delta t + f(1.2)\Delta t + \cdots + f(2)\Delta t$$
$$= \left(\frac{1}{1.1} + \frac{1}{1.2} + \cdots + \frac{1}{2}\right) 0.1$$
$$\approx 0.6688.$$

From Figure 3.13 you can see that the left-hand sum is larger than the area under the curve, and the right-hand sum is smaller, so

$$0.6688 < \int_1^2 \frac{1}{t}\, dt < 0.7188.$$

Notice that in this example the left- and right-hand sums trap the true value of the integral between them. As the subdivisions become finer, the left- and right-hand sums get closer together.

Convergence of Left and Right Sums

Let us continue the example

$$\int_1^2 \frac{1}{t}\, dt.$$

The left- and right-hand sums for $n = 2$, 10, 50, and 250 are listed in Table 3.7.

TABLE 3.7 *Left- and right-hand sums for $\int_1^2 \frac{1}{t}\, dt$*

n	Left-hand sum	Right-hand sum
2	0.8333	0.5833
10	0.7188	0.6688
50	0.6982	0.6882
250	0.6941	0.6921

Because the function $f(t) = 1/t$ is decreasing, the left-hand sums converge down on the integral from above, and the right-hand sums converge up from below. From the last row of the table we can deduce that

$$0.6921 < \int_1^2 \frac{1}{t}\, dt < 0.6941,$$

so $\int_1^2 \frac{1}{t}\, dt \approx 0.69$ to two decimal places. To approximate definite integrals accurately, we evaluate left- and right-hand sums with a large number of subdivisions, using a computer or calculator.

If a function is either increasing throughout an interval or decreasing throughout that interval, then the value of the definite integral of the function on that interval is between the values of the left- and right-hand sums. This is not true of functions in general. The left- and right-hand sums may both be larger (or smaller) than the definite integral. Nonetheless, for all functions we are likely to see, as n gets large, the left- and right-hand sums will approach the value of the definite integral.

Using a Calculator or Computer to Evaluate Definite Integrals

In Example 1, we approximated the integral $\int_{1}^{2} \frac{1}{t} dt$ using $n = 10$. Although 10 is a relatively small value for n, this calculation was a bit tedious. The approximations given in Table 3.7 would be very cumbersome to do by hand. In practice, we usually use a calculator or computer to compute Riemann sums.

Different calculators and computer packages may give slightly different values for definite integrals, due to a round-off error. In addition, on some calculators or computer packages, you have to specify the value of n desired. We can make the approximation more accurate by increasing the value of n, but the computation will take longer. We can use technology to find the value of the integral in Example 1. We see that

$$\int_{1}^{2} \frac{1}{t} dt = 0.69314718.$$

Example 2   Compute $\int_{1}^{3} t^2 dt$.

Solution   Using a calculator or computer, we see that $\int_{1}^{3} t^2 dt$ is about 8.667.

Example 3   A table of values for a function $P = f(t)$, where t is time measured in minutes, is given in Table 3.8. Estimate $\int_{20}^{30} f(t)dt$.

TABLE 3.8   *Estimate $\int_{20}^{30} f(t)dt$*

t	20	22	24	26	28	30
P	5	7	11	18	29	45

Solution   Since we only have a table of values, we must use Riemann sums to approximate the integral. We are given information at 6 times which divide the interval $20 \leq t \leq 30$ into 5 time intervals, so we use $n = 5$. We see that measurements are taken every 2 minutes, so $\Delta t = 2$. Our best estimate is obtained by calculating the left-hand and right-hand sums, and averaging the two.

$$\begin{aligned} \text{Left-hand sum} &= f(20)2 + f(22)2 + f(24)2 + f(26)2 + f(28)2 \\ &= 5 \cdot 2 + 7 \cdot 2 + 11 \cdot 2 + 18 \cdot 2 + 29 \cdot 2 \\ &= 10 + 14 + 22 + 36 + 58 \\ &= 140. \end{aligned}$$

$$\text{Right-hand sum} = f(22)2 + f(24)2 + f(26)2 + f(28)2 + f(30)2$$
$$= 7 \cdot 2 + 11 \cdot 2 + 18 \cdot 2 + 29 \cdot 2 + 45 \cdot 2$$
$$= 14 + 22 + 36 + 58 + 90$$
$$= 220.$$

We average the two to obtain our best estimate of the integral:

$$\int_{20}^{30} f(t)dt \approx \frac{140 + 220}{2} = 180.$$

The Definite Integral as Total Change and as Area

We have seen that the left- and right-hand sums used to calculate the definite integral are the same left- and right-hand sums used to calculate total change from rate of change. As n gets larger, the estimate of total change gets more accurate. The definite integral of rate of change gives total change.

If $f(t)$ is the rate of change of a quantity, then

$$\text{The total change in the quantity between } t = a \text{ and } t = b = \int_{a}^{b} f(t)\, dt.$$

Furthermore, since the left- and right-hand sums are represented graphically by the sums of areas of rectangles, the definite integral is represented graphically by an area.

When $f(t)$ is positive and $a < b$,

$$\text{Area under graph of } f \text{ between } a \text{ and } b = \int_{b}^{a} f(t)\, dt.$$

Example 4   The graph of a function $y = f(x)$ is given in Figure 3.14. Estimate $\int_{0}^{6} f(x)\, dx$.

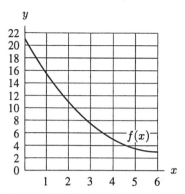

Figure 3.14: Estimate the area under
this function

Solution   We know that $\int_0^6 f(x)\, dx$ equals the area under the graph of $f(x)$ between $x = 0$ and $x = 6$. We can use Riemann sums to estimate this area. We can make n as large as we like depending on how accurate we want the estimate to be. Let's start with $n = 3$. Since $0 \le x \le 6$, we have $\Delta x = 2$, and we are partitioning the area into three parts: the part over $0 \le x \le 2$, the part over $2 \le x \le 4$, and the part over $4 \le x \le 6$. We will compute the left-hand sum first. Figure 3.15 shows the three rectangles. The height of the first rectangle is $f(0) = 21$ and the width is 2, so the area of the first rectangle is $(21)(2) = 42$.

$$\begin{aligned}
\text{Left-hand estimate} &= (f(0))(2) + (f(2))(2) + (f(4))(2)\\
&= (21)(2) + (11)(2) + (5)(2)\\
&= 42 + 22 + 10\\
&= 74 \text{ square units.}
\end{aligned}$$

Similarly, we compute a right-hand estimate. (See Figure 3.16.)

$$\begin{aligned}
\text{Right-hand estimate} &= (f(2))(2) + (f(4))(2) + (f(6))(2)\\
&= (11)(2) + (5)(2) + (3)(2)\\
&= 22 + 10 + 6\\
&= 38 \text{ square units.}
\end{aligned}$$

We can see from Figure 3.15 and Figure 3.16 that the left-hand sum is an overestimate and the right-hand sum is an underestimate, so the true area is between 38 and 74 square units.

To improve the accuracy, we use a larger number of rectangles. If $n = 6$, the width of each rectangle is 1, and we obtain the following estimates:

Left-hand estimate $= (21)(1) + (15.5)(1) + (11)(1) + (7.5)(1) + (5)(1) + (3.5)(1) = 63.5,$

and

Right-hand estimate $= (15.5)(1) + (11)(1) + (7.5)(1) + (5)(1) + (3.5)(1) + (3)(1) = 45.5.$

We have improved the estimate. Now we know that the true value of the integral is between 45.5 and 63.5. For even greater accuracy, we would take an even larger value of n.

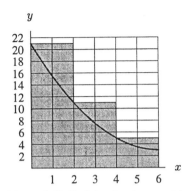

Figure 3.15: Left-hand estimate, with $n = 3$

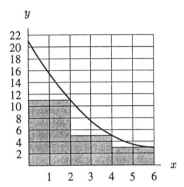

Figure 3.16: Right-hand estimate, with $n = 3$

Example 5   Suppose that the velocity of an object (in m/sec) is given by $v(t) = 10 + 8t - t^2$, where t is in seconds.

   (a)   Estimate the distance traveled by the object during the first 5 seconds (that is, between $t = 0$ and $t = 5$) using $n = 5$.

   (b)   Use technology to find the distance traveled.

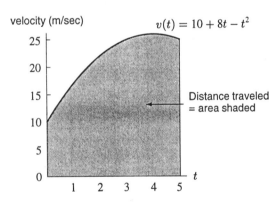

Figure 3.17: A velocity function

Solution   A graph of this velocity function is given in Figure 3.17. Finding the distance traveled is equivalent to finding the area under this curve between $t = 0$ and $t = 5$. We estimate that this area is about 100 since the average height appears to be about 20 and the width is 5.

   (a)   If we use $n = 5$, the width of each rectangle is 1 and, using $v(t) = 10 + 8t - t^2$, we compute the following estimates:

$$\text{Left-hand estimate} = (v(0))(1) + (v(1))(1) + (v(2))(1) + (v(3))(1) + (v(4))(1)$$
$$= (10)(1) + (17)(1) + (22)(1) + (25)(1) + (26)(1)$$
$$= 100 \text{ meters.}$$

$$\text{Right-hand estimate} = (v(1))(1) + (v(2))(1) + (v(3))(1) + (v(4))(1) + (v(5))(1)$$
$$= (17)(1) + (22)(1) + (25)(1) + (26)(1) + (25)(1)$$
$$= 115 \text{ meters.}$$

We average these to obtain a better estimate:

$$\text{Distance traveled} \approx \frac{100 + 115}{2} = 107.5 \text{ meters.}$$

(b)   We know that

$$\text{Distance traveled} = \text{Area under the velocity curve (See Figure 3.17)}$$
$$= \int_0^5 v(t) \, dt$$
$$= \int_0^5 (10 + 8t - t^2) \, dt$$
$$\approx 108.33333.$$

We used technology on the last step.

_____

Problems for Section 3.2

For Problems 1–2 estimate by hand the value of the definite integral by computing left-hand and right-hand sums with $n = 2$ and with $n = 4$. Draw a graph of the integrand $f(x)$ in each case, and illustrate (using rectangles) each of your approximations. Is each approximation an underestimate or an overestimate?

1. $\int_0^2 (x^2 + 1) \, dx$
    2. $\int_1^5 \left(\frac{2}{x}\right) dx$

In Problems 3–6,
(a)   Use a graph of the integrand to give a rough estimate of the integral. Explain how you arrived at your answer.
(b)   Use a computer or calculator to find the value of the definite integral.

3. $\int_0^1 x^3 \, dx$
    4. $\int_0^3 \sqrt{x} \, dx$

5. $\int_0^1 3^t \, dt$
    6. $\int_1^2 x^x \, dx$

In Problems 7–12, compute the definite integrals to one decimal place.

7. $\int_0^5 x^2 \, dx$
    8. $\int_1^4 \frac{1}{\sqrt{1 + x^2}} \, dx$

9. $\displaystyle\int_{1}^{5} (3x+1)^2\,dx$

10. $\displaystyle\int_{1.1}^{1.7} 10(0.85)^t\,dt$

11. $\displaystyle\int_{1}^{2} 2^x\,dx$

12. $\displaystyle\int_{1}^{2} (1.03)^t\,dt$

13.   A graph of the velocity function $v(t) = 10 + 8t - t^2$ is given in Figure 3.17. In Example 5, we calculated left- and right-hand estimates for distance traveled between $t = 0$ and $t = 5$. On a graph similar to Figure 3.17, shade in rectangles to represent each of these estimates.

14.   Values of a function $W(t)$ are given in Table 3.9.

   (a) Estimate $\displaystyle\int_{3}^{4} W(t)\,dt$.

   (b) For your estimate in part (a), what is n? What is Δt?

TABLE 3.9

t	3.0	3.2	3.4	3.6	3.8	4.0
$W(t)$	25	23	20	15	9	2

15.   Figure 3.18 shows the graph of the velocity, v, of an object (in ft/sec).

   (a) Estimate the distance the object traveled between $t = 0$ and $t = 8$, using a right-hand sum with $n = 4$.

   (b) On a graph similar to Figure 3.18, shade in rectangles to represent your estimate in part (a).

   (c) Is the estimate in part (a) an underestimate or an overestimate of the actual distance traveled? Explain.

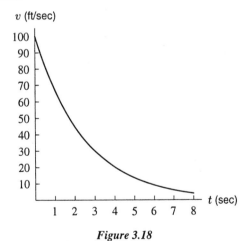

Figure 3.18

16.   The velocity of a car (in miles per hour) is given by $v(t) = 40t - 10t^2$, where t is measured in hours.

   (a) Write a definite integral for the distance the car travels during the first three hours.

   (b) Sketch a graph of velocity against time and represent the distance traveled during the first three hours as an area on your graph.

   (c) Use a computer or calculator to find this distance.

17. When you jump out of an airplane and your parachute fails to open, your downward velocity (in meters/second), t seconds after the jump, is approximated by

$$v(t) = 49(1 - (0.8187)^t).$$

(a) Write a definite integral to represent the distance you travel between time $t = 0$ and time $t = 5$.
(b) Graph $v(t)$ and represent this distance graphically.
(c) Use a computer or graphing calculator to estimate the distance you fall during the first five seconds.

18. As in Exercise 17, if you jump out of an airplane without a parachute, your downward velocity (in m/sec), t seconds after you jump, is given by

$$v(t) = 49(1 - (0.8187)^t).$$

Write a definite integral to represent the distance you fall during the first ten seconds and then find this distance. How far do you fall between $t = 5$ and $t = 10$?

19. Oil is leaking out of a tanker at a rate of $10(0.7)^t$ gallons per minute, with t in minutes.

(a) Sketch a graph of this rate against time, and use your graph to give a rough estimate of the total quantity of oil that leaks out between $t = 0$ and $t = 3$. Explain.
(b) Write a definite integral for the total quantity of oil that leaks out between $t = 0$ and $t = 3$. Use technology to find the value of the integral.
(c) Does more oil leak out during the first minute (between $t = 0$ and $t = 1$) or the third minute (between $t = 2$ and $t = 3$)? Justify your answer graphically.

20. Sketch the graph of a function f (you do not need to give a formula for f) on an interval $[a, b]$ with the property that with $n = 2$ subdivisions,

$$\int_a^b f(x)\, dx < \text{Left-hand sum} < \text{Right-hand sum}.$$

3.3 THE DEFINITE INTEGRAL AS AREA

The Definite Integral as an Area, when $f(x)$ is Positive

If $f(x)$ is positive we can interpret each term $f(x_0)\Delta x$, $f(x_1)\Delta x$, ... in a left- or right-hand Riemann sum as the area of a rectangle. As the width Δx of the rectangles approaches zero, the tops of the rectangles fit the curve of the graph more exactly, and the sum of their areas gets closer and closer to the area under the curve, shaded in Figure 3.19. Thus, we know that

When $f(x)$ is positive and $a < b$:

$$\frac{\text{Area under graph of } f}{\text{between } a \text{ and } b} = \int_a^b f(x)\, dx.$$

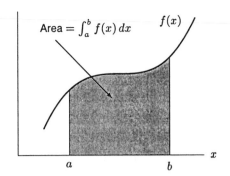

Figure 3.19: The definite integral $\int_a^b f(x)\,dx$

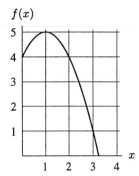

Figure 3.20: Estimate $\int_0^3 f(x)dx$

Example 1   A graph of $y = f(x)$ is shown in Fig 3.20. Estimate $\int_0^3 f(x)dx$.

Solution   We know that $\int_0^3 f(x)dx$ is equal to the area under the graph of f between $x = 0$ and $x = 3$, so we must estimate this area. (See Figure 3.21.) We can use left-hand and right-hand sums, choosing n depending on how ambitious we feel! Another excellent way to estimate area when a grid is given is simply to estimate the number of boxes in the area we are estimating. If we count boxes here, it appears that the shaded area in Figure 3.21 includes about 11.5 boxes. Since each box has area 1, we estimate that the shaded area is about 11.5 square units. We have

$$\int_0^3 f(x)dx \approx 11.5.$$

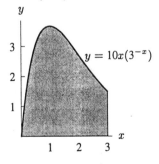

Figure 3.21: Area shaded = $\int_0^3 f(x)dx$

Example 2   Find the area under the graph of $y = 10x(3^{-x})$ between $x = 0$ and $x = 3$.

Figure 3.22

Solution   A graph of this function is shown in Figure 3.22. We see that a rough estimate of the area is about 9, since it appears to have about the same area as a rectangle of width 3 and height 3. To find the area as accurately as we wish, we realize that

$$\text{Area shaded} = \int_0^3 10x(3^{-x})dx,$$

and we use technology to evaluate the integral. We obtain

$$\int_0^3 10x(3^{-x})\,dx \approx 6.967.$$

The shaded area is about 7.0 square units.

Interpreting the Definite Integral as Area, when $f(x)$ is Not Positive

We have assumed in drawing Figure 3.19 that the graph of $f(x)$ lies above the x-axis. If the graph lies below the x-axis, then each value of $f(x)$ is negative, so each $f(x)\Delta x$ is negative, and the area gets counted negatively. In that case, the definite integral is the negative of the area between the graph of f and the horizontal axis.

Example 3   What is the relation between the definite integral $\int_{-1}^{1} (x^2 - 1)\, dx$ and the area between the parabola $y = x^2 - 1$ and the x-axis?

Solution   The parabola lies below the x-axis between $x = -1$ and $x = 1$. (See Figure 3.23.) So,

$$\int_{-1}^{1} (x^2 - 1)\, dx = -\text{Area} \approx -1.33.$$

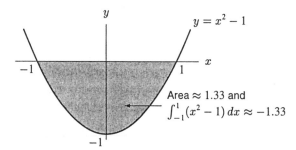

Figure 3.23: Integral $\int_{-1}^{1} (x^2 - 1)\, dx$ is negative of shaded area

---

When $f(x)$ is positive for some x-values and negative for others, and $a < b$:

$\int_a^b f(x)\, dx$ is the sum of the areas above the x-axis, counted positively, and the areas below the x-axis, counted negatively.

Example 4   Interpret the definite integral $\int_0^4 (x^3 - 7x^2 + 11x)\, dx$ in terms of areas.

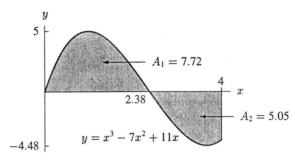

Figure 3.24: Integral $\int_0^4 (x^3 - 7x^2 + 11x)\, dx = A_1 - A_2$

Solution   The graph of $f(x) = x^3 - 7x^2 + 11x$ is shown in Figure 3.24. The graph crosses the x-axis at approximately $x = 2.38$. We see in Figure 3.24 that $f(x)$ is positive for $0 < x < 2.38$ and negative for $2.38 < x < 4$. The integral is the area above the x-axis, A_1, minus the area below the x-axis, A_2. Approximating the integral with technology shows

$$\int_0^4 (x^3 - 7x^2 + 11x)\, dx \approx 2.67.$$

Breaking the integral into two parts and calculating each one separately gives

$$\int_0^{2.38} (x^3 - 7x^2 + 11x)\, dx \approx 7.72 \quad \text{and} \quad \int_{2.38}^4 (x^3 - 7x^2 + 11x)\, dx \approx -5.05,$$

so $A_1 \approx 7.72$ and $A_2 \approx 5.05$. Then, as we would expect,

$$\int_0^4 (x^3 - 7x^2 + 11x)\, dx = A_1 - A_2 \approx 7.72 - 5.05 = 2.67.$$

Example 5   Find the total area of the shaded regions in Figure 3.24.

Solution   We saw in Example 4 that $A_1 \approx 7.72$ and $A_2 \approx 5.05$. Thus we have

$$\text{Total shaded area} = A_1 + A_2 \approx 7.72 + 5.05 = 12.77.$$

Example 6   Graphs of several functions are shown in Figure 3.25. In each case, indicate whether $\int_0^5 f(x)dx$ is positive, negative or approximately zero. Explain your answers.

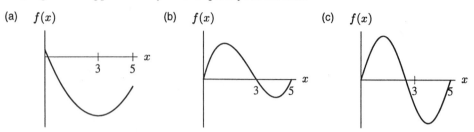

Figure 3.25: Is $\int_0^5 f(x)dx$ positive, negative or zero?

Solution   (a)   The graph lies almost entirely below the x-axis, so the integral is negative.
   (b)   The graph lies partly below the x-axis and partly above the x-axis. However, we see that the area above the x-axis is larger than the area below the x-axis, and so the integral is positive.
   (c)   The graph lies partly below the x-axis and partly above the x-axis. Since the areas above and below the x-axis appear to be approximately equal, they will cancel each other out. Therefore, the integral is approximately zero.

Example 7   Find the area between the graph of $y = x^2 - 2$ and the x-axis between $x = 0$ and $x = 3$.

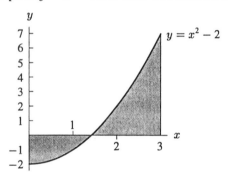

Figure 3.26

Solution   The graph of this function is shown in Figure 3.26, and the relevant area is shaded. If you compute the integral $\int_0^3 (x^2 - 2)dx$, you will find that

$$\int_0^3 (x^2 - 2)dx = 3.0.$$

However, since part of the area lies below the x-axis and part of it lies above the x-axis, this computation does not help us at all. (In fact, it is clear from the graph that the shaded area is more than 3.) We have to find the area above the x-axis and the area below the x-axis separately. We find that the graph crosses the x-axis at $x = 1.414$, and we compute the two areas separately:

$$\int_0^{1.414} (x^2 - 2)dx = -1.886 \quad \text{and} \quad \int_{1.414}^3 (x^2 - 2)dx = 4.886.$$

As we expect, we see that the integral between 0 and 1.414 is negative and the integral between 1.414 and 3 is positive. The total area shaded is the sum of the absolute values of the two integrals:

$$\text{Area shaded} = 1.886 + 4.886 = 6.772.$$

The shaded area is about 6.77 square units.

Area Between Two Curves

We can extend the use of rectangles to approximate the area between two curves. If $f(x) \geq g(x)$, as in Figure 3.27, we see that the height of a rectangle is given by $f(x) - g(x)$. The area of the rectangle is $(f(x) - g(x))\Delta x$, so the area between the two curves is given as follows.

If $g(x) \leq f(x)$ for $a \leq x \leq b$:

$$\begin{array}{c} \text{Area between graphs of } f(x) \text{ and } g(x) \\ \text{between } a \text{ and } b \end{array} = \int_a^b (f(x) - g(x))\, dx.$$

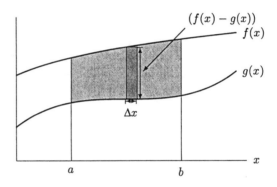

Figure 3.27: Area between two curves

Example 8   Sketch the graphs of $f(x) = 4x - x^2$ and $g(x) = \sqrt{x}$ for $x \geq 0$. Write a definite integral representing the area between the graphs of these two functions and use it to estimate the area.

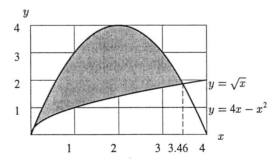

Figure 3.28: Area between $y = 4x - x^2$ and $y = \sqrt{x}$

Solution   The graphs of the two functions are shown in Figure 3.28; the area between them is shaded. We see that the two graphs cross at $x = 0$ and at $x \approx 3.46$. Between these values, the graph of $y = 4x - x^2$ lies above the graph of $y = \sqrt{x}$. Thus, the area between the two functions is given by

$$\text{Area} = \int_0^{3.46} ((4x - x^2) - (\sqrt{x})) \, dx.$$

Counting boxes on the graph shows that the shaded area is about 6 square units. Alternatively, we can use a calculator or computer to estimate the value of the definite integral. We see that the area is 5.845.

Problems for Section 3.3

1.   If the graph of f is in Figure 3.29, what is $\displaystyle\int_1^6 f(x) \, dx$?

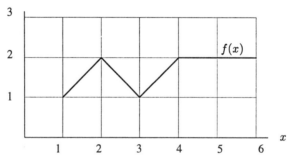

Figure 3.29

2.   A graph of $y = f(x)$ is given in Figure 3.30. Estimate $\displaystyle\int_0^3 f(x) \, dx$.

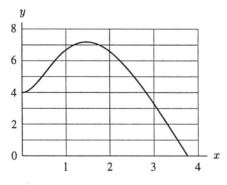

Figure 3.30

3.   Find the area under $y = x^3 + 2$ between $x = 0$ and $x = 2$.

4.   Find the area under $P = 100(0.6^t)$ between $t = 0$ and $t = 8$.

5. In Figure 3.31, use the grid to estimate the area of the region bounded by the curve, the horizontal axis and the vertical lines $x = 3$ and $x = -3$. Explain what you are doing.

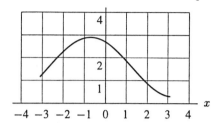

Figure 3.31

6. The graph of f is shown in Figure 3.32.

   (a) Estimate (by counting the squares) the total area shaded in Figure 3.32.

   (b) Estimate

   $$\int_0^8 f(x)\,dx.$$

   (c) Why are your answers to parts (a) and (b) different?

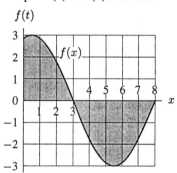

Figure 3.32

Without computing the integrals in Problems 7–10, decide if each is positive or negative, and explain your decision. [Hint: Sketch the graph of each. Compare the areas above and below the x-axis.]

7. $\displaystyle\int_0^3 (x^3 - 4x^2 + 2)\,dx$

8. $\displaystyle\int_0^2 (x^2 - x)\,dx$

9. $\displaystyle\int_0^4 (2 - x)3^x\,dx$

10. $\displaystyle\int_2^5 \frac{4 - x}{x}\,dx$

11. Find the area between the parabola $y = 4 - x^2$ and the x-axis.

12. Find the area between $y = x^2 - 9$ and the x-axis.

13. Find the area between $y = 3x$ and $y = x^2$.

14. Find the area between $y = x$ and $y = \sqrt{x}$.

15. (a) Sketch a graph of $f(x) = x(x + 2)(x - 1)$.

    (b) Find the total area between the graph and the x-axis between $x = -2$ and $x = 1$.

    (c) Find $\displaystyle\int_{-2}^1 f(x)\,dx$ and interpret it in terms of areas.

16. Compute the definite integral $\int_1^4 \dfrac{x^2 - 3}{x}\, dx$ and interpret the result in terms of areas.

17. Use a graph of $y = 2^{-x^2}$ to explain why $\int_{-1}^1 2^{-x^2}\, dx$ must be between 0 and 2.

18. (a) Sketch a graph of $x^3 - 5x^2 + 4x$ and mark on it points where $x = 1, 2, 3, 4, 5$.
    (b) Use your graph and area interpretation of the definite integral to decide which of the five numbers

$$I_n = \int_0^n (x^3 - 5x^2 + 4x)\, dx \quad \text{for } n = 1, 2, 3, 4, 5$$

is largest. Which is smallest? How many of the numbers are positive? (You do not need to calculate these integrals.)

19. For the function f graphed in Figure 3.33:
    (a) Suppose you know $\int_0^2 f(x)\, dx$. What is $\int_{-2}^2 f(x)\, dx$?

    (b) Suppose you know $\int_0^5 f(x)\, dx$ and $\int_2^5 f(x)\, dx$. What is $\int_0^2 f(x)\, dx$?

    (c) Suppose you know $\int_{-2}^5 f(x)\, dx$ and $\int_{-2}^2 f(x)\, dx$. What is $\int_0^5 f(x)\, dx$?

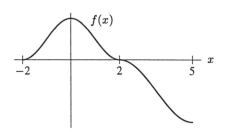

Figure 3.33

20. For the function f graphed in Figure 3.33:
    (a) Suppose you know $\int_{-2}^2 f(x)\, dx$ and $\int_0^5 f(x)\, dx$. What is $\int_2^5 f(x)\, dx$?
    (b) Suppose you know $\int_{-2}^5 f(x)\, dx$ and $\int_{-2}^0 f(x)\, dx$. What is $\int_2^5 f(x)\, dx$?
    (c) Suppose you know $\int_2^5 f(x)\, dx$ and $\int_{-2}^5 f(x)\, dx$. What is $\int_0^2 f(x)\, dx$?

3.4 THE DEFINITE INTEGRAL AS AVERAGE VALUE

The Average Value of a Function

We know how to find the average of n numbers: add them and divide by n. But how do we find the average value of a continuously varying function? Let us consider an example. Suppose $C = f(t)$ is the temperature at time t, measured in hours since midnight, and that we want to calculate the average temperature over a 24-hour period. One way to start would be to average the temperatures at n times, $t_1, t_2, \ldots, t_n$, during the day.

$$\text{Average temperature} \approx \frac{f(t_1) + f(t_2) + \cdots + f(t_n)}{n}.$$

The larger we make n, the better the approximation. We can rewrite this expression as a Riemann sum over the interval $0 \le t \le 24$ if we use the fact that $\Delta t = 24/n$, so $n = 24/\Delta t$:

$$\begin{aligned}
\text{Average temperature} &\approx \frac{f(t_1) + f(t_2) + \cdots + f(t_n)}{24/\Delta t} \\
&= \frac{f(t_1)\Delta t + f(t_2)\Delta t + \cdots + f(t_n)\Delta t}{24}
\end{aligned}$$

$$= \frac{1}{24} \sum_{i=1}^{n} f(t_i) \Delta t.$$

As $n \to \infty$, the Riemann sum tends towards an integral and also approximates the average temperature better. Thus, in the limit

$$\text{Average temperature} = \lim_{n \to \infty} \frac{1}{24} \sum_{i=1}^{n} f(t_i) \Delta t$$

$$= \frac{1}{24} \int_{0}^{24} f(t) \, dt.$$

Thus we have found a way of expressing the average temperature in terms of an integral. Generalizing for any function f, we define

$$\boxed{\begin{array}{l} \text{Average value of } f \\ \text{from } a \text{ to } b \end{array} = \frac{1}{b-a} \int_{a}^{b} f(x) \, dx}$$

How to Visualize the Average on a Graph

The definition of average value tells us that

$$(\text{Average value of } f) \cdot (b - a) = \int_{a}^{b} f(x) \, dx.$$

Thus, if we interpret the integral as the area under the graph of f (for positive f), then we can think of the average value of f as the height of the rectangle with the same area that is on the same base, $(b - a)$. (See Figure 3.34.)

If we think of the graph of the function as a wave in a fish tank, the average value of the function is the height of the water when the wave settles down and the water surface is level.

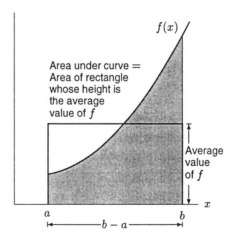

Figure 3.34: Area and average value

Example 1   Find the average value of $y = 5 + 6x - x^2$ on the interval $x = 0$ to $x = 5$.

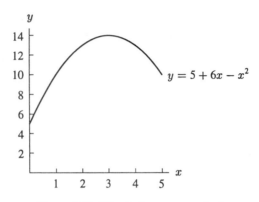

Figure 3.35: What is the average value?

Solution   A graph of this function is shown in Figure 3.35. We see that, on this interval, all the y-values are between 5 and 14. The average y-value must therefore be between 5 and 14. It appears in Figure 3.35 that the average value will be about 11. We use the formula for average value to calculate it exactly:

$$\text{Average value} = \frac{1}{b-a} \int_a^b f(x)\,dx$$
$$= \frac{1}{5-0} \int_0^5 (5 + 6x - x^2)\,dx$$
$$= \frac{1}{5}(58.333)$$
$$= 11.667.$$

The average value of y is 11.667. This matches what we see in the graph.

Example 2   The market research division of a fast food restaurant plans to track the effect of an extensive marketing campaign on business. In Figure 3.36, the number of customers per week, $N(t)$, is graphed against time, t (in weeks), for a period of six months (26 weeks). The marketing campaign starts at time $t = 0$ and lasts for 10 weeks.
   (a)   Describe the effect of the marketing campaign on the number of customers per week that come to the fast food restaurant during the six month period shown on the graph.
   (b)   From the graph, estimate visually the average number of customers per week over the six month period.
   (c)   Set up a definite integral representing the average number of customers per week over the six month period.
   (d)   Estimate the total number of customers to visit the store during this six month period.

customers per week

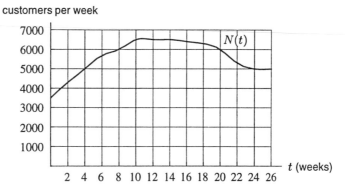

Figure 3.36: Number of customers at a fast food restaurant

Solution (a) While the marketing campaign lasts, the number of customers per week increases from about 3500 at the start of the campaign ($t = 0$) to about 6500 at the end of the campaign ($t = 10$). For the next 8 weeks, the number of customers stays pretty steady at about 6500. After this, the number of customers drops off again, and appears to be stabilizing at about 5000 customers per week.

(b) To approximate the average number of customers per week, we want to draw a horizontal line so that the area under this line is the same as the area under the curve. In other words, the area between the line and the curve that is below the line should be approximately equal to the area between the line and the curve that is above the line. We see in Figure 3.37 that a horizontal line at about 5500 satisfies this condition, so the average number of customers per week is about 5500.

(c)

$$\text{Average number of customers per week} \; = \; \frac{1}{26 - 0} \int_0^{26} N(t)\, dt$$

(d) The total number of customers is represented by the total area under the curve, so we estimate this area. Each grid box has an area of $(1,000 \text{ customers/week})(2 \text{ weeks}) = 2000$ customers, and there are about 74 square boxes under the curve (68 entire boxes and several pieces that we visually estimate to be about 6 more boxes), so we have

$$\text{Total number of customers} \; \approx \; (74)(2000) = 148,000.$$

customers per week

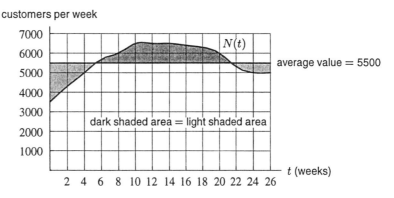

Figure 3.37: Average number of customers per week

Alternatively, we can estimate the total number of customers by multiplying our estimate of average number of customers per week (5500) by the number of weeks (26). We have

$$\text{Total number of customers} \approx (5500)(26) = 143,000.$$

Problems for Section 3.4

1. The graph of f is in figure 3.38. You calculated $\int_1^6 f(x)dx$ in Problem 1 of Section 3.3. What is the average value of f on the interval $1 \leq x \leq 6$?

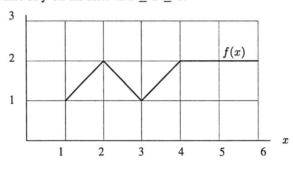

Figure 3.38

2. The graph of f is shown in Figure 3.39.

   (a) Estimate (by counting boxes)

   $$\int_0^5 f(x)\,dx.$$

   (b) Estimate the average value of f between $x = 0$ and $x = 5$ by estimating visually the average height.

   (c) Estimate the average value of f between $x = 0$ and $x = 5$ by using your answer to part (a) and the formula for average value. (Your answers to parts (b) and (c) should be about the same.)

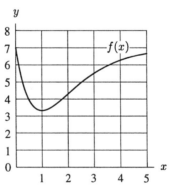

Figure 3.39

3. Find the average value of $g(t) = 1 + t$ over the interval $[0, 2]$.
4. Find the average value of $g(t) = 2^t$ over the interval $[0, 10]$.

5. Graphs of two functions are shown in Figure 3.40. In each case, give a rough estimate of the average value of the function between $x = 0$ and $x = 7$, and explain how you arrived at your answer.

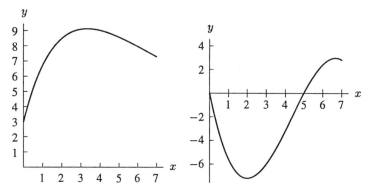

Figure 3.40: Estimate the average value

6. (a) What is the average value of $f(x) = \sqrt{1 - x^2}$ over the interval $0 \le x \le 1$?
   (b) How can you tell whether this average value is more or less than 0.5 without doing any calculations?

7. The warehouse for a mail-order company receives shipment for a certain item on June 1, and the stock of the item is steadily depleted over the next few months. Suppose the inventory function for this item is given by

$$I(t) = 5000e^{-0.1t},$$

where t is measured in days since June 1.

   (a) Find the average inventory in the warehouse during the 90 days after June 1.
   (b) Graph the function $I(t)$ and illustrate the average inventory graphically.

8. The value, V, of a Tiffany lamp, worth \$225 in 1965, increases at 15% per year. Its value in dollars t years after 1965 is given by

$$V = 225(1.15)^t.$$

Find the average value of the lamp over the period 1965–2000.

9. For the function f in Figure 3.41, write an expression involving one or more definite integrals that denotes:

   (a) The average value of f for $0 \le x \le 5$.
   (b) The average value of $|f|$ for $0 \le x \le 5$. (The notation $|f|$ stands for the absolute value of f.)

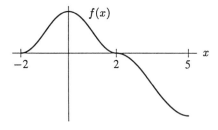

Figure 3.41

10. For the function f in Figure 3.41, consider the average value of f over the following intervals:

    (I) $0 \leq x \leq 1$ (II) $0 \leq x \leq 2$ (III) $0 \leq x \leq 5$ (IV) $-2 \leq x \leq 2$

    (a) For which interval is the average value of f least?
    (b) For which interval is the average value of f greatest?
    (c) For which pair of intervals are the average values equal?

11. The quantity of a certain radioactive substance at time t is given by

    $$Q = 4(0.96^t) \text{ grams.}$$

    (a) Find $Q(10)$ and $Q(20)$.
    (b) Find the average of $Q(10)$ and $Q(20)$.
    (c) Find the average value of Q over the interval $10 \leq t \leq 20$. (Use a calculator or computer to get a good estimate.)
    (d) Use what you know about the graph of Q to explain the relative sizes of your answers in parts (b) and (c).

12. Given the graph of f in Figure 3.42, arrange the following numbers in order from least to greatest:

    (a) $f'(1)$
    (b) the average value of f on $0 \leq x \leq 4$
    (c) $\int_0^1 f(x)dx$

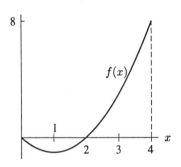

Figure 3.42

3.5 INTERPRETATIONS OF THE DEFINITE INTEGRAL

The Notation and Units for the Definite Integral

Just as the Leibniz notation dy/dx for the derivative reminds us that the derivative is the limit of a ratio of differences, the notation for the definite integral helps us recall the meaning of the integral. The symbol

$$\int_a^b f(x)\,dx$$

reminds us that an integral is a limit of sums (the integral sign is a misshapen S) of terms of the form "$f(x)$ times a small difference of x." Officially, dx is not a separate entity, but a part of the whole integral symbol. Thus, just as one thinks of d/dx as a single symbol meaning "the derivative with respect to x of...," one can think of $\int_a^b \ldots dx$ as a single symbol meaning "the integral of ... with respect to x."

However, most scientists and mathematicians informally think of dx as an "infinitesimally" small bit of x which in this context is multiplied by a function value $f(x)$. This viewpoint is often the key to interpreting the meaning of a definite integral. For example, if $f(t)$ is the velocity of a moving particle at time t, then $f(t)dt$ may by thought of informally as velocity × time, giving the distance traveled by the particle during a small bit of time dt. The integral $\int_a^b f(t)\,dt$ may then be thought of as the sum of all these small distances, giving us the net change in position of the particle between $t = a$ and $t = b$.

The notation for the integral also helps us determine what units should be used for the numerical value of the integral. Since the terms being added up are products of the form "$f(x)$ times a difference in x," the unit of measurement for $\int_a^b f(x)\,dx$ is the product of the units for $f(x)$ and the units for x. Thus if $f(t)$ is velocity measured in meters/second and t is time measured in seconds, then

$$\int_a^b f(t)\,dt$$

has for units (meters/sec)×(sec)=meters, which is what we expect since the value of the integral represents change in position. Similarly, if we graph $y = f(x)$ with the same units of measurement of length along the x- and y-axes, then $f(x)$ and x are measured in the same units so

$$\int_a^b f(x)\,dx$$

is measured in square units, say cm×cm=cm^2. Again this is what we would expect since in this context the integral represents an area. Finally for the average value,

$$\frac{1}{b-a}\int_a^b f(x)\,dx,$$

the units for dx inside the integral are canceled by the units for $1/(b-a)$ outside the integral, leaving only the units for $f(x)$. This is as it should be since the average value of f should be measured in the same units as $f(x)$.

Example 1   Suppose that $C(t)$ represents the cost per day to heat your house measured in dollars per day, where t is time measured in days and $t = 0$ corresponds to January 1, 1996. Interpret $\int_0^{90} C(t)\,dt$ and $\frac{1}{90-0}\int_0^{90} C(t)\,dt$.

Solution   The units for the integral $\int_0^{90} C(t)\,dt$ are (dollars/day)×(days)=dollars. The integral represents the total cost in dollars to heat your house for the first 90 days of 1996, namely the months of January, February, and March. The second expression is measured in (1/days)(dollars) or dollars per day, the same units as $C(t)$. It represents the average cost per day to heat your house during the first 90 days of 1996.

More Interpretations of the Definite Integral

Example 2 A bacteria colony has a population of 14 million bacteria at time $t = 0$. Suppose that the bacteria population is growing at a rate of $f(t) = 2^t$ million bacteria per hour.
  (a) Give a definite integral which represents the total change in the bacteria population during the two hours from $t = 0$ to $t = 2$.
  (b) Find the population at time $t = 2$.

Solution (a) Since $f(t) = 2^t$ gives the rate of change of population, the total change between $t = 0$ and $t = 2$ is given by

$$\int_0^2 2^t\, dt.$$

  (b) Using a computer or graphing calculator, we find $\int_0^2 2^t\, dt = 4.328$. The total change in the bacteria population during these two hours is 4.328 million bacteria. The bacteria population was 14 million at time $t = 0$, and increased 4.328 million between $t = 0$ and $t = 2$. Therefore, the population at time $t = 2$ is $14 + 4.328 = 18.328$ million bacteria.

Example 3 The graph of rate of change of the value of an investment with respect to time for a period of 5 months is shown in Figure 3.43.

  (a) When is the investment increasing in value and when is it decreasing?
  (b) Does the investment gain or lose value during the 5 months?

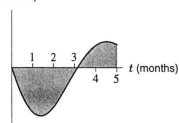

Figure 3.43: Did the company make money?

Solution (a) The investment decreases in value during the first 3 months, since the rate of change of value is negative then. The value is rising during the next 2 months.
  (b) We want to find the total change in the value of the investment between $t = 0$ and $t = 5$. Since the total change in value is the integral of the rate of change of value, we are looking for

$$\int_0^5 f(t)dt,$$

where $f(t)$ represents the rate of change of value, shown in Figure 3.43. The integral equals the shaded area above the t-axis minus the shaded area below the t-axis. Since the area below the axis is greater than the area above the axis, the integral is negative. Total change in value during this time is negative, so the investment decreased in value.

Example 4   A man takes a trip in a car, and his velocity (in mph) is given in Figure 3.44. At time $t = 0$, the man is 50 miles from his house. Positive velocities take him toward his house and negative velocities take him away from his house.

(a)   When is the man closest to his house, and approximately how far away is he then?
(b)   When is the man farthest from his house, and how far away is he then?

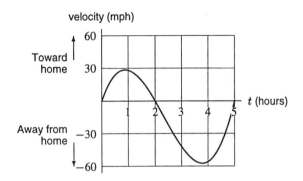

Figure 3.44: Describe this trip

Solution   What happens on this trip? The velocity function is positive the first two hours and negative between $t = 2$ and $t = 5$. Thus the man moves toward his house during the first two hours, then turns around at $t = 2$ and moves away from his house. The distance he travels is represented by area between the graph of velocity and the t-axis; since the area below the axis is greater than the area above the axis, we see that he ends up farther away from home than when he started. Thus he is closest to home at $t = 2$ and farthest from home at $t = 5$. We can estimate how far he went in each direction by estimating areas.

(a)   The man starts out 50 miles from home. During the first two hours, the distance the man travels is the area under the curve between $t = 0$ and $t = 2$. This area corresponds to about one box. Since each box has area (30 miles/hour)(1 hour) = 30 miles, the man travels about 30 miles toward home. He is closest to home after 2 hours, and he is about 20 miles away at that time.
(b)   Between $t = 2$ and $t = 5$, the man moves away from his house. Since this area is equal to about $3\frac{1}{2}$ boxes, which is $(3.5)(30) = 105$ miles, he has moved 105 miles farther from home. He was already 20 miles from home so at $t = 5$, he is about 125 miles from home. He is farthest from home at $t = 5$, and he is about 125 miles away at that time.

Notice that the man has covered a total distance of $30 + 105 = 135$ miles. However, he went toward his home for 30 miles and then away from his home for 105 miles. Thus, the *net* change in his position is 75 miles.

The Definite Integral of Marginal Cost Gives Total Cost

Recall that the derivative of the cost function is the marginal cost. A marginal cost function $C'(q)$ is given in Figure 3.45, where q is the quantity of items produced. We see, for example, that $C'(100) = 6$, which means that the additional cost to produce the 100th item is about \$6. How do we interpret

$$\int_a^b C'(q)\, dq?$$

Since marginal cost $C'(q)$ is the rate of change of the cost function, this definite integral represents the total change in the cost function between $q = a$ and $q = b$. In other words, this is the amount it will cost to increase production from a units to b units. Recall that the cost of producing 0 units is the fixed cost. The area under the marginal cost curve between 0 and q is the total increase in cost between a production of 0 and a production of q. This is called the *total variable cost*. If we add this to the fixed cost, we obtain the total cost to produce q units. In general,

---

If $C'(q)$ is the marginal cost function,

$$\int_a^b C'(q)\, dq = \text{Amount it will cost to increase production from } a \text{ units to } b \text{ units.}$$

$$\int_0^b C'(q)\, dq = \text{Total variable cost to produce } b \text{ units.}$$

To obtain the total cost of producing b units, add the fixed cost to the total variable cost.

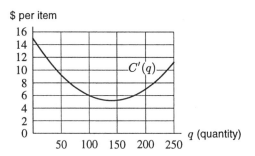

Figure 3.45: A marginal cost curve

Example 5   The marginal cost curve for a certain product is given in Figure 3.45. If the fixed costs for this production are $1000, estimate the total cost in producing 250 items.

Solution   The total cost of production equals fixed cost + variable cost. We know that the fixed cost is $1000 and the variable cost in producing 250 items is represented by the area under the marginal cost curve in Figure 3.45 between $q = 0$ and $q = 250$. Approximating this area as about 20 boxes each of area (2 dollars/item)(50 items) = 100 items, we get

$$\text{Total variable cost} = \int_0^{250} C'(q)\, dq \approx 20(100) = 2000.$$

We now find the total cost to produce 250 items:

$$
\begin{aligned}
\text{Total cost} &= \text{Fixed cost} + \text{Variable cost} \\
&= \$1000 + \$2000 \\
&= \$3000.
\end{aligned}
$$

A Pondwater Example

Example 6   Biological activity in a pond is reflected in the rate at which carbon dioxide, CO_2, is added to or withdrawn from the water. Plants take CO_2 out of the water during the day for photosynthesis and put CO_2 into the water at night. Animals put CO_2 into the water all the time as they breathe. Biologists are interested in how the net rate at which CO_2 enters or leaves a pond varies during the day. Figure 3.46 shows this rate as a function of time of day.[1] The rate is measured in millimoles (mmol) of CO_2 per liter of water per hour; time is measured in hours past dawn. At dawn, there were 2.600 mmol of CO_2 per liter of water.

(a)   What can be concluded from the fact that the rate is negative during the day and positive at night?

(b)   Some scientists have suggested that plants and animals respire (breathe) at a constant rate at night, and that plants photosynthesize at a constant rate during the day. Does Figure 3.46 support this view?

(c)   When was the CO_2 content of the water at its lowest? How low did it go?

(d)   How much CO_2 was released into the water during the 12 hours of darkness? Compare this quantity with the amount of CO_2 withdrawn from the water during the 12 hours of daylight. How can you tell by looking at the graph whether the CO_2 in the pond is in equilibrium?

(e)   Estimate the CO_2 content of the water at three hour intervals throughout the day. Use your estimates to plot a graph of CO_2 content throughout the day.

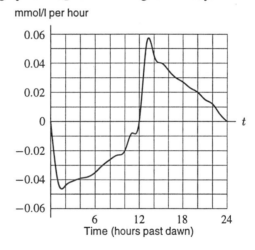

Figure 3.46: Rate at which CO_2 is entering or leaving water, $f'(t)$

Solution   (a) CO_2 is being taken out of the water during the day and returned at night. The pond must therefore contain some plants. (The data is in fact from pond water containing both plants and animals.)

(b)   Suppose t is the number of hours past dawn. The graph in Figure 3.46 shows that the CO_2 content changes at a greater rate for the first 6 hours of daylight, $0 < t < 6$, than it does for the final 6 hours of daylight, $6 < t < 12$. It turns out that plants photosynthesize more vigorously in the morning than in the afternoon. Similarly, CO_2 content changes more rapidly in the first half of the night, $12 < t < 18$, than in the 6 hours just before dawn, $18 < t < 24$. The reason seems to be that at night plants quickly use up most of the sugar that they synthesized during the day, and then their respiration rate is inhibited.

---

[1] Data from R. J. Beyers, *The Pattern of Photosynthesis and Respiration in Laboratory Microsystems* (Mem. 1st. Ital. Idrobiol., 1965).

(c) We are now being asked about the total quantity of CO_2 in the pond, rather than the rate at which it is changing. We will let $f(t)$ denote the CO_2 content of the pond water (in mmol/l) at t hours past dawn. Then Figure 3.46 is a graph of the derivative $f'(t)$. There are 2.600 mmol/l of CO_2 in the water at dawn, so $f(0) = 2.600$.

The CO_2 content $f(t)$ decreases during the 12 hours of daylight, $0 < t < 12$, when $f'(t) < 0$, and then $f(t)$ increases for the next 12 hours. Thus, $f(t)$ is at a minimum when $t = 12$, at dusk. The total quantity of CO_2 at time $t = 12$ is equal to the amount present at time $t = 0$ plus the total change between $t = 0$ and $t = 12$. We have

$$f(12) = f(0) + \int_0^{12} f'(t)\, dt = 2.600 + \int_0^{12} f'(t)\, dt.$$

We must approximate the definite integral by a Riemann sum. From the graph in Figure 3.46, we estimate the values of the function $f'(t)$ in Table 3.10.

TABLE 3.10 *Rate, $f'(t)$, at which CO_2 is entering or leaving water*

t	$f'(t)$	t	$f'(t)$	t	$f'(t)$	t	$f'(t)$	t	$f'(t)$	t	$f'(t)$
0	0.000	4	−0.039	8	−0.026	12	0.000	16	0.035	20	0.020
1	−0.042	5	−0.038	9	−0.023	13	0.054	17	0.030	21	0.015
2	−0.044	6	−0.035	10	−0.020	14	0.045	18	0.027	22	0.012
3	−0.041	7	−0.030	11	−0.008	15	0.040	19	0.023	23	0.005

The left Riemann sum with $n = 12$ terms, corresponding to $\Delta t = 1$, gives

$$\int_0^{12} f'(t)\, dt \approx (0.000)(1) + (-0.042)(1) + (-0.044)(1) + \cdots + (-0.008)(1) = -0.346.$$

At 12 hours past dawn, the CO_2 content of the pond water reaches its lowest level, which is approximately

$$2.600 - 0.346 = 2.254 \text{ mmol/l}.$$

(d) The increase in CO_2 during the 12 hours of darkness equals

$$f(24) - f(12) = \int_{12}^{24} f'(t)\, dt.$$

Using Riemann sums to estimate this integral, we find that about 0.306 mmol/l of CO_2 was released into the pond during the night. In part (c) we calculated that about 0.346 mmol/l of CO_2 was absorbed from the pond during the day. If the pond is in equilibrium, we would expect the daytime absorption to equal the nighttime release. These quantities are sufficiently close (0.346 and 0.306) that the difference could be due to measurement error.

If the pond is in equilibrium, the area between the rate curve in Figure 3.46 and the axis for $0 \le t \le 12$ will equal the area between the rate curve and the axis for $12 \le t \le 24$. (The axis is the horizontal line at 0.) In this experiment the areas do look approximately equal.

(e) We must evaluate

$$f(b) = f(0) + \int_0^b f'(t)\, dt = 2.600 + \int_0^b f'(t)\, dt$$

for the values $b = 0, 3, 6, 9, 12, 15, 18, 21, 24$. Left Riemann sums with $\Delta t = 1$ give the values for the CO_2 content in Table 3.11. The graph is shown in Figure 3.47.

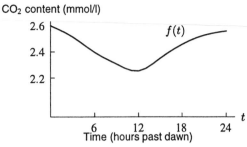

Figure 3.47: CO_2 content in pond water throughout the day

TABLE 3.11 CO_2 *content throughout the day*

b (hours after dawn)	0	3	6	9	12	15	18	21	24
$f(b)$ (CO_2 content)	2.600	2.514	2.396	2.305	2.254	2.353	2.458	2.528	2.560

Problems for Section 3.5

1. If $f(t)$ is measured in miles per hour and t is measured in hours, what are the units of measurement for

    $$\int_a^b f(t)\, dt?$$

2. If the marginal cost function $C'(q)$ is measured in dollars per ton, and q gives the quantity in tons, what are the units of measurement for the following definite integral and what does it represent?

    $$\int_{800}^{900} C'(q)\, dq$$

3. If $f(t)$ is measured in meters/second^2 and t is measured in seconds, what are the units of measurement for $\int_a^b f(t)\, dt$?

4. If $f(t)$ is measured in dollars per year and t is measured in years, what are the units of measurement for $\int_a^b f(t)\, dt$?

5. Oil is leaking out of a ruptured tanker at a rate of $r = f(t)$ gallons per minute. Write a definite integral expressing the total quantity of oil which leaks out of the tanker in the first hour.

6. The rate of growth of the net worth of a company is given by $r(t) = 2000 - 12t^2$, where t is measured in years since 1980. How does the net worth of the company change between 1980 and 1990? If the company is worth \$40,000 in 1980, what is it worth in 1990?

7. A news broadcast in early 1993 said the average American's annual income is changing at a rate given in dollars per month by $r(t) = 40(1.002)^t$ where t is in months from January 1, 1993. If this rate stayed constant throughout the year, what change in income did the average American receive during 1993?

8. A cup of coffee at 90°C is put into a 20°C room when $t = 0$. If the coffee's temperature is changing at a rate given in °C per minute by

$$r(t) = -7(0.9^t), \quad t \text{ in minutes,}$$

estimate, to one decimal place, the coffee's temperature when $t = 10$.

9. Water is leaking out of a tank at a rate of $R(t)$ gallons/hour, where t is measured in hours.

    (a) Write a definite integral that expresses the total amount of water that leaks out in the first two hours.

    (b) Figure 3.48 is a graph of $R(t)$. On a sketch, shade in the region whose area represents the total amount of water that leaks out in the first two hours.

    (c) Give an upper and lower estimate of the total amount of water that leaks out in the first two hours.

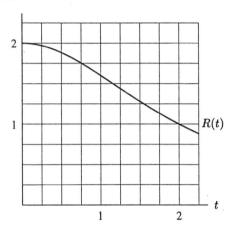

Figure 3.48

10. Filters at a water treatment plant become dirtier over time and thus become less effective. They are replaced every 30 days. During one 30-day period, the rate at which pollution passes through the filters into a nearby lake (in units of particulate matter per day) is measured every 6 days and is given in Table 3.12. Time t is measured in days since the filters were replaced.

    (a) Estimate the total amount of pollution entering the lake during the 30-day period.

    (b) Your answer to part (a) is only an estimate. Give bounds (lower and upper estimates) between which we know the true amount of pollution lies. (You may assume the rate of pollution is continually increasing.)

TABLE 3.12   *Rate of pollution*

day	0	6	12	18	24	30
units/day	7	8	10	13	18	35

11. The rate at which the world's oil is being consumed is continuously increasing. Suppose the rate (in billions of barrels per year) is given by the function $r = f(t)$, where t is measured in years and $t = 0$ is the start of 1990.

    (a) Write a definite integral which represents the total quantity of oil used between the start of 1990 and the start of 1995.

    (b) Suppose $r = 32(1.05^t)$. Using a left-hand sum with five subdivisions, find an approximate value for the total quantity of oil used between the start of 1990 and the start of 1995.

    (c) Interpret each of the five terms in the sum from part (b) in terms of oil consumption.

12. Two new salespeople have been hired by a company, and the number of sales per month is recorded for each in Figure 3.49, with the two curves labeled Salesperson A and Salesperson B. Which person has the most total sales after 6 months? After the first year? At approximately what times (if any) have they sold roughly equal total amounts? Approximately how many total sales has each made at the end of the first year?

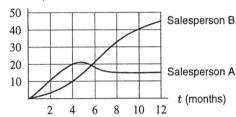

Figure 3.49: Which person sold the most items?

13. A marginal cost function $C'(q)$ is given in Figure 3.50. Assume fixed costs are \$10,000.

    (a) Estimate the total cost to produce 30 units.
    (b) Approximate the additional cost if the company increases productions from 30 units to 40 units.
    (c) Find the value of $C'(25)$. Interpret your answer in terms of costs of production.

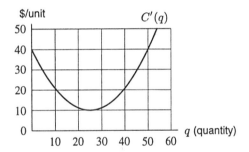

Figure 3.50: A marginal cost curve

14. The marginal cost function for a certain company is given by

$$C'(q) = q^2 - 16q + 70 \text{ dollars/unit,}$$

where q is the quantity produced. It is known that $C(0) = 500$. Estimate the total cost of producing 20 units. Explain how you arrived at your answer. What is the fixed cost and what is the total variable cost associated with this production size?

15. The marginal cost $C'(q)$ of producing q units is given in the table below.

    (a) If fixed cost is \$10,000, estimate the total cost of producing 400 units.
    (b) How much would the total cost increase if production were increased one unit, to 401 units?

TABLE 3.13 *Marginal cost in producing q units*

q	0	100	200	300	400	500	600
$C'(q)$	25	20	18	22	28	35	45

16. A car moves along a straight line with velocity, in feet/second, given by

$$v(t) = 6 - 2t \quad \text{for } t \geq 0.$$

(a) Describe the car's motion in words. (When is it moving forward, backward, and so on?)
(b) Suppose the car's position is measured from its starting point. When is it farthest forward? Backward?
(c) Find s, the car's position measured from its starting point, as a function of time.

17. Suppose that the graph in Figure 3.51 represents your velocity, in miles per hour, on a long bicycle trip. Suppose that you start out 10 miles from home, and positive velocities take you towards home and negative velocities take you away from home. Write a paragraph describing your trip: Do you start out going towards or away from home? How long do you continue in that direction and how far are you from home when you turn around? How many times do you change direction? Do you ever get home? Where are you at the end of the four-hour bike ride? Make up a story to go with this bicycle trip.

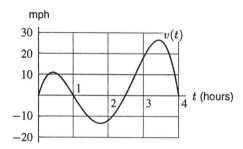

Figure 3.51: The velocity curve for a bicycle trip

18. A bicyclist is pedaling along a straight road with velocity, v, given in Figure 3.52. Suppose the cyclist starts 5 miles from a lake, and that positive velocities take her away from the lake and negative velocities towards the lake. When is the cyclist farthest from the lake, and how far away is she then?

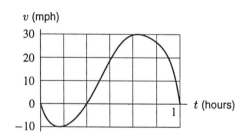

Figure 3.52

19. Suppose the rate of change of the price of stock in a certain company as a function of time, t, in weeks, is in Figure 3.53.

(a) At what time during this five-week period was the stock at its highest value?

(b)   At what time during this five-week period was the stock at its lowest value?

(c)   If the price of the stock as a function of time is given by $P(t)$, put the following quantities in increasing order:

$$P(0), \; P(1), \; P(2), \; P(3), \; P(4), \; P(5).$$

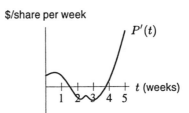

$/share per week

$P'(t)$

t (weeks)

Figure 3.53: Rate of change of the price of a company's stock

3.6 THE FUNDAMENTAL THEOREM OF CALCULUS

In Chapter 2, we learned that the rate of change of a quantity $F(t)$ is given by the derivative $F'(t)$. In Section 3.5, we learned that the definite integral of the rate of change $F'(t)$ is the total change in $F(t)$. We have the following result:

$$F(b) - F(a) = \begin{array}{c} \text{Total change in } F(t) \\ \text{between } t = a \text{ and } t = b \end{array} = \int_a^b F'(t)\, dt.$$

This result is one of the most important in calculus because it makes the connection between the derivative and the definite integral. It is called the Fundamental Theorem of Calculus and is often stated as follows:

The Fundamental Theorem of Calculus

If f is continuous and $f(t) = \dfrac{dF(t)}{dt}$, then

$$\int_a^b f(t)\, dt = F(b) - F(a).$$

In words:

     The definite integral of a rate gives total change.

The Fundamental Theorem provides a precise way of computing certain definite integrals.

Example 1   Compute $\displaystyle\int_1^3 2x\, dx$ by two different methods.

Solution   Using left- and right-hand sums, we can approximate this integral as accurately as we want. With $n = 100$, for example, the left-sum is 7.96 and the right sum is 8.04. Using $n = 500$, we learn that

$$7.992 < \int_1^3 2x \, dx < 8.008.$$

The Fundamental Theorem, on the other hand, allows us to compute the integral exactly. We take $f(x) = 2x$. By Example 4 in Section 2.3, we know that if $F(x) = x^2$, then $F'(x) = 2x$. So we take $f(x) = 2x$ and $F(x) = x^2$ and obtain

$$\int_1^3 2x \, dx = F(3) - F(1) = 3^2 - 1^2 = 8.$$

Example 2   If $F(t) = 3t^2$, it can be shown that $F'(t) = 6t$. Use this fact and the Fundamental Theorem of Calculus to find $\int_0^2 6t \, dt$.

Solution   Since $F'(t) = 6t$, we know that

$$\int_0^2 6t \, dt = F(2) - F(0)$$
$$= 3(2^2) - 3(0^2)$$
$$= 12 - 0$$
$$= 12.$$

The Fundamental Theorem can also be used when the rate, $F'(t)$, is known and we want to find the total change $F(b) - F(a)$. If we know $F(a)$, the theorem enables us to reconstruct the function F from knowledge about its derivative $F' = f$.

Example 3   Suppose you are given that $F'(t) = (1.8)^t$ and $F(0) = 2$. Find the value of $F(b)$ at the points $b = 0$, $0.1, 0.2, \ldots, 1.0$.

Solution   We apply the Fundamental Theorem with $f(t) = (1.8)^t$ and $a = 0$ to get values for $F(b)$. Since

$$F(b) - F(0) = \int_0^b F'(t) \, dt = \int_0^b (1.8)^t \, dt$$

and $F(0) = 2$, we have

$$F(b) = 2 + \int_0^b (1.8)^t \, dt.$$

We use Riemann sums to estimate the definite integral $\int_0^b (1.8)^t \, dt$ for each of the values of b. For example, when $b = 0.1$, we find that $\int_0^b (1.8)^t \, dt = 0.103$. Thus the change in F between 0 and 0.1 is 0.103. Since $F(0) = 2$, we see that $F(0.1) = 2.103$. Continuing in this way, we obtain the values given in Table 3.14.

TABLE 3.14   *Approximate values for F*

b	0	0.1	0.2	0.3	0.4	0.5	0.6	0.7	0.8	0.9	1.0
$F(b)$	2	2.103	2.212	2.328	2.451	2.581	2.719	2.866	3.021	3.186	3.361

Notice from the table that the function $F(b)$ is increasing between $b = 0$ and $b = 1$. This could have been predicted without the use of the Fundamental Theorem, from the fact that $(1.8)^t$, the derivative of $F(t)$, is positive for t between 0 and 1.

Example 4   The graph of the derivative F' of some function F is given in Figure 3.54. If you are told that $F(20) = 150$, estimate the maximum value attained by F.

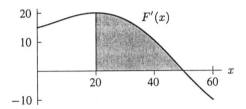

Figure 3.54: Graph of the derivative F' of some function F

Solution   Let's begin by getting a rough idea of how F behaves. We know that $F(x)$ increases for $x < 50$ because the derivative of F is positive for $x < 50$. Similarly, $F(x)$ decreases for $x > 50$ because $F'(x)$ is negative for $x > 50$. Therefore, the graph of F rises until the point at which $x = 50$, and then it begins to fall. Evidently, the highest point on the graph of F is at $x = 50$, and so the maximum value attained by F is $F(50)$. To evaluate $F(50)$, we use the Fundamental Theorem:

$$F(50) - F(20) = \int_{20}^{50} F'(x)\,dx,$$

which gives

$$F(50) = F(20) + \int_{20}^{50} F'(x)\,dx = 150 + \int_{20}^{50} F'(x)\,dx.$$

The definite integral equals the area of the shaded region under the graph of F', which we can estimate is roughly 300. Therefore, the greatest value attained by F is $F(50) \approx 150 + 300 = 450$.

Example 5   The graph of the derivative F' of some function F is given in Figure 3.55. Assume $F(0) = 0$. Of the four numbers $F(1)$, $F(2)$, $F(3)$, and $F(4)$, which is largest? Which is smallest? How many of these four numbers are negative?

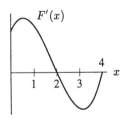

Figure 3.55: A graph of the derivative F'

Solution   For every number b, the Fundamental Theorem tells us that

$$\int_0^b F'(x)\,dx = F(b) - F(0) = F(b) - 0 = F(b).$$

Therefore, the values of $F(1)$, $F(2)$, $F(3)$, and $F(4)$ are values of definite integrals. Recall that the value of a definite integral is equal to the area of the regions under the graph above the x-axis minus the area of the regions below the x-axis above the graph. Let A_1 represent the area of the region between $x = 0$ and $x = 1$, let A_2 represent the area of the region between $x = 1$ and $x = 2$, let A_3 represent the area of the region between $x = 2$ and $x = 3$, and let A_4 represent the area of the region between $x = 3$ and $x = 4$, as shown in Figure 3.56.

We see that the region between $x = 0$ and $x = 1$ lies above the x-axis, and so $F(1)$ is positive, and we have

$$F(1) = \int_0^1 F'(x)\,dx = A_1.$$

We see that the region between $x = 0$ and $x = 2$ also lies entirely above the x-axis, and so $F(2)$ is positive. We have

$$F(2) = \int_0^2 F'(x)\,dx = A_1 + A_2.$$

We see that $F(2) > F(1)$. The region between $x = 0$ and $x = 3$ includes parts above and below the x-axis. We have

$$F(3) = \int_0^3 F'(x)\,dx = (A_1 + A_2) - A_3.$$

Since the area A_3 appears to be approximately the same as the area A_2, we have $F(3) \approx F(1)$. Finally, we see that

$$F(4) = \int_0^4 F'(x)\,dx = (A_1 + A_2) - (A_3 + A_4).$$

Since the area $A_1 + A_2$ appears to be larger than the area $A_3 + A_4$, we see that $F(4)$ is still positive, but it is smaller than all the others.

We see that the largest value is $F(2)$ and the smallest value is $F(4)$. None of the numbers is negative.

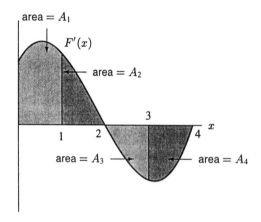

Figure 3.56: Using areas to understand definite integrals

Problems for Section 3.6

1. Figure 3.57 shows the graph of f. If $F' = f$ and $F(0) = 0$, find $F(b)$ for $b = 1, 2, 3, 4, 5, 6$.

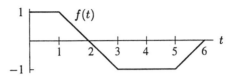

Figure 3.57

2. The graph of $y = f(x)$ is given in Figure 3.58.

    (a) What is $\displaystyle\int_{-3}^{0} f(x)\,dx$?

    (b) If the area of the shaded region is A, what is $\displaystyle\int_{-3}^{4} f(x)\,dx$?

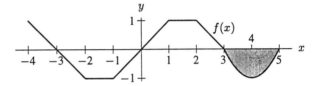

Figure 3.58

For Problems 3–5, suppose $F(0) = 0$ and $F'(x) = f(x) = 4 - x^2$ for $0 \le x \le 2.5$.

3. Approximate $F(b)$ for $b = 0,\ 0.5,\ 1,\ 1.5,\ 2,\ 2.5$.

4. Using a graph of F', decide where F is increasing and where F is decreasing.

5. Does F have a maximum value for $0 \le x \le 2.5$? If so, at what value of x does it occur, and approximately what is that maximum value?

6. The graph of a derivative $F'(t)$ is shown in Figure 3.59, and some areas have been labeled. If $F(0) = 3$, draw a careful graph of $F(t)$. What is the change in F between $t = 0$ and $t = 6$? Evaluate $F(6)$.

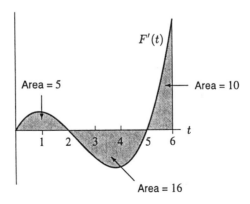

Figure 3.59

7. If $F(t) = t^2 + t$, it can be shown that $f(t) = F'(t) = 2t + 1$. Find $\int_1^4 (2t + 1)\, dt$ two ways:

    (a) Using technology.
    (b) Using the Fundamental Theorem of Calculus.

8. (a) Use a computer or graphing calculator to approximate the definite integral

    $$\int_1^4 3x^2\, dx.$$

    (b) Use the facts that $F(x) = x^3$ and therefore $F'(x) = 3x^2$, and the Fundamental Theorem of Calculus, to find the value of the integral in part (a) exactly.

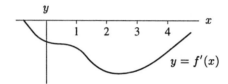

Figure 3.60: Note: This is the graph of f', not f.

Problems 9–10 concern the graph of f' in Figure 3.60.

9. Which is greater, $f(0)$ or $f(1)$?

10. List the following quantities in increasing order:

    $$\frac{f(4) - f(2)}{2}, \quad f(3) - f(2), \quad f(4) - f(3).$$

For Problems 11–14, mark the following quantities on a copy of the graph of f in Figure 3.61.

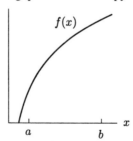

Figure 3.61

11. A length representing $f(b) - f(a)$.

12. A slope representing $\dfrac{f(b) - f(a)}{b - a}$.

13. An area representing $F(b) - F(a)$, where $F' = f$.

14. A length roughly approximating

    $$\frac{F(b) - F(a)}{b - a}, \text{ where } F' = f.$$

REVIEW PROBLEMS FOR CHAPTER THREE

For Problems 1–6, find the integrals to one decimal place. In each case say how many subdivisions you used.

1. $\displaystyle\int_0^{10} 2^{-x}\,dx$

2. $\displaystyle\int_1^5 x^2 + 1\,dx$

3. $\displaystyle\int_0^1 \sqrt{1 + t^2}\,dt$

4. $\displaystyle\int_{-1}^1 \frac{x^2 + 1}{x^2 - 4}\,dx$

5. $\displaystyle\int_2^3 \frac{-1}{(r + 1)^2}\,dr$

6. $\displaystyle\int_1^3 \frac{z^2 + 1}{z}\,dz$

7. If $F(x) = \sqrt{x}$, then $f(x) = F'(x) = \dfrac{1}{2\sqrt{x}}$. Find $\displaystyle\int_1^4 \frac{1}{2\sqrt{x}}\,dx$ two ways:

    (a) Using technology.
    (b) Using the Fundamental Theorem of Calculus.

8. If $F(t) = t^2 + 3t$, it can be shown that $f(t) = F'(t) = 2t + 3$. Find $\displaystyle\int_1^4 (2t + 3)\,dt$ two ways:

    (a) Using technology.
    (b) Using the Fundamental Theorem of Calculus.

9. A car accelerates smoothly from 0 to 60 mph in 10 seconds. Suppose the car's velocity as a function of time is given in Figure 3.62. Estimate how far the car travels during the 10-second period.

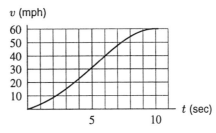

Figure 3.62

10. The Environmental Protection Agency was recently asked to investigate a spill of radioactive iodine. Measurements showed the ambient radiation levels at the site to be four times the maximum acceptable limit, so the EPA ordered an evacuation of the surrounding neighborhood.

    It is known that the level of radiation from an iodine source decreases according to the formula
    $$R(t) = R_0(0.996)^t$$
    where R is the radiation level (in millirems/hour) at time t, R_0 is the initial radiation level (at $t = 0$), and t is the time measured in hours.

    (a) If the maximum acceptable limit is 0.6 millirems/hour, use a graph of $R(t)$ to determine how long it will take for the site to reach an acceptable level of radiation.
    (b) How much total radiation (in millirems) will have been emitted by that time?

11. The graph of a derivative $F'(t)$ is shown in Figure 3.63. Make a table of values of $F(t)$ for $t = 0, 1, 2, 3, 4, 5$, given that $F(0) = 5$.

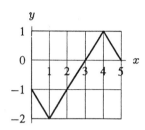

Figure 3.63

12. Assume the population, P, of Mexico (in millions), is given by

$$P = 67.38(1.026)^t,$$

where t is the number of years since 1980.

(a) What was the average population of Mexico between 1980 and 1990?
(b) What is the average of the population of Mexico in 1980 and the population in 1990?
(c) Explain, in terms of the concavity of the graph of P (see Figure 1.54 on page 48), why your answer to part (b) is larger or smaller than your answer to part (a).

13. Suppose the cost function $C(q)$ represents the total cost to produce a quantity q units of a certain product. The fixed costs for the production are $20,000. The marginal cost function is given by

$$C'(q) = 0.005q^2 - q + 56.$$

(a) On a graph of $C'(q)$, illustrate graphically the total variable cost of producing 150 units.
(b) Estimate $C(150)$, the total cost to produce 150 units.
(c) Find the value of $C'(150)$ and interpret your answer in terms of costs of production.
(d) Use your answers to parts (b) and (c) to estimate $C(151)$.

14. A table of values for the function $y = f(x)$ is given in Table 3.15. Estimate $\int_0^{25} f(x)dx$.

TABLE 3.15

x	0	5	10	15	10	25
$f(x)$	100	82	69	60	53	49

15. Let $F(b) = \int_0^b 2^x \, dx$.

(a) What is $F(0)$?
(b) Does the value of F increase or decrease as b increases? (Assume $b \geq 0$.)
(c) Estimate $F(1)$, $F(2)$, and $F(3)$.

16. The Montgolfier brothers (Joseph and Etienne) were eighteenth-century pioneers in the field of hot-air ballooning. Had they had the appropriate instruments, they might have left us a record of one of their early experiments, like that shown in Figure 3.64. The graph shows their vertical velocity, v, with upward as positive.

(a) Over what intervals was the acceleration positive? Negative? Zero?

(b) What was the greatest altitude achieved, and at what time?
(c) At what time was the upward acceleration greatest?
(d) At what time was the deceleration greatest?
(e) What might have happened during this flight to explain the answer to part (d)?
(f) This particular flight ended on top of a hill. How do you know that it did, and what was the height of the hill above the starting point?

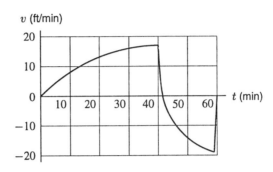

Figure 3.64

17. A mouse moves back and forth in a tunnel, attracted to bits of cheddar cheese alternately introduced to and removed from the ends (right and left) of the tunnel. The graph of the mouse's velocity, v, is given in Figure 3.65, with velocity being positive moving towards the right end of the tunnel, and negative towards the left end. Assuming that the mouse starts $(t = 0)$ at the center of the tunnel, use the graph to estimate the time(s) at which

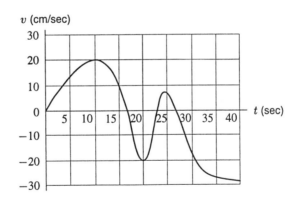

Figure 3.65

(a) The mouse changes direction.
(b) The mouse is moving most rapidly to the right; to the left.
(c) The mouse is farthest to the right of center; farthest to the left.
(d) The mouse's speed (i.e., the magnitude of its velocity) is decreasing.
(e) The mouse is at the center of the tunnel.

18.   The graph of some function f is given in Figure 3.66. List, from *least* to *greatest*,

    (a) $f'(1)$.

    (b)   The average value of $f(x)$, $0 \le x \le a$.

    (c)   The average value of the rate of change in $f(x)$, for $0 \le x \le a$.

    (d) $\displaystyle\int_0^a f(x)\,dx$.

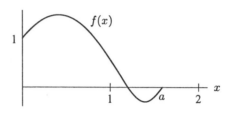

Figure 3.66

19.   If you jump out of an airplane and your parachute fails to open, your downward velocity t seconds after the jump is approximated by

$$v(t) = 49(1 - (0.8187)^t).$$

    (a)   Write an expression for the distance you fall in T seconds.

    (b)   If you jump from 5000 meters above the ground, estimate, using left- and right-hand sums, how many seconds you fall before hitting the ground.

CHAPTER FOUR

A LIBRARY OF FUNCTIONS

In Chapter 1, we learned about linear functions, exponential functions, and power functions. In this chapter, we extend our library of functions to include exponential functions to the base e, logarithmic functions, polynomials, and periodic functions. We examine the effect of parameters on graphs of functions. The emphasis is on understanding families of functions, including the logistic function and the surge function.

We explore the graphs, tables, and formulas that represent these functions, and we examine applications of these functions to business, social sciences, and life sciences.

4.1   THE NUMBER e

When we studied exponential functions in Chapter 1, we saw that the graphs of exponential functions such as $y = 2^t$ or $y = 3^t$ have the shape shown in Figure 4.1. Recall that in the exponential function $y = 2^t$, the number 2 is called the base. The most frequently used base is the famous number $e = 2.71828\cdots$. This base is used so often that you will find an $\boxed{e^x}$ button on most calculators[1] The graph of $y = e^t$ is shown in Figure 4.1. Since the number e is between 2 and 3, it should not surprise you that the graph of $y = e^t$ lies between the graphs of $y = 2^t$ and $y = 3^t$.

You are probably wondering why the strange (and irrational) number e is so important. One of the reasons is that many calculus formulas come out neater when e is used as the base rather than any other base, as we will see in chapter 5.

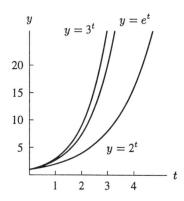

Figure 4.1: The graph of $y = e^t$

Compound Interest

Suppose that we deposit $100 in a bank paying interest at an 8% annual percentage rate. How much is in the account at the end of the year? This depends on how often the interest is compounded. If the interest is paid into the account only at the end of the year, then the balance in the account after one year will be $108. However, if the interest is paid quarterly (four times per year), then 2% interest is paid after the first quarter, 2% after the next quarter, and so on. Slightly more money is earned this way, since the interest paid early in the year will earn interest during the rest of the year. This effect is called *compounding*. If interest is compounded monthly (or weekly or daily), even more money is earned is shown in Table 4.1. (See Appendix B for a full discussion of compounding.)

TABLE 4.1   *The effect of compounding*

Frequency of Compounding	Balance in account after one year
annually	$108.00
quarterly	$108.24
monthly	$108.30
weekly	$108.32

---

[1] An alternative notation for e^x is $\exp(x)$.

In general, the more often interest is compounded, the more money will be earned (although the increase may not be very large). What do you think will happen if interest is compounded even more frequently, such as every minute or every second? The benefit of increasing the frequency of compounding becomes negligible beyond a certain point. When that point is reached, we say that the interest is *compounded continuously*. To find the balance in the account at the end of one year when the interest is compounded continuously, we use the number e. If we have deposited $100 in an account paying 8% interest compounded continuously, we multiply 100 by $e^{0.08}$. If the interest is compounded continuously, the balance after one year $= 100e^{0.08} = \$108.33$.
In general,

If an initial amount of $\$P_0$ is deposited in an account paying an annual interest rate of r, then the balance P in the account after t years is given by

$$P = P_0(1 + r)^t, \text{ if the interest is compounded annually, and}$$

$$P = P_0 e^{rt}, \text{ if the interest is compounded continuously.}$$

We use the notation P_0 for the initial amount deposited since it is the value of P when $t = 0$.

Example 1   If $1000 is deposited in a bank account paying 5% annual interest compounded continuously, how much will be in the account 10 years later?

Solution   We use the formula $P = P_0 e^{rt}$. The annual rate is 5% so $r = 0.05$, the length of time is $t = 10$, and the initial deposit is $P_0 = 1000$. We have

$$P = P_0 e^{rt} = 1000 e^{(0.05)(10)} = 1000 e^{0.5} = 1648.72.$$

The amount in the account after 10 years is $1648.72.

Example 2   Suppose that a bank advertises an annual rate of 8% interest. If you deposit $5000, how much will be in the account 3 years later if the interest is compounded
(a)   Annually?   (b)   Continuously?

Solution   (a)   For interest compounded annually, $P = P_0(1 + r)^t = 5000(1.08)^3 = 6298.56$. If interest is compounded annually, the amount in the account after 3 years is $6298.56.
(b)   For interest compounded continuously, $P = P_0 e^{rt} = 5000 e^{(0.08)(3)} = 6356.25$. As expected, the amount in the account 3 years later is larger if the interest is compounded continuously ($6356.25) than if the interest is compounded annually ($6298.56).

Example 3   Suppose you want to invest money in a certificate of deposit (CD) for your child's education. You want it to be worth $12,000 in 10 years. How much should you invest if the CD pays interest at a 9% annual rate compounded (a)   Annually?   (b)   Continuously?

Solution   (a)   If the CD pays 9% interest compounded annually over a 10 year period, then the interest rate is $r = 0.09$, and the length of time is $t = 10$. We must find the initial amount P_0 if the balance, P, after 10 years is $12,000. We have

$$P = P_0(1 + r)^t$$
$$12,000 = P_0(1.09)^{10}$$

Solve for P_0:

$$P_0 = \frac{12,000}{(1.09)^{10}} \approx \frac{12,000}{2.36736} = 5068.93.$$

The initial deposit must be \$5068.93 if interest is compounded annually.

(b)   On the other hand, if the CD pays 9% interest compounded continuously, we have

$$P = P_0 e^{rt}$$
$$12,000 = P_0 e^{(0.09)(10)}$$

Solve for P_0:

$$P_0 = \frac{12,000}{e^{(0.09)(10)}} = \frac{12,000}{e^{0.9}} \approx \frac{12,000}{2.45960} = 4878.84.$$

The initial deposit must be \$4878.84 if the interest is compounded continuously. Notice that to achieve the same result, continuous compounding requires a smaller initial investment than annual compounding. This is to be expected, since continuous compounding yields more money.

Using the Effective Annual Yield

We can measure the effect of compounding by introducing the notion of *effective annual yield*. Since \$100 invested at 8% compounded quarterly grows to \$108.24 by the end of one year, we say that the *effective annual yield* in this case is 8.24%. We now have two interest rates which describe the same investment: the 8% compounded quarterly and the 8.24% effective annual yield. Banks call the 8% the *annual percentage rate*, or *APR*. We may also call the 8% the *nominal rate* (nominal means "in name only"). However, it is the effective yield which tells you exactly how much interest the investment really pays. Thus, to compare two bank accounts, simply compare the effective annual yields. The next time that you walk by a bank, look at the advertisements, which should (by law) include both the APR, or nominal rate, and the effective annual yield. We will often abbreviate *annual percentage rate* to *annual rate*.

Example 4   Find the effective annual yield of a 6% annual rate, compounded continuously.

Solution   In one year, an investment of P_0 becomes $P_0 e^{0.06}$. Using a calculator, we see that

$$P_0 e^{0.06} = P_0(1.0618365)$$

So the effective annual yield is about 6.18%.

Exponential Growth and Decay

In Section 1.5, we saw the family of exponential functions

$$P = P_0 a^t,$$

where P_0 is the initial value of P and a is the growth factor. The case $a > 1$ represents exponential growth; $0 < a < 1$ represents exponential decay. For any positive number a, we can write $a = e^r$ for some r. If $a > 1$, r is positive, and if $0 < a < 1$, r is negative. Thus, the function representing an exponentially growing population can be rewritten as

$$P = P_0 a^t = P_0(e^r)^t = P_0 e^{rt}$$

with r positive. In the case when $0 < a < 1$, we can use another positive constant, r, and write

$$a = e^{-r}.$$

Thus if Q is a quantity that is decaying exponentially and Q_0 is the initial quantity, at time t we will have

$$Q = Q_0 a^t = Q_0(e^{-r})^t = Q_0 e^{-rt} = \frac{Q_0}{e^{rt}}.$$

Since e^{rt} is now in the denominator, Q will decrease as time goes on—as you'd expect if Q is decaying.

Any **exponential growth** function can be written in either of the two forms

$$P = P_0 a^t \qquad \text{or} \qquad P = P_0 e^{rt}$$

and any **exponential decay** function can be written as either

$$Q = Q_0 b^t \qquad \text{or} \qquad Q = Q_0 e^{-rt}$$

where P_0 and Q_0 are the initial quantities, r is positive, $a > 1$, and $0 < b < 1$.

We say that P and Q are growing or decaying at a *continuous rate* of r. (Note that, for example, $r = 0.02$ corresponds to a continuous growth rate of 2%.)

Example 5   Sketch the graphs of $P = e^{0.5t}$ and $Q = e^{-0.2t}$.

Solution

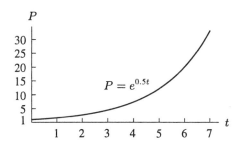

Figure 4.2: An exponential growth function

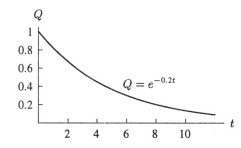

Figure 4.3: An exponential decay function

The graph of $P = e^{0.5t}$ is in Figure 4.2. Notice that the graph has the same shape as the previous exponential growth curves: increasing and concave up. Indeed, the families a^t with $a > 1$ and e^{rt} with $r > 0$ are one and the same. The graph of $Q = e^{-0.2t}$ is in Figure 4.3; it too has the same shape as other exponential decay functions. The families a^t with $0 < a < 1$ and e^{rt} with $r < 0$ are also one and the same.

Example 6   The population of a city is 50,000 in 1990 and is growing at a continuous yearly rate of 4.5%.
  (a) Give the population of the city as a function of the number of years since 1990.
  (b) What will be the population of the city in the year 2000?
  (c) Sketch a graph of the population of the city as a function of time.
  (d) Use your graph to estimate the doubling time for the population of the city.

Solution   (a)   We use the formula $P = P_0 e^{rt}$, so we have

$$P = 50{,}000 e^{0.045t},$$

where t is the number of years since 1990.

(b)   The year 2000 is when $t = 10$, so we have

$$P = 50{,}000 e^{(0.045)(10)} \approx 78{,}416.$$

We can predict that the population will be about 78,000 in the year 2000.

(c)   See Figure 4.4.

(d)   The doubling time is the time it takes for the population to double, from 50,000 to 100,000. We see in Figure 4.4 that $P = 100{,}000$ occurs approximately when $t = 15$, so the doubling time will be about 15 years if the population continues to grow at the same rate.

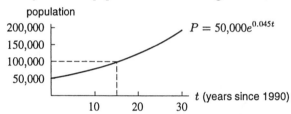

Figure 4.4: What is the doubling time?

Example 7   The Environmental Protection Agency was recently asked to investigate a spill of radioactive iodine. Initial measurements showed the ambient radiation level at the site to be about 2.4 millirems/hour (four times the maximum acceptable limit of 0.6 millirems/hour), so the EPA ordered an evacuation of the surrounding area. The level of radiation from an iodine source is known to decay at a continuous hourly rate of $r = -0.004$.

(a)   What is the level of radiation after 24 hours?

(b)   Use a graph to estimate the number of hours until the level of radiation will reach the maximum acceptable limit, so that the inhabitants can be allowed to return.

Solution   (a)   The level of radiation, R, in millirems/hour, at time t, in hours since the initial measurement, is given by

$$R = 2.4 e^{-0.004t},$$

so the level of radiation after 24 hours is

$$R = 2.4 e^{(-0.004)(24)} = 2.18.$$

After 24 hours, the level of radiation has been reduced to 2.18 millirems per hour.

(b)   A graph of $R = 2.4 e^{-0.004t}$ is shown in Figure 4.5. We are told that the maximum acceptable value of R is 0.6 millirems per hour, and we see that this occurs at approximately $t = 350$. The inhabitants will not be able to return for about 350 hours (or about 15 days).

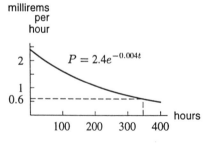

Figure 4.5: The level of radiation from radioactive iodine

In an exponential growth or decay problem, it is not always clear whether the growth rate given is a continuous rate or an annual rate. For example, if we are told that "the city's population is increasing at a rate of 5% per year," it is not obvious whether we should use $P_0(1.05^t)$ or $P_0 e^{0.05t}$ for the population of the city at time t. Either formula would be acceptable, and the difference in the results for short time periods will usually be insignificant. In this text, we will generally use the continuous formula $(P = P_0 e^{rt})$ for exponential growth and decay problems.

Problems for Section 4.1

1. If you deposit $10,000 in an account earning interest at an 8% annual rate compounded continuously, how much money is in the account after five years?

2. Use the number e to find the effective annual yield of a 6% annual rate, compounded continuously.

3. Suppose $1000 is invested in an account paying a nominal annual rate of 5.5%. How much is in the account after 8 years if the interest is compounded (a) Annually?   (b) Continuously?

4. Suppose $1000 is invested at 6% annual interest compounded continuously. Use trial and error, or a graph, to determine how long it will take for the balance to double.

5. Each of the curves in Figure 4.6 represents the balance in a bank account into which a single deposit was made at time zero. Assuming interest is compounded continuously, find:

   (a) The curve representing the largest initial deposit.
   (b) The curve representing the largest interest rate.
   (c) Two curves representing the same initial deposit.
   (d) Two curves representing the same interest rate.

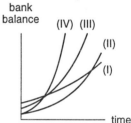

Figure 4.6

6. What is the effective annual yield of an investment paying a 12% annual rate, compounded continuously?

7. If you need $10,000 in your account 3 years from now and the annual interest rate on your account is 8% compounded continuously, how much should you deposit now?

8. Explain how you can match the interest rates (a)–(e) with the effective annual yields (I)–(V) without doing any calculations.

   (a) 5.5% annual rate, compounded continuously.   (I)   5%
   (b) 5.5% annual rate, compounded quarterly.   (II) 5.06%
   (c) 5.5% annual rate, compounded weekly.   (III) 5.61%
   (d) 5% annual rate, compounded yearly.   (IV) 5.651%
   (e) 5% annual rate, compounded twice a year.   (V)   5.654%

9. When you rent an apartment, you are often required to give the landlord a security deposit which is returned if you leave the apartment undamaged. In Massachusetts the landlord is required to pay the tenant interest on the deposit once a year, at a 5% annual rate, compounded annually. The landlord, however, may invest the money at a higher (or lower) interest rate. Suppose the landlord invests a $1000 deposit at an annual rate of
   (a) 6%, compounded continuously   (b) 4%, compounded continuously.

   In each case, determine the net gain or loss by the landlord at the end of the first year. (Give your answer to the nearest cent.)

10. Figure 4.7 shows the balances in two bank accounts over time; both accounts pay the same nominal interest rate, but one compounds continuously and the other compounds annually. Which curve corresponds to which compounding method? What is the initial deposit in each case?

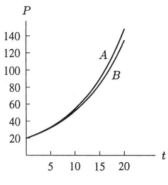

Figure 4.7

11. In 1980, there were about 170 million vehicles (cars and trucks) and about 227 million people in the United States. If the number of vehicles was growing at 4% a year, while the population was growing at 1% a year, in what year was there, on average, one vehicle per person?

12. A fishery stocks a pond with 1000 young trout. The number of the original trout still alive after t years is given by $P(t) = 1000e^{-0.5t}$.

    (a) How many trout are left after six months? After 1 year?
    (b) Find $P(3)$ and interpret it in terms of trout.
    (c) At what time will there be 100 of the original trout left?
    (d) Sketch a graph of the number of trout against time, and describe how the population is changing. What might be causing this?

13. Air pressure, P, decreases exponentially with the height above the surface of the earth, h:

$$P = P_0 e^{-0.00012h}$$

where P_0 is the air pressure at sea level and h is in meters.

    (a) If you go to the top of Mount McKinley, height 6198 meters (about 20,320 feet), what is the air pressure, as a percent of the pressure at sea level?
    (b) The maximum cruising altitude of an ordinary commercial jet is around 12,000 meters (about 40,000 feet). At that height, what is the air pressure, as a percent of the sea level value?

14. One of the main contaminants of a nuclear accident, such as that at Chernobyl, is strontium-90, which decays exponentially at a continuous rate of approximately 2.47% per year. Preliminary estimates after the Chernobyl disaster suggested that it would be about 100 years before the region would again be safe for human habitation. What percent of the original strontium-90 would still remain by this time?

15. Sketch a graph of $f(x) = e^{-x^2}$. Be sure to use a window that includes both positive and negative values of x.

    (a) For what values of x is f increasing? For what values is it decreasing?
    (b) Is the graph of f concave up or concave down at $x = 0$?
    (c) Discuss the end behavior of f: As $x \to \infty$, what happens to $f(x)$? As $x \to -\infty$, what happens to $f(x)$?

16. It is believed by some that the earth's population cannot exceed 40 billion people. If this is true, then the population, P, in billions, t years after 1990, can be modeled by the function

$$P = \frac{40}{1 + 11e^{-0.08t}}.$$

   (a) Sketch the graph of P against t.
   (b) According to this model, approximately when will the earth's population reach 20 billion people?
   (c) According to this model, approximately when will the earth's population reach 39.9 billion people?
   (d) According to this model, by how many people will the earth's population increase between 1990 and 2000?

17. Under certain circumstances, the velocity, V, of a falling raindrop is given by $V = V_0(1 - e^{-t})$, where t is time and V_0 is a positive constant.

   (a) Sketch a rough graph of V against t, for $t \geq 0$.
   (b) What does V_0 represent?

18. Sketch a rough graph of $f(x) = e^x$, and use the graph to decide whether the derivative of $f(x)$ at $x = 1$ is positive or negative. Give reasons for your decision.

19. Estimate $f'(0)$ if $f(x) = e^{2x}$.

20. Use technology to compute the following definite integral to one decimal place: $\int_{-3}^{3} e^{-t^2}\, dt.$

21. Find the average value of $g(t) = e^t$ over the interval $[0, 10]$.

22. The newspaper article below is from *The New York Times*, May 27, 1990. Fill in the three blanks. (For the first blank, assume that daily compounding is essentially the same as continuous compounding. For the last blank, assume the interest has been compounded yearly, and give your answer in dollars. Exclude the occurrence of leap years.)

213 Years After Loan, Uncle Sam Is Dunned
By LISA BELKIN
Special to The New York Times

SAN ANTONIO, May 26 — More than 200 years ago, a wealthy Pennsylvania merchant named Jacob DeHaven lent $450,000 to the Continental Congress to rescue the troops at Valley Forge. That loan was apparently never repaid.

So Mr. DeHaven's descendants are taking the United States Government to court to collect what they believe they are owed. The total: ____ in today's dollars if the interest is compounded daily at 6 percent, the going rate at the time. If compounded yearly, the bill is only ____.

Family Is Flexible

The descendants say that they are willing to be flexible about the amount of a settlement and that they might even accept a heartfelt thank you or perhaps a DeHaven statue. But they also note that interest is accumulating at ____ a second.

4.2 THE NATURAL LOGARITHM

What do the following three questions have in common?

1. If a population of 10,000 grows at an annual rate of 2% per year, how long will it take for the population to reach 100,000?
2. If stocks worth $5000 increase in value to $8000 in three years, what is the (continuous) yearly growth rate?
3. What value of t satisfies the equation $2^t = 7$?

To answer question 1 we must find the value of t which satisfies the equation $100{,}000 = 10{,}000(1.02)^t$. Question 2 translates to: What value of t satisfies the equation $8000 = 5000e^{3t}$? Thus, in order to answer each of the three questions, we need to solve for a variable that appears in an exponent. We can approximate the answer to each of them using trial and error or using a graph. However, if we want to find the exact answer analytically, we need to use *logarithms*. Logarithms enable us to solve many equations when the unknown quantity is in the exponent.

A logarithm can be defined for any positive base. Logarithms to base 10 are used for such things as the Richter scale, which measures the energy released by an earthquake, and for determining pH, a measure of the acidity of a chemical solution. However, the most frequently used base is the famous number $e = 2.71828\dots$. In fact, this base is used so often that the logarithm to base e is called the *natural logarithm* and is denoted by "ln". You will find an $\boxed{\ln}$ button on most scientific calculators as well as an $\boxed{e^x}$ button; that should be some indication of how important the base e is. At first glance, this is all somewhat mysterious. What can possibly be natural about using logarithms to the base 2.71828? The full answer to that question must wait until Chapter 5, where we show that many calculus formulas come out neater when e is used as the base rather than any other base. In this text, we will only use the natural logarithm.

Definition and Properties of the Natural Logarithm

The natural logarithm of x, written $\ln x$, is defined as follows:

$$\ln x = \log_e x = c \quad \text{means} \quad e^c = x$$

so

$$\ln x \text{ is the power of } e \text{ needed to get } x.$$

Thus, for example, $\ln e^3 = 3$ since 3 is the power of e needed to get e^3. Similarly, $\ln \frac{1}{e^2} = -2$ since $\frac{1}{e^2} = e^{-2}$. Use your calculator to find $\ln 5$. You will see that $\ln 5 \approx 1.6094$. This is because $5 \approx e^{1.6094}$. (Use the $\boxed{e^x}$ button on your calculator to check!) Since no power of e yields a negative number or zero, it follows that

$$\ln x \text{ is not defined if } x \text{ is negative or zero.}$$

In working with logarithms, you will need to use the following properties:

Rules For Computing Using Natural Logarithms

1. $\ln(AB) = \ln A + \ln B$

2. $\ln\left(\dfrac{A}{B}\right) = \ln A - \ln B$

3. $\ln(A^p) = p \ln A$

4. $\ln e^x = x$

5. $e^{\ln x} = x$

In addition, $\ln 1 = 0$ because $e^0 = 1$.

Solving Equations Using Natural Logarithms

Logarithms are useful when we have to solve for unknown exponents, as in the following examples.

Example 1   Find t such that $3^t = 10$.

Solution   First, notice that we expect t to be between 2 and 3, because $3^2 = 9$ and $3^3 = 27$. To find t exactly, we take the natural logarithm of both sides and then use the rules of logarithms to solve for t:

$$\ln(3^t) = \ln 10.$$

Then using the third log rule, we have

$$t \ln 3 = \ln 10$$
$$t = \frac{\ln 10}{\ln 3}.$$

Using a calculator to find the natural logs gives

$$t \approx \frac{2.3026}{1.0986} \approx 2.096.$$

Example 2   Find t such that $12 = 5e^{3t}$.

Solution   Since t is in the exponent, in order to solve for t we will use logarithms. It is easiest to begin by isolating the exponential, so we divide both sides of the equation by 5:

$$2.4 = e^{3t}.$$

Now take the natural logarithm of both sides:

$$\ln 2.4 = \ln(e^{3t}).$$

Using the fourth log rule gives

$$\ln 2.4 = 3t,$$

so

$$t = \frac{\ln 2.4}{3}.$$

Using a calculator, we get $t \approx \frac{0.8755}{3} \approx 0.2918$.

Example 3   Suppose that $5000 is deposited in an account paying 8% annual interest. How long will it take for the money to double if the interest is compounded (a)   Annually?   (b)   Continuously?

Solution   (a)   We use the formula $P = P_0(1 + r)^t$, where t is the number of years since the deposit. We have $P = 5000(1.08)^t$, and we wish to find the value of t for which $P = 10{,}000$. We have

$$10{,}000 = 5000(1.08)^t.$$

Dividing both sides by 5000 gives

$$2 = (1.08)^t.$$

Taking the natural log of both sides, we get

$$\ln 2 = \ln(1.08^t).$$

Using the third log rule gives

$$\ln 2 = t \ln 1.08.$$

so

$$t = \frac{\ln 2}{\ln 1.08}.$$

Using a calculator gives

$$t \approx \frac{0.6931}{0.07696} \approx 9.006.$$

Thus, if the interest is compounded annually, it takes about 9 years for the balance to double.

(b)   Since the interest is compounded continuously, we use the formula

$$P = P_0 e^{rt}.$$

We have

$$P = 5000 e^{0.08t}.$$

We wish to find the value of t for which $P = 10{,}000$. We have

$$10{,}000 = 5000 e^{0.08t}.$$

Dividing both sides by 5000 gives

$$2 = e^{0.08t}.$$

Taking the natural log of both sides, we get

$$\ln 2 = \ln(e^{0.08t}).$$

Using the fourth log rule, we have

$$\ln 2 = 0.08t,$$

so

$$t = \frac{\ln 2}{0.08}.$$

Using a calculator gives

$$t \approx \frac{0.6931}{0.08} \approx 8.664.$$

Hence, if the interest is compounded continuously, it takes about eight years and eight months for the balance to double. As we would expect, the balance is doubled faster when the interest is compounded continuously.

Example 4 The release of chlorofluorocarbons used in air conditioners and, to a lesser extent, in household sprays (hair spray, shaving cream, etc.) destroys the ozone in the upper atmosphere. At the present time, the amount of ozone in the atmosphere, Q, is decaying exponentially at a continuous yearly rate of 0.25%. What is the half-life of ozone? In other words, at this rate, how long will it take for half the ozone to disappear?

Solution If Q_0 is the initial quantity of ozone, then

$$Q = Q_0 e^{-0.0025t},$$

where t is in years. We want to find the value of t such that $Q = Q_0/2$, so

$$\frac{Q_0}{2} = Q_0 e^{-0.0025t}.$$

Divide by Q_0 to obtain

$$\frac{1}{2} = e^{-0.0025t}.$$

Taking natural logs yields

$$-0.0025t = \ln\left(\frac{1}{2}\right) = -0.6931,$$

so

$$t \approx 277 \text{ years.}$$

Half the present atmospheric ozone will be gone in 277 years.

In Example 4 the decay rate was given. However, in many situations where we expect to find exponential growth or decay, the rate is not given. To find it, we must know the quantity at two different times and then solve for the growth rate, as in the next example.

Example 5 The population of Kenya was 19.5 million in 1984 and 21.2 million in 1986. Assuming it increases exponentially, find a formula for the population of Kenya as a function of time.

Solution If we measure the population, P, in millions and the time, t, in years since 1984, we can say that

$$P = P_0 e^{kt} = 19.5 e^{kt}$$

where $P_0 = 19.5$ is the initial value of P. We find k by using the fact that $P = 21.2$ when $t = 2$, so

$$21.2 = 19.5 e^{k \cdot 2}.$$

We divide both sides by 19.5, giving

$$\frac{21.2}{19.5} = 1.0872 = e^{2k}.$$

Now take natural logs of both sides:

$$\ln(1.0872) = \ln(e^{2k}).$$

Using a calculator and the fact that $\ln(e^{2k}) = 2k$, this becomes

$$0.0836 = 2k.$$

So

$$k \approx 0.042,$$

and therefore

$$P = 19.5 e^{0.042t}.$$

Since $k = 0.042 = 4.2\%$, we say that the population of Kenya was growing at a continuous rate of 4.2% a year.

Example 6   If $10,000 is deposited in an account paying an annual interest rate of 5%, compounded continuously, how long will it take for the balance in the account to reach $15,000?

Solution   Since the interest is being compounded continuously, we use the formula $P = P_0 e^{rt}$, where the interest rate is $r = 0.05$, and the initial amount is $P_0 = 10,000$. We wish to find the value of t for which $P = 15,000$. Our equation is

$$15,000 = 10,000e^{0.05t}.$$

Since we want to solve for t, and t is in the exponent, we need to use logarithms. First, we divide both sides of the equation by 10,000 to isolate the exponential, and then we take the natural logarithm of both sides and solve for t.

$$15,000 = 10,000e^{0.05t}$$
$$1.5 = e^{0.05t}$$
$$\ln(1.5) = \ln(e^{0.05t})$$
$$\ln(1.5) = 0.05t$$
$$t = \frac{\ln(1.5)}{0.05} \approx 8.1093.$$

It will take about 8.1 years for the balance in the account to reach $15,000.

Example 7   (a)   Find the doubling time, D, for annual growth rates of 2%, 3%, 4%, and 5%.
(b)   Since D decreases as the growth rate increases, we might guess that D is inversely proportional to the growth rate. Use your answers to part (a) to confirm that a growth rate of i% gives a doubling time of

$$D = \frac{70}{i} \text{ years.}$$

This is the "rule of 70" used by bankers. To compute the approximate doubling time of an investment, the banker divides 70 by the annual interest rate.

Solution   (a)   We begin by finding the doubling time for an annual growth rate of 2%. We use the formula $P = P_0(1.02)^t$. We wish to find the value of t for which $P = 2P_0$, so we are solving

$$2P_0 = P_0(1.02)^t$$
$$2 = (1.02)^t$$
$$\ln 2 = \ln(1.02)^t$$
$$\ln 2 = t\ln(1.02) \quad \text{(using a property of logarithms)}$$
$$t = \frac{\ln 2}{\ln 1.02} \approx 35.003.$$

We see that if the annual interest rate is 2%, it will take about 35 years for an investment to double in value. Similarly, we can find the doubling times for 3%, 4%, and 5%. The results are given in Table 4.2.

TABLE 4.2 *Doubling times for investments*

i (% annual growth rate)	2	3	4	5
D (doubling time in years)	35.003	23.450	17.673	14.207

(b) We compute $(70/i)$ for $i = 2, 3, 4, 5$. The results are shown in Table 4.3.

TABLE 4.3 *Rule of 70*

i (% annual interest)	2	3	4	5
$(70/i)$ (approximate doubling time in years)	35.000	23.333	17.500	14.000

Comparing the values in Table 4.2 and Table 4.3, we see that the approximation $(70/i)$ given by the "rule of 70" gives a reasonably accurate approximation to the exact doubling time D for interest rates in the range considered.

Relationship between a^t and e^{kt}

In Example 5 we chose to use e for the base of the exponential function representing Kenya's population, making clear that the continuous growth rate was 4.2%. If, however, we had wanted to emphasize the annual growth rate, we could have expressed the exponential function in the form

$$P = P_0 a^t.$$

Since the population grew from 19.5 to 21.2 million in 2 years, we know that

$$21.2 = 19.5a^2$$

so

$$a = \left(\frac{21.2}{19.5}\right)^{(1/2)} \approx 1.043.$$

Thus,

$$P = 19.5(1.043)^t.$$

Notice that the base $a \approx 1.043$ corresponds to an annual growth rate of about 4.3%, which is just slightly more than the continuous growth rate of 4.2% found in Example 5.

In general, an exponential function of the form $P_0 e^{kt}$ can always be written in the form $P_0 a^t$, and vice versa, simply by letting $e^k = a$ or $k = \ln a$. The two different formulas, $P = P_0 e^{kt}$ and $P = P_0 a^t$, have the same graph and represent the same function.

> To convert between a^t and e^{kt} use
>
> $$a = e^k \quad \text{or} \quad k = \ln a$$

If a comes from a percentage growth rate r, that is, $a = 1 + r$, then the continuous growth rate $k = \ln(1 + r)$ will be slightly less than, but very close to, r, provided r is small.

The Graph of ln x

Using the $\boxed{\text{LN}}$ button on a calculator to plot a graph of $f(x) = \ln x$ for $0 < x \le 10$, we get Figure 4.8. Because no power of e gives 0, ln 0 is undefined. Also, no power of e gives a negative number, so ln x is only defined for $x > 0$, as we see in the graph. The log function has a vertical asymptote at $x = 0$. As x gets closer and closer to 0 from the right, the natural logarithm graph drops toward $-\infty$. Check this on a calculator by finding ln(0.1), ln(0.01), ln(0.001), and so on. In addition, you should notice that the natural logarithm function crosses the x-axis at $x = 1$. This is because $e^0 = 1$ and so ln 1 = 0.

We know that the exponential function grows extremely quickly. The logarithm function, on the other hand, grows extremely slowly. This is because ln x is the power of e you need to get x and you only need a relatively small power of e to get a pretty large x. In fact, to make ln x large, you will need a gigantic value of x, as we see in Table 4.4. However, though it grows slowly, the ln function does go to infinity as x increases.

Comparison Between Logarithms and Power Functions

The graph of ln x in Figure 4.8 might remind you of the graph of $y = \sqrt{x}$. Both are concave down and increasing, and both go to infinity. We see in Table 4.4 that ln x increases very slowly. But how does it compare to $y = \sqrt{x}$? Figure 4.9 compares the graphs of $\sqrt{x}$ and ln x, and we see that indeed ln x grows substantially slower than $\sqrt{x}$. Recall that the exponential function climbs faster than any positive power of x. The logarithm function is at the other extreme: it climbs slower than any positive power of x.

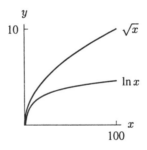

Figure 4.9: Comparison of $y = \sqrt{x}$ and $y = \ln x$.

In Section 1.7, we discussed fitting formulas to data. If we have a data set that appears to be increasing and concave down (such as the graph in Figure 4.8), often the best formula to fit the

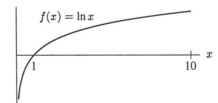

Figure 4.8: Graph of the natural logarithm

TABLE 4.4   *Values for* ln x

x	1	10	100	1000	10,000	100,000	1,000,000
ln x	0	2.3	4.6	6.9	9.2	11.5	13.8

data will be a logarithm function. When we fit a logarithm function to data, the process is called *logarithmic regression*. Many calculators or computers will do logarithmic regression and will find the best logarithm function to model a set of data.

Problems for Section 4.2

For Problems 1–2, plot a graph of the given function on a calculator or computer. Describe and explain what you see.

1. $y = \ln e^x$

2. $y = e^{\ln x}$

3. Answer the three questions posed in the first paragraph of this section.

For Problems 4–18, solve for t using natural logarithms.

4. $5^t = 7$ 7. $10 = 2^t$ 10. $a = b^t$
13. $e^{3t} = 100$ 16. $2P = Pe^{0.3t}$

5. $130 = 10^t$ 8. $100 = 25(1.5)^t$ 11. $10 = e^t$
14. $10 = 6e^{0.5t}$ 17. $7(3^t) = 5(2^t)$

6. $2 = (1.02)^t$ 9. $50 = 10(3^t)$ 12. $5 = 2e^t$
15. $B = Pe^{rt}$ 18. $5e^{3t} = 8e^{2t}$

19. If \$12,000 is deposited in an account paying 8% interest per year, compounded continuously, how long will it take for the balance to reach \$20,000?

20. If an investment of \$5000 grows to \$8080 in four years, what was the annual rate of return on the investment? (Assume continuous compounding.)

21. In 1994, the world's population was 5.6 billion, and the population was projected to reach 8.5 billion by the year 2030. What annual rate of growth is assumed in this prediction?

22. Find the doubling time of a quantity that is increasing by 7% per year.

23. If the quantity of a certain substance decreases by 4% in 10 hours, find the half-life of the substance.

24. You invest \$5000 in an account which pays interest compounded continuously.

    (a) How much money is in the account after 8 years, if the annual interest rate is 4%?
    (b) If you want the account to contain \$8000 after 8 years, what yearly interest rate is needed?

25. The half-life of a certain radioactive substance is 12 days. If there are 10.32 grams initially

    (a) Write an equation to determine the amount, A, of the substance as a function of time.
    (b) When will the substance be reduced to 1 gram?

26. Owing to an innovative rural public health program, infant mortality in Senegal, West Africa, is being reduced at a rate of 10% per year. How long will it take for infant mortality to be reduced by 50%?

27. Assume that the rate of inflation continues at the 1991 rate of 4.6% a year. If the price of postage stamps goes up, on average, at the rate of inflation, when will it cost \$1 to mail a letter? [Note: That cost rose to 29 cents in 1990.]

28. (a) Use the "rule of 70" to predict the doubling time of an investment which is earning 8% interest per year.
    (b) Find the doubling time exactly, and compare your answer to part (a).

29. What is the doubling time of prices which are increasing by 5% a year?

30. The population of a region is growing exponentially. If there were 40,000,000 people in 1980 ($t = 0$) and 56,000,000 in 1990, find an expression for the population at any time t. What would you predict for the year 2000? What is the doubling time?

31. A standard cup of coffee contains about 100 mg of caffeine. The half-life of caffeine in the body is about four hours. How long after you drink a cup of coffee will the level of caffeine in your body be reduced to 5 mg?

32. Table 4.5 gives the amount of rain forest destroyed for agriculture and development, for several years. [2]

    (a) Plot these data.
    (b) Are the data increasing or decreasing? Concave up or concave down? In each case, indicate whether your answer is good news or bad news to someone who hopes to save rain forests.
    (c) Use technology to do logarithmic regression on this data. If you do not have the technology, draw a logarithmic curve through the data.
    (d) Use the curve you found in part (c) to predict how much rain forest will be destroyed in the year 2000 for agriculture and development.

TABLE 4.5 *Rain forest destroyed for agriculture and development*

Year	1960	1970	1980	1988
Hectares $\times 10^6$	2.21	3.79	4.92	5.77

Convert the functions in Problems 33–37 into the form $P = P_0 a^t$. Which represent exponential growth and which represent exponential decay?

33. $P = P_0 e^{0.2t}$

34. $6\ P = 10e^{0.917t}$

35. $P = P_0 e^{-0.73t}$

36. $P = 79e^{-2.5t}$

37. $P = 7e^{-\pi t}$

Convert the functions in Problems 38–41 into the form $P = P_0 e^{kt}$.

38. $P = P_0 2^t$

39. $P = 10(1.7)^t$

40. $P = 5.23(0.2)^t$

41. $P = 174(0.9)^t$

42. The island of Manhattan was sold for \$24 in 1626. Suppose the money had been invested in an account which compounded interest continuously.

    (a) How much money would be in the account in the year 2000 if the yearly interest rate was 5%?
    (b) How much money would be in the account in the year 2000 if the yearly interest rate was 7%?
    (c) If the yearly interest rate was 6%, in what year would the account be worth one million dollars?

43. The solid waste generated in cities in the US is increasing. The solid waste generated, in millions of tons, was 82.3 in 1960 and 139.1 in 1980. In Example 3 in Section 1.3, we found

---

[2]C. Schaufele and N. Zumoff, *Earth Algebra, Preliminary Version*, (New York: Harper Collins, 1993) 131.

the equation of a line through these two points and used this line to predict the amount of municipal solid waste in the year 2000. Since we only have the two points, it is not clear that the line gives the correct relationship between solid waste and year. In this problem, we assume that solid waste is increasing exponentially rather than linearly, and we find an exponential curve to fit the data.

(a)   Using the information about municipal solid waste in the years 1960 and 1980, find the equation of an exponential growth curve through the two points.

(b)   Use your answer to part (a) to predict US municipal solid waste in the year 2000. Compare your answer to the earlier prediction of 195.9 million tons, which we found using the assumption that the growth was linear.

44.   The air in a factory is being filtered so that the quantity of a pollutant, P (measured in mg/liter), is decreasing according to the equation $P = P_0 e^{-kt}$, where t represents time in hours . If 10% of the pollution is removed in the first five hours:

(a)   What percentage of the pollution is left after 10 hours?

(b)   How long will it take before the pollution is reduced by 50%?

(c)   Plot a graph of pollution against time. Show the results of your calculations on the graph.

(d)   Explain why the quantity of pollutant might decrease in this way.

45.   The population, P, in millions, of Nicaragua was 3.6 million in 1990 and growing at 3.4% per year. Let t be time in years since 1990.

(a)   Express P as a function in the form $P = P_0 a^t$.

(b)   Express P as an exponential function using the base e.

(c)   Compare the annual and continuous growth rates.

46.   The quantity, Q, of radioactive carbon-14 remaining t years after an organism dies is given by the formula

$$Q = Q_0 e^{-0.000121t},$$

where Q_0 is the initial quantity.

(a)   A skull uncovered at an archeological dig has 15% of the original amount of carbon-14 present. Estimate its age.

(b)   Show how you can calculate the half-life of carbon-14 from this equation.

47.   A picture supposedly painted by Vermeer (1632–1675) contains 99.5% of its carbon-14 (half-life 5730 years). From this information can you determine whether or not the picture is a fake? Explain your reasoning.

48.   Sketch a graph of $f(x) = \ln(x^2 + 1)$. Be sure to use a window that includes both positive and negative values of x.

(a)   For what values of x is f increasing? For what values is it decreasing?

(b)   Is f concave up or concave down at $x = 0$?

(c)   Discuss the end behavior of f: As $x \to \infty$, what happens to $f(x)$? As $x \to -\infty$, what happens to $f(x)$?

49.   Let $f(x) = \ln x$. Estimate $f'(2)$ by two different methods.

(a)   Using a graph.

(b)   By using difference quotients for smaller and smaller intervals around 2. Give your answer accurate to one decimal place.

50.   Compute the definite integral $\int_1^4 (x - 3 \ln x)\, dx$ and interpret the result in terms of areas.

4.3 NEW FUNCTIONS FROM OLD; POLYNOMIALS

New Functions from Old

We have studied power functions, exponential functions, and logarithm functions. We know that a parabola opening up, for example, might come from a function containing x^2 with a positive coefficient. But what if the vertex of the parabola is not at the origin? And what if the sides of the parabola are not as steep as they are for the function $y = x^2$? In this section, we learn how to shift, stretch, and add functions we already know about to create new functions. We end by discussing polynomials, which are obtained by shifting, stretching, and adding power functions.

Shifts

Consider the function $y = x^2 + 4$. The y-coordinates for this function will all be exactly 4 units larger than the y-coordinates of the function $y = x^2$ for the same values of x. It follows that the graph of $y = x^2 + 4$ should be obtained from the graph of $y = x^2$ by adding 4 to the y-coordinate of each point. We see in Figure 4.10 that the graph of $y = x^2 + 4$ is the graph of $y = x^2$ moved up four units.

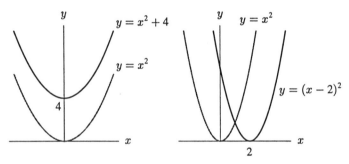

Figure 4.10: Graphs of $y = x^2$ with $y = x^2 + 4$ and $y = (x - 2)^2$

A graph can also be shifted to the left or to the right. In Figure 4.10, we see that the graph of $y = (x - 2)^2$ is the graph of $y = x^2$ shifted to the right 2 units. In general,

---

- The graph of $y = f(x) + k$ is the graph of $y = f(x)$ moved up k units (down if k is negative).
- The graph of $y = f(x - k)$ is the graph of $y = f(x)$ moved to the right k units (to the left if k is negative).

Example 1   Use your knowledge of graphs of functions to sketch a rough graph of each of the following:

(a) $y = x^3 + 1$

(b) $y = e^x - 3$

(c) $y = (x + 2)^6$

(d) $y = (x - 2)^2 - 1$

Solution   The four graphs are shown in Figure 4.11.

(a)   The graph of $y = x^3 + 1$ is the graph of $y = x^3$ moved up 1 unit.

(b)   The graph of $y = e^x - 3$ is the graph of $y = e^x$ moved down 3 units.

(c)   The graph of $y = (x + 2)^6$ is the graph of $y = x^6$ moved to the left 2 units.

(d)   The graph of $y = (x - 2)^2 - 1$ is the graph of $y = x^2$ moved to the right 2 units and down 1 unit.

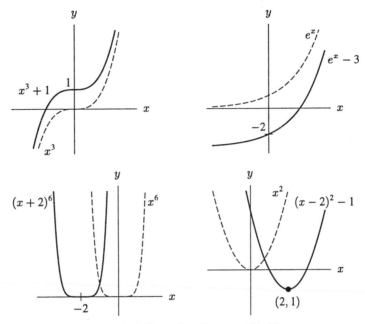

Figure 4.11: Examples of graphs shifted by a constant

Example 2   (a)   A cost function, $C(q)$, for a company is shown in Figure 4.12. If the fixed cost suddenly increases by \$1000, sketch a graph of the new cost function.

(b)   A supply function for a product is given in Figure 4.13. A new factory opens and will produce 100 units of the product no matter what the price. Sketch a graph of the new supply function.

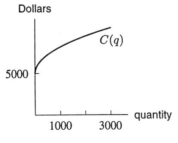

Figure 4.12: A cost function

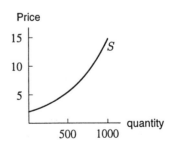

Figure 4.13: A supply function

Solution   (a)   For any given quantity, the new cost is $1000 more than the old cost. The new cost function is $C(q) + 1000$, and the graph is shifted vertically up 1000 units. See Figure 4.14.

(b)   We see that at a price of 10 dollars, approximately 800 units are currently produced. When the new factory opens, this amount will increase by 100 units, so the new amount produced at a price of 10 dollars will be 900 units. In fact, at any price, the quantity produced will increase by 100, and so the new supply curve is shifted horizontally to the right by 100 units. See Figure 4.15.

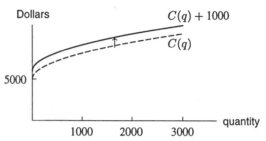

Figure 4.14: A vertical shift

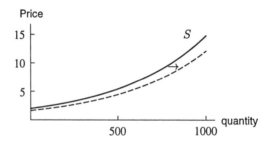

Figure 4.15: A horizontal shift

Stretches

The graph of a constant multiple of a given function is easy to visualize: each y-value is stretched or shrunk by that multiple. For example, consider the function $f(x)$ and its multiples $y = 3f(x)$ and $y = -2f(x)$, whose graphs are in Figure 4.16. The factor 3 in the function $y = 3f(x)$ stretches each $f(x)$ value by multiplying it by 3; the factor -2 in the function $y = -2f(x)$ stretches $f(x)$ by multiplying by 2 and reflecting it about the x-axis. You can think of the multiples of a given function as a family of functions.

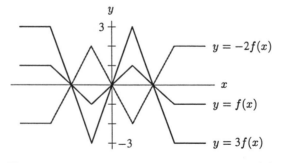

Figure 4.16: The family of multiples of the function $f(x)$

When we looked at the effect of coefficients on power functions in Section 1.6, we were studying the effect of multiplying a function by a constant. In general,

> Multiplying a function by a constant stretches or shrinks its graph vertically; a negative sign reflects the graph about the x-axis.

Sums of Functions

The graph on the right in Figure 4.17 shows the number of US students majoring in science and engineering, broken down into men and women.[3] The top curve represents the total and the bottom curve represents the men. It is really a picture of three functions: the number of men majoring in science and engineering, $m(t)$, the number of women majoring in science and engineering, $w(t)$, and the total number of students, $n(t)$, where t is the year. The men's and women's graphs are shown separately to the left.

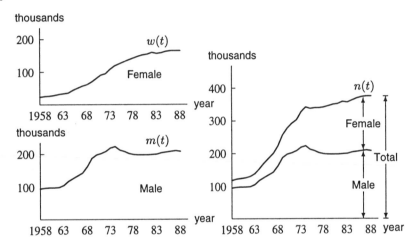

Figure 4.17: Numbers of US students majoring in science and engineering (1958–1988)

Now Total students = Number of men + Number of women, so, in functional notation,

$$n(t) = m(t) + w(t).$$

Figure 4.17 makes it clear that the graph of the sum function is obtained by stacking the other graphs one on top of the other.

Polynomials

Sums, stretches, and shifts of power functions produce *polynomials*. Some of the best-known functions for which there are formulas are the polynomials:

$$y = p(x) = a_n x^n + a_{n-1} x^{n-1} + \cdots + a_1 x + a_0,$$

where n is a positive integer, called the *degree* of the polynomial, and a_n is a non-zero number called the *leading coefficient*. We call $a_n x^n$ the *leading term*. If $n = 2$, then the polynomial has the form $ax^2 + bx + c$ and is often called a quadratic polynomial. If $n = 3$, the polynomial has the form $ax^3 + bx^2 + cx + d$ and is frequently referred to as a cubic polynomial.

---

[3]Data from the National Science Foundation, Washington D.C.

Graphs of Polynomials

The shape of the graph of a polynomial depends on its degree, as shown in Figure 4.18. The graph of a quadratic polynomial is always a parabola. It will open up (as shown in Figure 4.18) if the leading coefficient is positive and will open down if the leading coefficient is negative. A cubic polynomial can have several different shapes: it may have approximately the same shape as the graph of $y = x^3$, it may have the shape shown in Figure 4.18, or it may be a reflection of one of these if the leading coefficient is negative. Use a calculator or computer to graph the three cubic polynomials

$$y = x^3 - x^2 - 2x + 2,$$
$$y = x^3 - 3x^2 + 3x + 2,$$
$$\text{and} \quad y = -x^3 - x^2 - 2x + 2.$$

Which ones look like the one shown in Figure 4.18 (or a reflection of this)? Which ones have the same general shape as $y = x^3$ (or a reflection of this)?

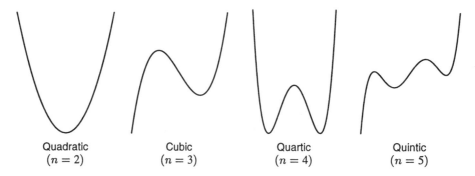

| Quadratic | Cubic | Quartic | Quintic |
| $(n = 2)$ | $(n = 3)$ | $(n = 4)$ | $(n = 5)$ |

Figure 4.18: Graphs of typical polynomials of degree n

Notice in Figure 4.18 that the graph of the quadratic "turns around" once, the cubic "turns around" twice, and the quartic (fourth degree) "turns around" three times. An nth degree polynomial "turns around" at most $n - 1$ times (where n is a positive integer), but there may be fewer turns.

Example 3   Using a calculator or computer, sketch graphs of $y = x^4$ and $y = x^4 - 15x^2 - 15x$ for $-4 \leq x \leq 4$ and for $-20 \leq x \leq 20$. Set the y range to $-100 \leq y \leq 100$ for the first domain, and to $-100 \leq y \leq 200{,}000$ for the second. What do you observe?

Solution   From the graphs in Figure 4.19 you can see that close up (for $-4 \leq x \leq 4$) the graphs look different; from far away, however, they are almost indistinguishable. The reason is that the leading term of each polynomial (the one with the highest power of x) is the same, namely x^4, and for large values of x, the leading term dominates the other terms.

TABLE 4.6   *Numerical values of $y = x^4$ and $y = x^4 - 15x^2 - 15x$*

x	$y = x^4$	$y = x^4 - 15x^2 - 15x$	Difference
-20	160,000	154,300	5700
-15	50,625	47,475	3150
15	50,625	47,025	3600
20	160,000	153,700	6300

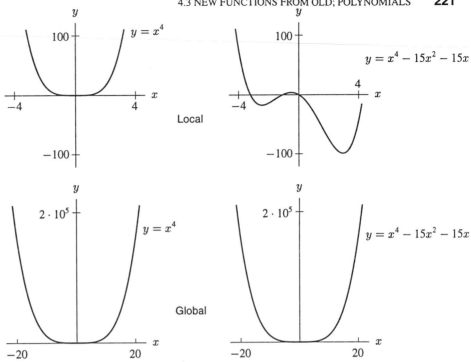

Figure 4.19: Local and global views of $y = x^4$ and $y = x^4 - 15x^2 - 15x$

Looked at numerically in Table 4.6, the differences in the values of the two functions when $x = \pm 20$, although large, are tiny compared with the vertical scale (-100 to $200,000$) and so cannot be seen on the graph.

We saw in Example 3 that if we look at the degree four polynomial $y = x^4 - 15x^2 - 15x$ from a distance, it looks just like the power function $y = x^4$. This phenomenon happens with all polynomials.

> If the graph of a polynomial of degree n
> $$y = a_n x^n + a_{n-1} x^{n-1} + \cdots + a_1 x + a_0$$
> is viewed in a large enough window, it will have essentially the same shape as the graph of the power function given by the leading term:
> $$y = a_n x^n.$$

Example 4   Sketch by hand the global shape of the graph of the polynomial $f(x) = -4x^3 + 2x^2 - 12$ as it would appear in a very large window. Use your graph to explain what happens to the function as $x \to \infty$ and $x \to -\infty$.

Solution   The leading term of $f(x)$ is $-4x^3$, so this polynomial will have the global shape of $y = -4x^3$, which looks approximately like the graph of $y = x^3$ stretched vertically and reflected over the x-axis. See Figure 4.20. In this graph, we see that as $x \to \infty$ (off to the right), $y \to -\infty$. As $x \to -\infty$ (off to the left), $y \to +\infty$.

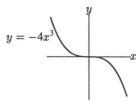

Figure 4.20: A graph of $y = -4x^3$

Using Polynomials

Example 5   The average miles per gallon of gasoline used by US automobiles declined until the 1960s and then started to rise as manufacturers started to make cars more fuel efficient. Table 4.7 shows average miles per gallon for selected years between 1940 and 1986.[4]

(a)   Plot this data. What family of functions should be used to model the data: linear, exponential, logarithmic, power function, or a polynomial? If a polynomial, state what degree and whether the leading coefficient will be positive or negative.

(b)   If you have technology that will do quadratic regression, find the best quadratic polynomial to model this data and graph it with the data.

TABLE 4.7   *What function fits this data?*

Year	1940	1950	1960	1970	1980	1986
MPG	14.8	13.9	13.4	13.5	15.5	18.3

Solution   (a)   The data is shown in Figure 4.21, with time t given in years since 1940. Miles per gallon decreases and then increases, so the best function to model this data is a quadratic (degree 2) polynomial. Since the parabola opens up, the leading coefficient will be positive.

(b)   If we do quadratic regression on this data, we see that the best quadratic polynomial to model the data is

$$f(t) = 0.00617t^2 - 0.225t + 15.10,$$

where $f(t)$ is average miles per gallon, and t is given in years since 1940. A graph of this function with the data is given in Figure 4.22, and we see that it matches the data very well.

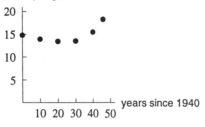

Figure 4.21: Average miles per gallon

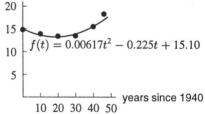

Figure 4.22: Using quadratic regression

[4]C. Schaufele and N. Zumoff, *Earth Algebra, Preliminary Version*, Harper Collins, New York, New York, 1993, p.91.

Example 6   A company offers dinner cruises on the St. Lawrence River. The company has found that the average number of passengers per night is 75 if the price is $50 per person. At a price of $35 per person, the average number of passengers per night is 120.

(a)   Assume that the demand function is linear and write the demand, q, as a function of price, p.

(b)   Since revenue is given by $R = pq$, use your answer to part (a) to write the revenue function R as a function of price, p.

(c)   Use a graph of the revenue function to determine what price should be charged to obtain the greatest revenue.

Solution   (a)   The two points given are $(p, q) = (50, 75)$ and $(p, q) = (35, 120)$. The slope of the line between them is

$$m = \frac{120 - 75}{35 - 50} = \frac{45}{-15} = -3 \frac{\text{passengers}}{\text{dollar}}.$$

To find the vertical intercept of the line, we use the slope and one of the points:

$$75 = b + (-3)(50)$$
$$225 = b$$

The demand function is $q = 225 - 3p$.

(b)   We see that $R = pq = p(225 - 3p) = 225p - 3p^2$.

(c)   The graph of the revenue function is given in Figure 4.23. We see that the maximum revenue is obtained when $p = 37.5$. To maximize revenue, the company should charge $37.50. The company's profit, of course, will depend on its costs as well as its revenue. The best price for the company to charge (to reap the greatest profit) cannot be determined without additional information about its costs.

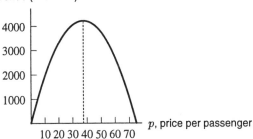

Figure 4.23: Revenue function for dinner cruises

Problems for Section 4.3

For each of the functions $y = f(x)$ given in Problems 1–4, sketch graphs of

(a) $y = f(x) + 2$ (b) $y = f(x - 1)$ (c) $y = 3f(x)$ (d) $y = -f(x)$

1.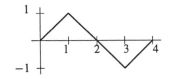

Figure 4.24

2.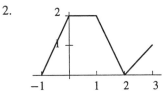

Figure 4.25

3.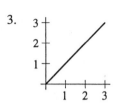

Figure 4.26

4.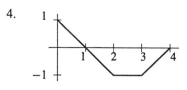

Figure 4.27

5. (a) Write an equation for a graph obtained by vertically stretching the graph of $y = x^2$ by a factor of 2, followed by a vertical upward shift of 1 unit. Sketch the graph.
    (b) What is the equation if the order of the transformations (stretching and shifting) in part (a) is interchanged?
    (c) Are the two graphs the same? Explain the effect of reversing the order of transformations.

6. Some values for the functions $y = f(x)$ and $y = g(x)$ are given in Table 4.8 and Table 4.9. Make a table of values for each of the following functions:
    (a) $y = f(x) + 3$ (b) $y = f(x - 2)$ (c) $y = 5g(x)$
    (d) $y = -f(x) + 2$ (e) $y = g(x - 3)$ (f) $y = f(x) + g(x)$

TABLE 4.8

x	0	1	2	3	4	5
$y = f(x)$	10	6	3	4	7	11

TABLE 4.9

x	0	1	2	3	4	5
$y = g(x)$	2	3	5	8	12	15

For Problems 7–12, use the graph of $y = f(x)$ in Figure 4.28 to sketch the graph indicated:

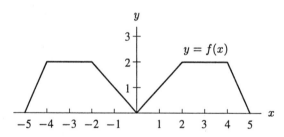

Figure 4.28

7. $y = 2f(x)$ 8. $y = f(x) + 2$ 9. $y = -3f(x)$
10. $y = f(x - 1)$ 11. $y = 2 - f(x)$ 12. $y = 2f(x) - 1$

13. (a) Sketch a graph of a possible demand curve, with price on the vertical axis and quantity on the horizontal axis.
    (b) If a fixed tax is added to the product, the demand curve will change. If the tax is $2, the consumer will pay $12 when the price is $10. Thus, the demand with the tax at a price of $10 is the same as the demand without the tax at a price of $12. Draw the demand curve with the tax on the same axes as the demand curve without tax drawn in part (a), and clearly label which is which. Indicate whether the new demand curve represents a vertical shift, a horizontal shift, or a stretch of the old demand curve.

14. The Heaviside step function, H, is graphed in Figure 4.29. Sketch graphs of the following functions.
    (a) $2H(x)$
    (b) $H(x) + 1$
    (c) $H(x + 1)$
    (d) $-H(x)$
    (e) $H(-x)$

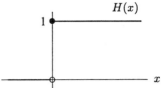

Figure 4.29

15. Assume that each of the graphs in Figure 4.30 is of a polynomial. For each graph:
    (a) What is the minimum possible degree of the polynomial?
    (b) Is the *leading coefficient* of the polynomial positive or negative? (You may assume that the graphs are in windows large enough to show the global behavior.)

(I) 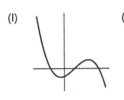 (II) 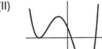 (III)   (IV) 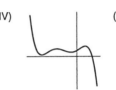 (V)

Figure 4.30

For each of the functions $f(x)$ given in Problems 16–23, answer the following questions:
(a) What is the degree of the polynomial? Is the leading coefficient positive or negative?
(b) Without using a calculator or computer, sketch a very global picture of the graph of the function.
(c) Use your answer to part (b) to explain what happens to $f(x)$ as $x \to \infty$ and as $x \to -\infty$.
(d) Using a calculator or computer, sketch a graph of the function. How many turning points does the function have? How does the number of turning points compare to the degree of the polynomial?

16. $x^2 + 10x - 5$

17. $5x^3 - 17x^2 + 9x + 50$

18. $8x - 3x^2$

19. $17 + 8x - 2x^3$

20. $-9x^5 + 82x^3 + 12x^2$

21. $0.01x^4 + 2.3x^2 - 7$

22. $100 + 5x - 12x^2 + 3x^3 - x^4$

23. $0.2x^7 + 1.5x^4 - 3x^3 + 9x - 15$

24. A dog food company finds that its profit function (in dollars) is given by

$$\pi(p) = -p^2 + 130p - 225,$$

where p is the price per pound of the dog food (in cents).
    (a) Sketch a graph of the profit function.
    (b) What price should be charged to maximize the profit? What is the profit at this price?
    (c) For what prices is the profit function positive?

25. A sporting goods wholesaler finds that when the price of one of the products is $25, the company sells, on average, 500 units per week. When the price is $30, the number sold per week decreases to 460 units.
    (a) Assume that the demand function for this product is linear and find the demand, q, as a function of price, p.

    (b)   Use your answer to part (a) to write revenue as a function of price.

    (c)   Sketch a graph of the revenue function found in part (b), and find the price that should be charged to maximize revenues. What is the revenue at this price?

26.   A private health club has determined that its cost and revenue functions are given by $C = 10{,}000 + 35q$ and $R = pq$ respectively, where q represents the number of annual club members and p represents the price of a one-year membership. The demand function for the health club is known to be $q = 3000 - 20p$.

    (a)   Use the demand function to write cost and revenue as functions of the price of membership in the club.

    (b)   Sketch graphs of cost and revenue as a function of price, on the same axes. (In order to get a good view of the graphs, it may help to know that price will not go above \$170 and that the annual costs of running the club can reach \$120,000.)

    (c)   Explain why the graph of the revenue function has the shape it does.

    (d)   For what prices does the club make a profit?

    (e)   Estimate the annual membership fee that will maximize profit for the club. Illustrate this point on your graph.

27.   A pomegranate is thrown from ground level straight up into the air at time $t = 0$ with velocity 64 feet per second. Its height at time t will be $f(t) = -16t^2 + 64t$. Find the time it hits the ground and the instant that it reaches its highest point. What is the maximum height?

28.   Evidence has linked ozone depletion in the atmosphere to chlorofluorocarbons (CFCs). Global CFC production increased from 42 thousand tons in 1950 to a high of 1260 thousand tons in 1988. International efforts to protect stratospheric ozone began in October 1987 with the Montreal Protocol. This document recommended cutting CFC production in half over 10 years. Additional evidence has spurred the world to further action, and efforts are underway to achieve a more rapid and complete phaseout of CFCs. The goal to cut the high of 1260 in half was in fact achieved by 1992.

    (a)   Draw a graph of world CFC production (in thousands of tons) as a function of year since 1950. Include scales on both your axes.

    (b)   What family of functions should be used to model data on CFC production since 1987?

29.   Figure 4.31 shows oil production in the Middle East, in millions of tons, as a function of year.[5] If you were to model this function with a polynomial, what degree would it have, and would the leading coefficient be positive or negative?

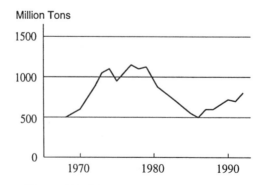

Figure 4.31: Oil production in the Middle East

---

[5]Lester R. Brown, et.al., *Vital Signs*, (New York: W.W. Norton and Co., 1994) 49.

30. World energy consumption (in quadrillion Btu) is given in Table 4.10.

    (a) Describe how world energy consumption has changed.
    (b) Find the average rate of change of energy consumption between 1985 and 1992. Give units with your answer and interpret your answer in terms of energy consumption.
    (c) Plot the data, with energy consumption as a function of years since 1980.
    (d) What family of functions could be used to model these data? If you suggest a polynomial, state what degree and whether the leading coefficient will be positive or negative.
    (e) If you have technology to do quadratic regression, find the best quadratic polynomial to model this data, and graph it with the data.

    TABLE 4.10   *World energy consumption*

Year	1980	1982	1985	1987	1988	1990	1992
Energy consumption (quadrillion Btu)	76.0	70.8	74.0	76.8	79.9	81.3	82.2

31. The rate, R, at which a population in a confined space increases is proportional to the product of the current population, P, and the difference between the *carrying capacity*, L, and the current population. (The carrying capacity is the maximum population the environment can sustain.)

    (a) Write R as a function of P.
    (b) Sketch R as a function of P.

32. For $y = f(x) = 3x^{3/2} - x$, use your calculator to construct a graph of $y = f(x)$, for $0 \le x \le 2$. From your graph, estimate $f'(0)$ and $f'(1)$.

4.4 PERIODIC FUNCTIONS

What Are Periodic Functions?

Figure 4.32 shows a graph of the number of new housing construction starts in the United States, where t is time measured in quarter-years. The data is from 1977, 1978, and 1979. Notice that very few new homes begin construction during the first quarter of a year (January, February, and March), whereas a great many new homes are begun in the second quarter (April, May, and June). Presumably, if we continued through the current year this pattern of oscillating up and down would continue.

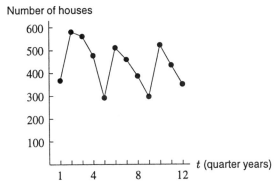

Figure 4.32: Housing construction starts, 1977–1979

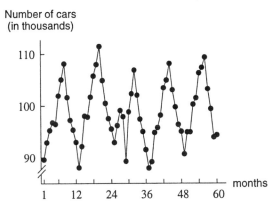

Figure 4.33: Traffic on the Golden Gate Bridge, 1976-80

Many functions have graphs that oscillate up and down, somewhat resembling a wave. Let's look at another example. Figure 4.33 is a graph of the number of cars (in thousands) traveling across the Golden Gate Bridge per month, from 1976-1980. Notice that traffic is at its minimum in January of each year (except 1978) and reaches its maximum in August of each year. Again, the graph looks like a wave. Functions that exactly repeat their values at regular intervals are called *periodic*, or repeating. Many processes, such as the number of housing starts or the number of cars that cross the bridge, are approximately periodic. The water level in a tidal basin, the blood pressure in a heart, retail sales in the United States, and the position of air molecules transmitting a musical note are also all periodic functions of time.

Amplitude and Period

We know that periodic functions are oscillating functions that repeat forever. If we know one cycle of the graph, we know the entire graph. The time to complete one cycle is called the *period* of the function.

The **amplitude** of a periodic function is half the difference between its maximum and minimum values.

The **period** of a periodic function is the time needed for the function to execute one complete cycle.

Example 1   Estimate the amplitude and period of the function for new housing starts shown in Figure 4.32.

Solution   Figure 4.32 is not a perfect periodic function since the maximum and minimum are not the same for each wave. Nonetheless, we approximate the minimum to be 300 and the maximum to be about 550. The difference between them is 250, so the amplitude is $\frac{1}{2}(250) = 125$.

The wave completes a cycle between $t = 1$ and $t = 5$, so the period is $t = 4$ quarter-years. The business cycle for new housing construction is 4 quarter-years, or one year.

Example 2   Figure 4.34 shows the temperature in an unopened freezer. Estimate the temperature in the freezer at 12:30 and at 2:45.

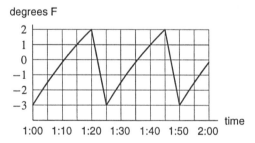

degrees F

Figure 4.34: What is the temperature at 2:45?

Solution   The maximum and minimum values occur every 25 minutes, so the period is 25 minutes. The temperature at 12:30 should be the same as at 12:55 and at 1:20. We see that the temperature at these times is 2°F. Similarly, the temperature at 2:45 should be the same as at 2:20 and 1:55, about −1.5°F.

The Sine and Cosine

Many periodic functions can be represented using the functions called *sine* and *cosine*. You will find keys for the sine and cosine on your calculator, usually abbreviated as [sin] and [cos].

Warning: Your calculator can be in either "degree" mode or "radian" mode. When you are working the sine and cosine problems in this book, you must make sure your calculator is in "radian" mode.

Graphs of the Sine and Cosine

The graphs of the sine and the cosine are shown in Figures 4.35 and 4.36. We see that both functions are periodic. Notice that the graph of the cosine function is identical to the graph of the sine function, except that the graph of the cosine function is shifted $\pi/2$ to the left. We will focus our attention primarily on the sine function.

The maximum and minimum values of $\sin t$ are $+1$ and -1, so the amplitude of the sine function is 1. The graph of $y = \sin t$ moves through a complete cycle between $t = 0$ and $t = 2\pi$; all the rest of the graph is just a repeat of this portion. The period of the sine function is 2π.

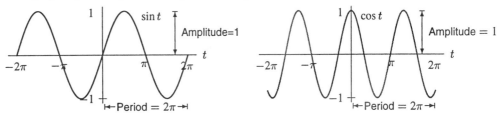

Figure 4.35: Graph of $\sin t$ *Figure 4.36:* Graph of $\cos t$

Example 3   Sketch a graph of $y = 3 \sin 2t$ and use the graph to determine the amplitude and period.

Solution

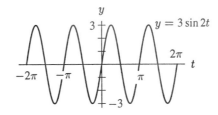

Figure 4.37: What are the amplitude and period?

The waves have a maximum of $+3$ and a minimum of -3, and so the amplitude is 3. The graph completes one complete cycle between $t = 0$ and $t = \pi$, so the period is π.

Example 4   How do the graphs of each of the following functions differ from the graph of $y = \sin t$? Explain your answer in terms of what you learned about shifts and stretches in the previous section.

(a) $y = 3 \sin t$ (b) $y = 5 + \sin t$ (c) $y = \sin(t + \frac{\pi}{2})$

Solution   (a)   The graph of $y = 3 \sin t$ is shown in Figure 4.38. The maximum and minimum values of this function are $+3$ and -3, and so the amplitude of the function is 3. This is the graph of $y = \sin t$ stretched vertically by 3 units.

(b)   The graph of $y = 5 + \sin t$ is shown in Figure 4.39. The maximum and minimum values of this function are 6 and 4, and so the amplitude is $(6 - 4)/2$, which is 1. The amplitude (or size of the wave) is the same as it is for $y = \sin t$, since this is a graph of $y = \sin t$ shifted up 5 units.

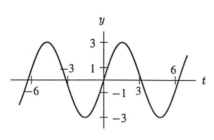

Figure 4.38: Graph of $y = 3\sin t$

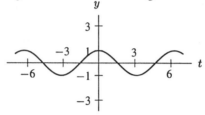

Figure 4.39: Graph of $y = 5 + \sin t$

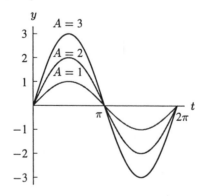

Figure 4.40: Graph of $y = \sin(t + \frac{\pi}{2})$

(c) The graph of $y = \sin(t + \frac{\pi}{2})$ is shown in Figure 4.40. This has the same amplitude (1) and period (2π) as the graph of $y = \sin t$. It is the graph of $y = \sin t$ shifted $\pi/2$ units to the left. (In fact, this is the graph of $y = \cos t$.)

Families of Curves: Understanding the graph of $y = A \sin Bt$

The constants A and B in the expression $y = A \sin Bt$ are called *parameters*. We can study families of curves by varying one parameter at a time and studying the result.

Example 5   (a) Let $y = A \sin t$. Substitute different values for A and sketch the graphs. Explain the effect of A on the graph.

       (b) Let $y = \sin Bt$. Substitute different values for B and sketch the graphs. Explain the effect of B on the graph.

Solution   (a) Graphs of $y = A \sin t$ for different values of A are shown in Figure 4.41. We see that A is the amplitude of the function, for positive A.

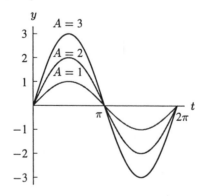

Figure 4.41: Graphs of $y = A \sin t$

(b) The graphs of $y = \sin Bt$ for $B = \frac{1}{2}, B = 1$, and $B = 2$ are shown in Figure 4.42. When $B = 1$, the period is 2π; when $B = 2$, the period is π; and when $B = \frac{1}{2}$, the period is 4π. The parameter B affects the period of the function. We see that the larger B is, the "sooner" the wave repeats itself (and hence the shorter the period).

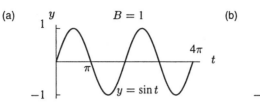
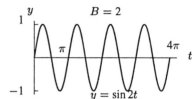

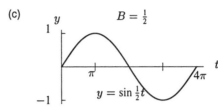

Figure 4.42: Graphs of $y = \sin Bt$

Example 6 On February 10, 1990, high tide in Boston was at midnight. The height of the water in the harbor is a periodic function, since it oscillates between high and low tide. A reasonable formula to model the height of the water (in feet) is

$$y = 5 + 4.9 \cos\left(\frac{\pi}{6}t\right),$$

where t is time measured in hours since midnight on February 10, 1990.
(a) Sketch a graph of this function on February 10, 1990 (from $t = 0$ to $t = 24$).
(b) What was the water level at high tide?
(c) When was low tide, and what was the water level at that time?
(d) What is the period of this function, and what does it represent in terms of tides?
(e) What is the amplitude of this function, and what does it represent in terms of tides?

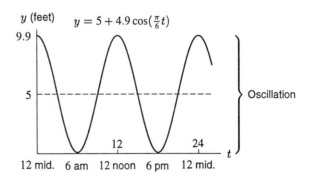

Figure 4.43: Function approximating the tide in Boston on February 10, 1990

Solution   (a)   See Figure 4.43.

       (b)   The water level at high tide was 9.9 feet.

       (c)   Low tide occurs at $t = 6$ (6 a.m.) and at $t = 18$ (6 p.m.). The water level at this time is 0.1 feet.

       (d)   The period is 12 hours, and it represents the interval between successive high tides or successive low tides. Of course, there is something wrong with the assumption in the model that the period is 12 hours. If so, the high tide would always be at noon or midnight, instead of progressing slowly through the day, as it in fact does. The interval between successive high tides actually averages about 12 hours 24 minutes, which could be taken into account in a more precise mathematical model.

       (e)   The maximum is 9.9, and the minimum is 0.1, so the amplitude is $(9.9 - 0.1)/2$, which is 4.9 feet. This represents half the difference between the depths at high and low tide.

Problems for Section 4.4

1. Sketch a possible graph of sales of sunscreen in the Northeast over a 3-year period, as a function of months since January 1 of the first year. Explain why your graph should be periodic. What is the period?

2. Figure 4.44 shows the levels of the hormones estrogen and progesterone during the monthly ovarian cycles in females.[6] Is the level of both hormones periodic? What is the period in each case? Approximately when in the monthly cycle is estrogen at a peak? Approximately when in the monthly cycle is progesterone at a peak?

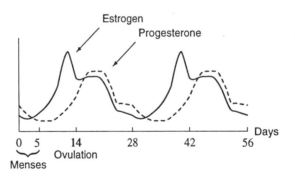

Figure 4.44: Levels of hormones in ovulating females

3. Figure 4.45 shows the number of reported cases of mumps by month, in the United States, as a function of the number of months since January 1, 1972.[7]

       (a)   Find the period and amplitude of this function, and interpret each in terms of the number of cases of mumps.

       (b)   Predict the number of cases of mumps 30 months and 45 months after January 1, 1972. Explain your reasoning.

---

[6] Robert M. Julien, *A Primer of Drug Action*, Seventh Edition, W.H. Freeman and Co., New York, 1995., p. 360.

[7] Center for Disease Control, 1974, *Reported Morbidity and Mortality in the United States 1973*, Vol. 22, No. 53.

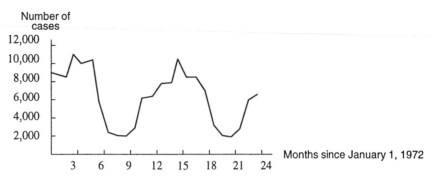

Figure 4.45: Number of reported cases of mumps

For Problems 4–9, sketch graphs of the functions. What are their amplitudes and periods?

4. $y = 3 \sin x$

5. $y = 3 \sin 2x$

6. $y = -3 \sin 2\theta$

7. $y = 4 \cos 2x$

8. $y = 4 \cos(\frac{1}{2}t)$

9. $y = 5 - \sin 2t$

10. Use the solution to Example 6 on page 231 to estimate the water level in Boston Harbor at 3:00 am, 4:00 am, and 5:00 pm on February 10, 1990.

11. A population of animals varies periodically between a low of 700 on January 1 and a high of 900 on July 1. Graph the population against time.

12. Use a graphing calculator or computer to find the period of $2 \sin 3t + 3 \cos t$.

13. A table of data for a function is given below. Explain why this function appears to be periodic. Approximately what are the period and amplitude of the function? If we assume that the function is periodic, estimate the value of the function at $t = 15$, at $t = 75$, and at $t = 135$.

TABLE 4.11

t	20	25	30	35	40	45	50	55	60
$f(t)$	1.8	1.4	1.7	2.3	2.0	1.8	1.4	1.7	2.3

14. Give the amplitude and period of the function $y = \cos t$.

15. The population of a herd of deer is modeled by

$$P(t) = 4000 + 400 \sin\left(\frac{\pi}{6}t\right) + 180 \sin\left(\frac{\pi}{3}t\right),$$

where t is measured in months from the first of April.

(a) Use a calculator or computer to sketch a graph showing how this population varies with time.
    Use the graph to answer the following questions.

(b) When is the herd largest? How many deer are in it at that time?

(c) When is the herd smallest? How many deer are in it then?

(d) When is the herd growing the fastest? When is it shrinking the fastest?

(e) How fast is the herd growing on April 1?

16. The number of hours, H, of daylight in Madrid as a function of date is approximated by the formula

$$H = 12 + 2.4 \sin[0.0172(t - 80)],$$

where t is the number of days since the start of the year. Using the formula for average value of a function over an interval given in Section 3.4, find the average number of hours of daylight in Madrid:

(a) in January (b) in June (c) over a whole year

(d) Comment on the relative magnitudes of your answers to parts (a), (b), and (c). Why are they reasonable?

17. The Bay of Fundy in Canada is reputed to have the largest tides in the world, with the difference between low and high water level being as much as 15 meters (nearly 50 feet). Suppose at a particular point in the Bay of Fundy, the depth of the water, y meters, as a function of time, t, in hours since midnight on January 1, 1994, is given by

$$y = y_0 + A \cos[B(t - t_0)].$$

(a) What is the physical meaning of y_0?
(b) What is the value of A?
(c) What aspect of the tides does B reflect?
(d) What is the physical meaning of t_0?

18. (a) Match the functions f, g, h, k, whose values are given in the table, with the functions whose formulas are

(i) $\omega = 1.5 + \sin t$ (ii) $\omega = 0.5 + \sin t$ (iii) $\omega = -0.5 + \sin t$ (iv) $\omega = -1.5 + \sin t$.

t	$\omega = f(t)$	t	$\omega = g(t)$	t	$\omega = h(t)$	t	$\omega = k(t)$
6.0	−0.78	3.0	1.64	5.0	−2.46	3.0	0.64
6.5	−0.28	3.5	1.15	5.1	−2.43	3.5	0.15
7.0	0.16	4.0	0.74	5.2	−2.38	4.0	−0.26
7.5	0.44	4.5	0.52	5.3	−2.33	4.5	−0.48
8.0	0.49	5.0	0.54	5.4	−2.27	5.0	−0.46

(b) Based on the table, what is the relationship between the values of $g(t)$ and $k(t)$? Explain this relationship using the formulas you chose for g and k.

(c) Using the formulas you chose for g and h, explain why all the values of g are positive, whereas all the values of h are negative.

4.5 THE LOGISTIC CURVE

Suppose that several rabbits are released on an island previously free of rabbits. We expect that the population of rabbits on the island will increase. If the population of rabbits grows exponentially, the graph of the population of rabbits against time is shown in Figure 4.46. Is this a reasonable model? Since there is only a finite amount of space on the island, it is not realistic to assume that the population of rabbits will continue to grow without bound forever. Instead we expect that there is a maximum rabbit population that the island can sustain. As the population approaches this upper limit, we expect that the rate of growth will slow down as in Figure 4.47. Population growth with an upper bound such as this can be modeled with a *logistic* or *inhibited growth model*, which, as we will see in this section, has many applications in business as well as in the life and social sciences.

Number of rabbits

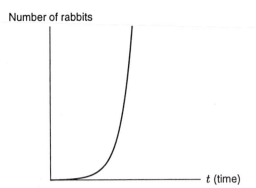

t (time)

Figure 4.46: An exponential model for the population of rabbits

Number of rabbits

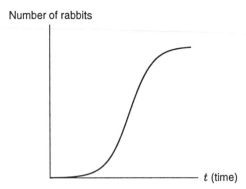

t (time)

Figure 4.47: A logistic model for the population of rabbits

Modeling the US Population

Population projections first became important to political philosophers in the late eighteenth century. As concern for scarce resources has grown, so has the interest in accurate population projections. Here we examine models for population growth in the United States. Every ten years the population of the United States is recorded by a census. The first such census was in 1790. Table 4.12 contains the census data from 1790 to 1940.

The US Population: 1790-1860

We begin by finding a model for the US population for the years 1790–1860. You can check that the data for these years is approximately exponential. The data is plotted in Figure 4.48, and we see that an exponential function would appear to fit this data well. We use a calculator or computer to do exponential regression on the data, and we see that the best exponential function to model this data is approximately

$$P = 3.93(1.03)^t,$$

where t is the number of years since 1790. Between 1790 and 1860, the US population was growing at an annual rate of about 3% per year.

The function $P = 3.93(1.03)^t$ is plotted in Figure 4.49 with the data, and we see that it fits the data very well. If we put $t = 0, 10, 20, \cdots, 70$ into this function, we get the populations predicted by our model for the years 1790, 1800, $\cdots$, 1860. Table 4.13 contains the comparison to the actual population.

TABLE 4.12   *US Population in millions, 1790–1940*

Year	Population	Year	Population	Year	Population
1790	3.9	1850	23.2	1910	92.0
1800	5.3	1860	31.4	1920	105.7
1810	7.2	1870	38.6	1930	122.8
1820	9.6	1880	50.2	1940	131.7
1830	12.9	1890	62.9		
1840	17.1	1900	76.0		

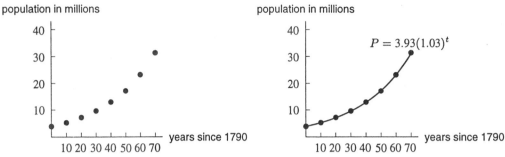

Figure 4.48: US Population 1790–1860

Figure 4.49: An exponential model for US population 1790–1860

TABLE 4.13 *Predicted versus actual US population 1790–1860, in millions. (exponential model)*

Year	Actual	Predicted	Year	Actual	Predicted
1790	3.9	3.9	1830	12.9	12.8
1800	5.3	5.3	1840	17.1	17.2
1810	7.2	7.1	1850	23.2	23.2
1820	9.6	9.5	1860	31.4	31.1

The agreement is remarkable. Of course, since we used the data from throughout the 70 year period, we should expect good agreement throughout that period. What is surprising is that if we had used only the populations in 1790 and 1800 to create our exponential function, the predictions would still be very accurate. It is amazing that a person in 1800 could predict the population 60 years later so accurately, especially when one considers all the wars, recessions, epidemics, additions of new territory, and immigration that took place from 1800 to 1860.

The US Population: 1790-1940

How well does our exponential function fit the US population beyond 1860? Figure 4.50 shows a graph of the US population from 1790 until 1940, and this data is shown with the exponential function $P = 3.93(1.03)^t$ in Figure 4.51. The exponential function which fit the data so well for the years 1790–1860 does not fit very well beyond 1860, and we must look for a better way to model this data.

The graph of the function given by the data in Figure 4.50 is concave up for small values of t, but then appears to become concave down and to be leveling off. This kind of growth is modeled

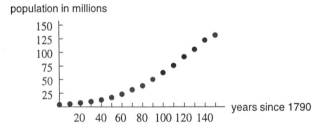

Figure 4.50: US Population 1790–1940

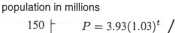

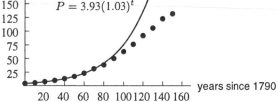

Figure 4.51: Not a good fit

with a *logistic function*. If we use a calculator or computer to do logistic regression on this data, we obtain the model

$$P = \frac{184.8177}{1 + 48.35e^{-0.032169t}},$$

where t is years since 1790. A graph of this function with the data is shown in Figure 4.52, and we see that this logistic function agrees very well with the actual population values up to 1940.

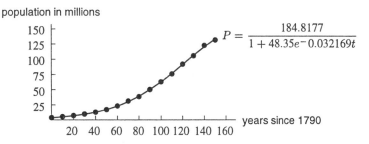

Figure 4.52: A logistic model for US population 1790–1940

The Logistic Function

We saw one example of a logistic function when we modeled the US population.

A **logistic function** has the form

$$P = f(t) = \frac{L}{1 + Ce^{-kt}},$$

where L, C, and k are positive constants.

A logistic function is used to model many situations, from sales of a new product to the spread of a virus. As we have seen, the graph of a logistic function is everywhere increasing. It is concave up at first, and then becomes concave down and levels off at a horizontal asymptote. As we saw when we examined US population growth, the logistic function is approximately exponential for small enough values of t[8].

Notice that our general logistic function has three parameters: L, C, and k. In Example 1, we investigate the effect of two of these parameters on the graph. (You are asked to investigate the third in the problems.)

[8]Just how small is small enough depends on the values of the parameters C and k.

Example 1   Let $P = \dfrac{L}{1 + 100e^{-kt}}$.

    (a)   Let $k = 1$. Sketch the graph of P using different values for L, and then explain the effect of the parameter L.

    (b)   Now let $L = 1$. Sketch a graph of P using different values for k, and then explain the effect of the parameter k.

Solution   (a)   See Figure 4.53. What do you observe in the graph? The graph levels off at the value of L. The parameter L determines where the horizontal asymptote is and is the maximum potential value of P.

    (b)   See Figure 4.54. What effect does the parameter k have on the graph? We notice that as k increases, the curve levels off at the asymptote much more rapidly. The parameter k affects the steepness of the curve.

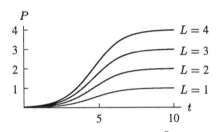   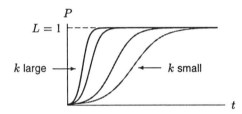

Figure 4.53: Graph of $P = \dfrac{L}{1 + 100e^{-t}}$ for various values of L

Figure 4.54: Graph of $P = \dfrac{1}{1 + 100e^{-kt}}$ with various values of k

The Carrying Capacity and the Point of Diminishing Returns

We saw in Example 1 that the numerator L of the logistic function

$$P = \frac{L}{1 + Ce^{-kt}}$$

is the maximum potential value of P. This value L is called the *carrying capacity* of the environment and represents the largest population the environment can support (or the maximum potential sales in a market).

    One way to estimate the carrying capacity is to look for the point where the concavity changes. We know that the graph of a logistic curve is concave up at first and then is concave down. At the point where the concavity changes, the slope is steepest. This is the point at which the population is growing the fastest. To the left of this point, the graph is concave up and the rate of growth is increasing. To the right of this point, the graph is concave down and the rate of growth is diminishing. For this reason, the point where the concavity changes is called the point of *diminishing returns*. It can be shown that this point always occurs where $P = \frac{L}{2}$. See Figure 4.55. Finding this point can often help us estimate the limiting value L. Many companies will watch for this concavity change in sales of a new product and use it to predict what the maximum potential sales will be.

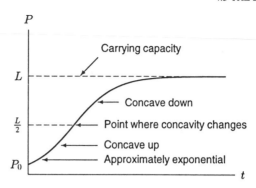

Figure 4.55: Logistic growth

> For the logistic curve $P = \frac{L}{1+Ce^{-kt}}$
>
> - The limiting value L represents the carrying capacity or maximum possible value of P.
> - The point of diminishing returns is the point where the concavity changes. This is the point where the population is growing the fastest, and it occurs where $P = \frac{L}{2}$.
> - The logistic function is approximately exponential for small enough values of t, with rate of growth roughly equal to k.

Example 2 Total sales of a new product often follow a logistic model. For example, if a new CD appears on the market, we expect sales to increase more and more rapidly as word of the CD spreads. Eventually, however, most of the people who want the CD will have already bought it and sales will slow down. The graph of total sales against time will be concave up at first and then concave down, with the limiting value L equal to the maximum potential sales. Suppose that total sales (in thousands) of a new CD are given in Table 4.14, with t given in months since the CD was introduced.

(a) Find the point where concavity changes in this function and use it to estimate maximum potential sales, L.

(b) If you have technology that will do logistic regression, find the best logistic function to model this data. What maximum potential sales does this function predict?

TABLE 4.14 *Total sales of a new CD*

t (months)	0	1	2	3	4	5	6	7
P (total sales in 1000s)	0.5	2	8	33	95	258	403	496

Solution (a) We see that the rate of change of total sales is increasing until $t = 5$ and is decreasing after $t = 5$, so the inflection point is approximately at $t = 5$. The vertical coordinate at $t = 5$ is 258, so we have $\frac{L}{2} = 258$, and thus $L = 516$. We estimate that the maximum potential sales for this CD are 516,000. Notice that we could have predicted maximum potential sales as soon as we saw the concavity change, which in this case happened after 6 months.

(b) Logistic regression gives the following function:

$$P = \frac{531.6327}{1 + 869.06e^{-1.33t}}.$$

Maximum potential sales predicted by this function are $L = 531.6327$, or about 530,000 CDs.

This function is graphed with the data in Figure 4.56.

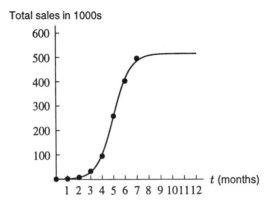

Figure 4.56: Total sales of CD

Example 3 Compact Disc (CD) players were first introduced in 1982, and sales have risen steadily since then. Total cumulative worldwide sales are shown in Figure 4.57.[9] Do sales appear to have reached the point of diminishing returns (that is, the point at which concavity changes from concave up to concave down)? What does this tell you about maximum potential sales?

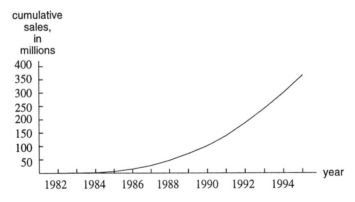

Figure 4.57: Cumulative sales of CD players

Solution It appears that the graph in Figure 4.57 continues to be concave up, so the point of diminishing returns has not yet been reached. Since the concavity has not yet changed, the point $L/2$ must lie above the value in 1995, which appears to be about 365 million. We predict the maximum potential sales of CD players will be greater than 2 times 365 million, or more than 730 million CD players.

The US Population: 1790-1990

We used a logistic function to model the US population between 1790 and 1940. How well does this model fit the US population since 1940? We take one more look at the US population, using all the data from 1790 to 1990.

[9] Adapted from Alan E. Bell, "Next-generation Compact Discs", *Scientific American* (July 1996) 45.

Example 4   Earlier in this section, we used the following logistic function to model the US population between 1790 and 1940:

$$P = \frac{184.8177}{1 + 48.35e^{-0.032169t}},$$

where t is in years since 1790, and P is in millions. According to this function, what is the maximum US population? Is this prediction accurate? How well does this logistic model fit the growth of the US population since 1940? (Table 4.15 shows the actual US population between 1790 and 1990 and the predicted values using this logistic model.)

TABLE 4.15   *Predicted versus actual US population, in millions, 1790–1990 (logistic model)*

Year	Actual	Predicted	Year	Actual	Predicted	Year	Actual	Predicted
1790	3.9	3.7	1860	31.4	30.4	1930	122.8	120.4
1800	5.3	5.1	1870	38.6	39.4	1940	131.7	133.2
1810	7.2	7.0	1880	50.2	50.3	1950	150.7	144.2
1820	9.6	9.5	1890	62.9	62.9	1960	179.3	153.5
1830	12.9	12.9	1900	76.0	76.9	1970	203.3	161.0
1840	17.1	17.3	1910	92.0	91.6	1980	226.5	166.9
1850	23.2	23.1	1920	105.7	106.3	1990	248.7	171.5

Solution   According to the function given, the maximum potential population is $L = 184.8177$ million people. The model predicts the US population will never go above 185 million people. Is this accurate? No. In fact, we see in Table 4.15 that the actual US population was above this figure by 1970. The fit between the logistic function and the actual population values is obviously not a good one beyond 1940.

Despite World War II, which undoubtedly depressed population growth between 1942 and 1945, in the last half of the 1940s the US population surged, wiping out in five years a deficit caused by 15 years of depression and war. The 1950s saw a population growth of 28 million, leaving our logistic model in the dust. This surge in population is referred to as the baby boom. All one can say is that based on 150 years of data, what happened in the US in the 20 years after World War II was completely without precedent. The baby boom could well end up being for the United States one of the most important sociological events of the twentieth century, and its consequences will be felt for many years to come.

Once again we have reached a point where our model is no longer useful. This should not lead you to believe that a reasonable mathematical model cannot be found; rather it should serve to point out that no model is perfect and that when one model fails, we seek a better one. Just as we abandoned the exponential model in favor of the logistic model for the US population, we could look further.

Dose-Response Curves

A *dose-response curve* plots the intensity of physiological response to a drug against the dose administered. We expect that as the dose increases, the intensity of the response will increase, and thus a dose-response curve is increasing. The intensity of the response is generally scaled as a

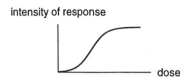

Figure 4.58: A dose-response curve

percentage of the maximum response. The curve cannot go above the maximum response (or 100%) and so the curve will level off at a horizontal asymptote. Dose-response curves generally start out concave up for low doses and become concave down for high doses. Notice that the independent variable for a dose-response curve is the dose of the drug, not time. A dose-response curve is shown in Figure 4.58. We see that the dose-response curve can be modeled with a logistic function.

All dose-response curves demonstrate important characteristics. We can read a dose-response curve to determine the amount of drug needed to produce the desired effect, or to determine the maximum effect attainable, or to determine the dose required to obtain the maximum effect. In addition, the derivative (or slope) of the dose-response curve has practical significance.

The goal in administering drugs is to administer a dose which is large enough to be effective but not so large as to be dangerous. What does the slope of the dose-response curve tell us about the range of safe but effective doses? Figure 4.59 shows two different dose-response curves: one with a large slope and one with a smaller slope. On each graph, we have labeled the vertical axis with the minimum desired response and the response beyond which the drug is dangerous. Notice that in Figure 4.59(a), there is a broad range of dosages at which the drug is both safe and effective. In Figure 4.59(b), however, when the slope of the curve is steep, the range of dosages at which the drug is both safe and effective is very small. When the slope of the dose-response curve is steep, a small mistake in the dose of the drug could have dangerous results. Administration of such a drug is difficult.

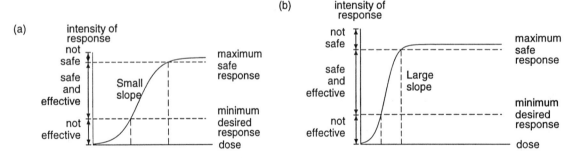

Figure 4.59: What does the slope of the dose-response curve tell us?

Example 5   Figure 4.60 shows dose-response curves for three different drugs used for the same purpose. Discuss the advantages and disadvantages of the three drugs.

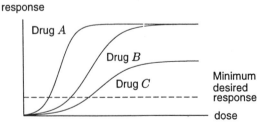

Figure 4.60: What are the advantages and disadvantages of each of these drugs?

Solution   Drugs A and B exhibit the same maximum response, while the maximum response of Drug C is significantly less. The potency of Drugs B and C (dose required to reach desired effect) is significantly less than the potency of Drug A. (Potency, however, is a relatively unimportant characteristic of a drug, since it makes little difference what the effective dose is as long as the drug is administered in an appropriate dose with no undue toxicity.) Drug A has a steeper slope than either of the other two. Thus, Drug C may be the preferred drug despite its lower maximum effect, since all three drugs reach the minimum desired response, and Drug C is the easiest to administer.

Problems for Section 4.5

1. A rumor spreads among a group of 400 people in an enclosed region. Sociologists who study the propagation of rumors have found that $N(t)$, the number of people who have heard the rumor by time t (where t is the time in hours since the rumor started to spread) can be approximated by a function of the form

$$N(t) = \frac{400}{1 + 399e^{-0.4t}}.$$

   (a) Find $N(0)$ and interpret it.
   (b) How many people will have heard the rumor after 2 hours? After 10 hours?
   (c) Sketch a graph of $N(t)$.
   (d) Approximately how long will it take until half the people in the region have heard the rumor?
   (e) Approximately how long will it take until virtually everyone in the region has heard the rumor?
   (f) Approximately when is the rumor spreading fastest?

2. One model for the population of the world, P, in billions, is

$$P = \frac{40}{1 + 11e^{-0.08t}},$$

   where t is years since 1990.

   (a) What does this model assume is the maximum sustainable population of the world?
   (b) Sketch the graph of P against t.
   (c) According to this model, approximately when will the earth's population reach 20 billion people?
   (d) According to this model, approximately when will the earth's population reach 39.9 billion people?

3. Table 4.16 gives the percentage, P, of households with a television set that also have a VCR, as a function of year.[10]

   TABLE 4.16   *Percent of households with VCRs*

Year	1978	1979	1980	1981	1982	1983	1984
Percent	0.3	0.5	1.1	1.8	3.1	5.5	10.6

Year	1985	1986	1987	1988	1989	1990	1991
Percent	20.8	36.0	48.7	58.0	64.6	71.9	71.9

[10]*Statistical Abstract of the United States*, 1992.

(a)   Explain why a logistic model is a reasonable one to use for this data.

(b)   Use the data to estimate the point of diminishing returns. What limiting value L does this point of concavity change predict? Does this limiting value appear to be accurate given the percentages for 1990 and 1991?

(c)   The best logistic equation for this data turns out to be

$$P = \frac{75}{1 + 316.75e^{-0.669t}},$$

where t is years since 1978. What limiting value does this model predict?

(d)   Explain in terms of percentages of households with VCRs what the limiting value is telling you. Do you think your answer to part (c) is an accurate prediction? What do you think will ultimately be the percentage of households with a VCR?

4.   Write a paragraph explaining why sales of a new product will often follow a logistic curve, and explain why it is beneficial to the company to watch for the point of diminishing returns.

5.   A new game is put on the market, and the company keeps a close watch on total sales, as shown in the table below.

(a)   Plot this data and mark on your plot the point of diminishing returns.

(b)   Predict total possible sales of this game, using the point of diminishing returns.

(c)   If you have technology to do logistic regression, find the best logistic function to fit this data, and graph it with the data. What total possible sales does the function predict for this game?

TABLE 4.17

months since game was introduced	0	2	4	6	8	10	12	14
total sales (1000's)	0	2.3	5.5	9.6	18.2	31.8	42.0	50.8

6.   Figure 4.61 shows world electrical generating capacity of nuclear power plants between 1960 and 1993. The data appear to follow a logistic curve. [11]

(a)   What is the concavity of the curve between 1960 and 1975? What does this tell you about the rate of increase of nuclear power during this time?

(b)   Approximately when does the concavity change? How many gigawatts of nuclear power is being generated at the time when the concavity changes? What limiting value does this predict, and does your answer match what you see in the graph?

(c)   If you use a logistic model for these data, what is the prediction for the generating capacity of nuclear power plants beyond 1993? Does the model predict that nuclear power use will rise, fall, or stay about the same?

---

[11]Lester R. Brown, et.al., *Vital Signs 1994*, (New York: W.W. Norton and Co., 1994) 53.

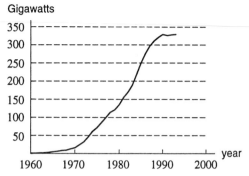

Figure 4.61: World electrical generating capacity of
nuclear power plants

7. The Tojolobal Mayan Indian community in Southern Mexico has available a fixed amount of land. The proportion, P, of available arable land in use for farming t years after 1935 is modeled with the logistic function

$$P = \frac{1}{1 + 2.968e^{-0.0275t}}.$$

See Figure 4.62.[12]

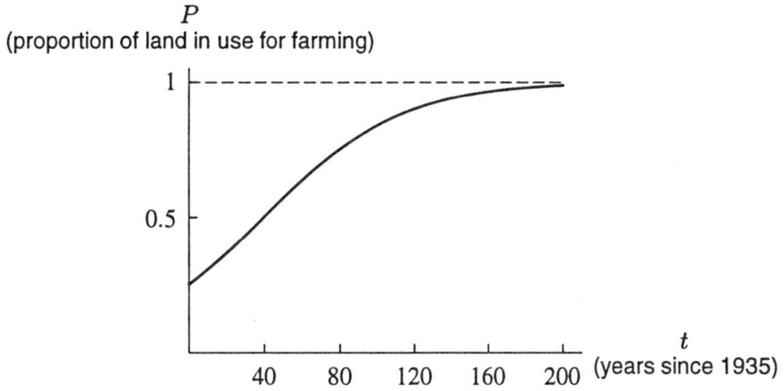

P
(proportion of land in use for farming)

Figure 4.62: Land use by Tojolobal Mayans

    (a) What proportion of the land was in use for farming in 1935?
    (b) What is the long run prediction of this model?
    (c) When was half the land in use for farming?
    (d) When is the proportion of arable land used for farming increasing most rapidly?.

8. Investigate the effect of the parameter C on the logistic curve $P = \frac{10}{1+Ce^{-t}}$. Substitute several values for C and explain, with a graph and with words, the effect of C on the graph.

---

[12]From J. S. Thomas and M. C. Robbins, The limits to growth in a Tojolobal Maya Ejido, *Geoscience and Man 26*, (Baton Rouge: Geoscience Publications, 1988) 9-16.

9. The total number of people infected with a virus often has the shape of a logistic curve. Suppose that 10 people originally have the virus, and that in the early stages of the virus (with time t measured in weeks) the number of people infected is increasing approximately exponentially, with a continuous growth rate of 1.78. It is estimated that, in the long run, approximately 5000 people will become infected.

   (a) In the logistic function $P = \frac{L}{1+Ce^{-kt}}$, we can use $k = 1.78$ as a rough approximation. What should we use for the parameter L, and why?
   (b) Use the fact that when $t = 0$, we have $P = 10$, to find C.
   (c) Now that you have estimated L, C, and k, what is the logistic function you are using to model the data?
   (d) Sketch a graph of your answer to part (c).
   (e) Estimate the length of time until the rate at which people are becoming infected starts to decrease. What is the vertical coordinate at this point?

10. (a) Draw a logistic curve. Label the carrying capacity L and the point of diminishing returns t_0.
    (b) Draw the derivative of the logistic curve. Mark the point t_o on the horizontal axis.
    (c) Suppose a company keeps track of the rate of sales (for example, sales per week) rather than total sales. Explain how the company can tell on a graph of rate of sales when the point of diminishing returns is reached.

11. Rate of sales, in sales per month, of a new automobile anti-theft device are given in Table 4.18. We see that there were 140 sales during the first month, 520 during the second month, etc.

    (a) When is the point of diminishing returns reached?
    (b) What are the total sales at this point?
    (c) Use your answer to part (b) to estimate total potential sales of the device.

TABLE 4.18 *Sales per month*

Sales per month	140	520	680	750	700	550
Months	1	2	3	4	5	6

12. Suppose a dose-response curve is given by $R = f(x)$ where R is percent of maximum response and x is the dose of the drug in mg. Assume that the curve has the general shape shown in Figure 4.58 on page 242, that the inflection point for the curve is at $(15, 50)$ and that $f'(15) = 11$.

    (a) Explain what $f'(15)$ tells you in terms of dose and response for this drug.
    (b) Is $f'(10)$ greater than or less than 11? Is $f'(20)$ greater than or less than 11? Explain.

13. A dose-response curve for a certain drug is given by

$$R = \frac{100}{1 + 100e^{-0.1x}},$$

where R = percent of maximum response and x is the dose in mg. Sketch a graph of this curve.

    (a) What dose corresponds to a response of 50% of the maximum? This is the concavity change — the point at which the response is increasing the fastest.
    (b) For this drug, the minimum desired response is 20% and the maximum safe response is 70%. What range of doses is both safe and effective for this drug?

14. Dose-response curves for three different products are given in Figure 4.63.

    (a) For the desired response, which drug requires the largest dose? The smallest dose?
    (b) Which drug has the largest maximum response? The smallest?
    (c) Which drug is the safest to administer? Explain.

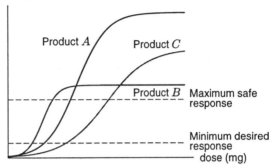

Figure 4.63: Which product is best?

15. Clearly explain why it is safer to use a drug for which the derivative of the dose-response curve is smaller.

16. There are two kinds of dose-response curves. One type, discussed in this section, plots the intensity of response against the dose of the drug. In this problem, we consider a dose-response curve in which the percentage of subjects showing a specific response is plotted against the dose of the drug. Two such dose-response curves are plotted together in Figure 4.64 and in Figure 4.65. In each case, the curve on the left shows the percentage of subjects exhibiting the desired response at the given dose, and the curve on the right shows the percentage of subjects for which the given dose is lethal.

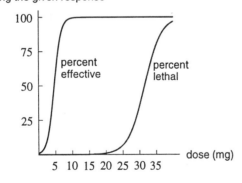

Figure 4.64: What doses are safe and effective?

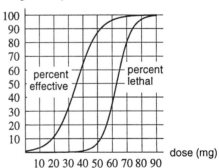

Figure 4.65: What happens when the dose is 50 mg?

    (a) In Figure 4.64, what range of doses appears to be both safe and effective for 99% of all patients?
    (b) In Figure 4.65, discuss the possible outcomes (and what percent of patients fall in each outcome) when 50 mg of the drug is administered.

4.6 THE SURGE FUNCTION

Measuring the Concentration of Nicotine in the Blood

When a person smokes a cigarette, the nicotine from the cigarette enters the person's body through the lungs and is then absorbed into the blood and spread throughout the body. Most cigarettes contain between 0.5 and 2.0 mg of nicotine and approximately 20% (between 0.1 and 0.4 mg) is actually inhaled and absorbed into the person's bloodstream. The half-life of nicotine in the bloodstream is about two hours, which explains the need for regular doses of the drug to avoid withdrawal symptoms in chronic smokers. The lethal dose of the drug is considered to be about 60 mg.

The nicotine level in the blood rises dramatically as a person smokes, and then, when the smoking ceases, begins to taper off. Table 4.19 shows blood nicotine concentration (in ng/ml) during and after the use of cigarettes. (Smoking occurred during the first ten minutes, and the experimental data shown represent average values for 10 people.)[13]

The points in Table 4.19 are plotted in Figure 4.66, enabling us to visualize a graph of blood nicotine level against time. Notice that the concentration seems to rise rapidly at first and then decays. Functions with this behavior are often called *surge functions*. We can model surge functions with a function of the form $y = ate^{-bt}$, where a and b are positive constants.

TABLE 4.19   *Blood nicotine concentrations during and after the use of cigarettes*

t (minutes)	0	5	10	15	20	25	30	45	60	75	90	105	120
C (ng/ml)	4	12	17	14	13	12	11	9	8	7.5	7	6.5	6

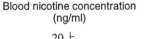

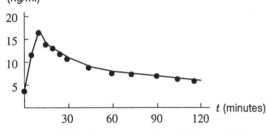

Figure 4.66: Blood nicotine concentrations during and after the use of cigarettes

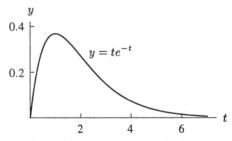

Figure 4.67: One member of the family $y = ate^{-bt}$

Investigating the Family of Functions $y = ate^{-bt}$

We begin our investigation of functions of the form $y = ate^{-bt}$ by looking at a graph of one member of this family. The graph of $y = te^{-t}$ (with $a = 1$ and $b = 1$) is shown in Figure 4.67. Such a graph represents a quantity that increases rapidly at first, and then decreases toward zero. For example, the number of bacteria in a person during the course of an infection, from onset to cure, might be described in this way. What effect do the parameters a and b have on the shape of this graph? You are asked to investigate the effect of parameter a on the graph of $y = ate^{-bt}$ in the problems; we will consider parameter b now.

---

[13]Benowitz, Porched, Skeiner, Jacog, "Nicotine Absorption and Cardiovascular Effects with Smokeless Tobacco Use: Comparison with Cigarettes and Nicotine Gum," *Clinical Pharmacology and Therapeutics* 44 (1988): 24.

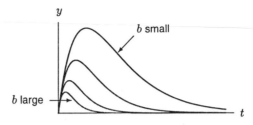

Figure 4.68: Graph of $y = te^{-bt}$, with b varying

The Effect of the Parameter b on $y = te^{-bt}$

Consider the family of curves $y = te^{-bt}$ for different positive values of b. The results are shown in Figure 4.68. We see that the general shape of the curve does not change, but as b gets smaller, the curve rises for a longer period of time and to a higher value before beginning to fall back toward the t-axis.

Certainly the point at which the function changes from increasing to decreasing is important. We see in Figure 4.69 that, when $b = 1$, the turning point occurs at $t = 1$. When $b = 2$, it occurs at $t = \frac{1}{2}$, and when $b = 3$, it occurs at $t = \frac{1}{3}$. We will see in Chapter 5 that, indeed, the point where the function $y = te^{-bt}$ changes from increasing to decreasing always occurs at $t = \frac{1}{b}$. See Figure 4.69.

> The surge function $y = ate^{-bt}$ increases rapidly and then decreases toward zero. The function changes from increasing to decreasing at $t = \frac{1}{b}$.

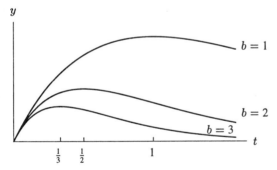

Figure 4.69: How does the time to the largest value depend on b?

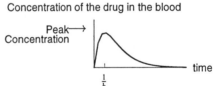

Figure 4.70: A drug concentration curve

Drug Concentration Curves

When the concentration of a drug in the body is plotted against the time since the drug is administered, the curve generally has the shape shown in Figure 4.70. This is called a *drug concentration curve*, and is modeled using a function of the form $C = ate^{-bt}$.

Two of the things we look for in a drug concentration curve are the peak concentration (the concentration of the drug in the body when it is at its highest level) and the length of time until peak concentration is reached. See Figure 4.70.

A drug concentration curve generally has the shape of the graph

$$C = ate^{-bt},$$

where C is the concentration of the drug in the blood and t is time since the drug was administered.

The concentration increases quickly and then decreases toward zero. Peak concentration occurs when $t = \frac{1}{b}$.

The Effect of Other Factors on Drug Absorption

Many factors can affect the rate of absorption of a drug, and we examine some of them here. In the next two examples, we see that drug interactions can affect the drug concentration curve, and that the age of the patient may also play a role. In the problems, we will see that food intake can affect the rate of absorption of some drugs, and (perhaps most surprising) that drug concentration curves can vary markedly between different commercial products of the same drug.

Example 1   Paracetamol (acetaminophen) is often used as a model for drug absorption since its rate of absorption in humans is directly related to the gastric emptying rate. Figure 4.71 shows the normal drug concentration curves for paracetamol alone and for paracetamol taken in conjunction with propantheline. Figure 4.72 shows drug concentration curves for patients known to be slow absorbers of the drug, for paracetamol alone and for paracetamol in conjunction with metoclopramide. In all cases, 1.5 g of paracetamol was administered and the graphs represent the concentration of paracetamol in the blood. Discuss the effects of the additional drugs on peak concentration and the time to reach peak concentration.[14]

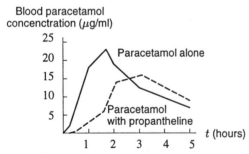

Figure 4.71: Drug concentration curves, normal patients

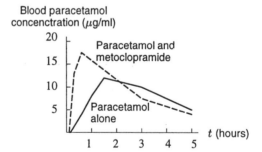

Figure 4.72: Drug concentration curves, patients with slow absorption

Solution   In Figure 4.71, we see that it takes about 1.5 hours for the paracetamol to reach its peak concentration, and that the maximum concentration reached is about 23μg of paracetamol per ml of blood. However, if propanthalene is administered with the paracetamol, it takes much longer to reach the peak concentration (in fact, the time to reach peak concentration is approximately doubled, to about three

[14]G.S. Avery, ed. *Drug Treatment: Principle and Practice of Clinical Pharmacology and Therapeutics*, (LOCATION: Adis Press, 1976).

hours) and the peak concentration is much lower, at about $16\mu g/ml$. The drug propanthalene is known to decrease gastric-intestinal motility and therefore to slow the rate of absorption.

Figure 4.72 shows drug concentration curves for patients who have a slow rate of absorption, and if we compare the paracetamol curves of Figure 4.71 and Figure 4.72, we see that in these patients, the time to reach peak concentration is unchanged (it is still about 1.5 hours), but the maximum concentration is much lower. The drug metoclopramide accelerates the rate of gastric emptying and increases the rate of absorption. We see that when metoclopramide is given with the paracetamol, the peak concentration is reached much faster, and the peak concentration is much higher.

Example 2   Figure 4.73 shows drug concentration curves for anhydrous ampicillin for newborn babies and adults.[15] Discuss the differences between newborns and adults in the processing of this drug.

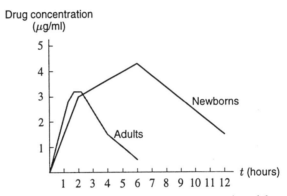

Figure 4.73: Drug concentration curves, comparing adults and newborns

Solution   Newborn babies are generally not as effective as adults in eliminating drugs from the body, and this is apparent in the drug concentration curves for this drug. The level of the drug in the body begins to decline after two hours in an adult, but not until six hours have elapsed in a newborn. In addition, the concentration in the blood reaches a higher level in the newborn than it does in the adult. Graphs such as this have taught us to modify the doses for drugs that must be given to pregnant women, women in labor, and women who are breastfeeding.

Minimum Effective Concentration

The minimum effective concentration of a drug is the blood concentration necessary to achieve a pharmacological response. The time taken to reach the minimum effective concentration is referred to as the onset of pharmacological response, and termination of drug effectiveness will occur when the drug concentration falls below the minimum effective concentration. See Figure 4.74.

---

[15]*Pediatrics*, 1973, 51, 578.

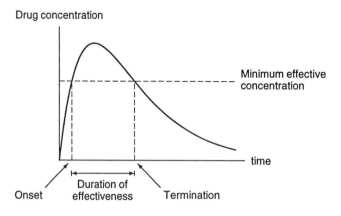

Figure 4.74: When is the drug effective?

Example 3   Depo-Provera is a progestin approved for use in the United States in 1992 after being used as an effective contraceptive. The drug concentration curve is shown in Figure 4.75 for a dose of 150 mg given intramuscularly.[16] The minimum effective concentration is about 4ng/ml. How often should the drug be administered?

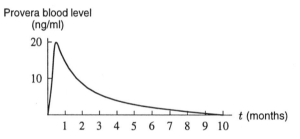

Figure 4.75: Drug concentration curve for Depo-Provera

Solution   We plot the minimum effective concentration on the drug concentration curve as a dotted horizontal line. See Figure 4.76. We see that the drug becomes effective almost immediately and ceases to be effective after about four months. Doses should be given about every four months.

   Although the dosage interval is four months, notice that up to ten months may be required for Depo-Provera to be virtually eliminated from the body after injections are discontinued. Thus, fertility during that period is unpredictable.

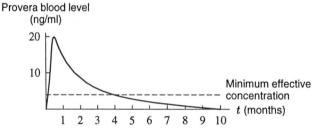

Figure 4.76: When should the next dose be administered?

---

[16]Robert M. Julien, *A Primer of Drug Action*, 1995. W.H. Freeman and Co.

Problems for Section 4.6

1. Investigate the effect of the parameter a on the function $C = ate^{-bt}$. Let $b = 1$, and sketch the graph of this function using different positive values for a. Explain what you observe.

2. Suppose that the drug concentration curve for a certain drug is given by

$$C = 12.4te^{-0.2t},$$

   where t is in hours and the concentration is in ng/ml.

   (a) Sketch a graph of this curve.
   (b) How many hours does it take for the drug to reach its peak concentration? What is the concentration at that time?
   (c) If the minimum effective concentration is 10 ng/ml, during what time period is the drug effective?
   (d) Complications can arise when this drug is used in conjunction with other drugs. These negative side effects are possible whenever the level of the drug is above 4 ng/ml. How long must a patient wait before it is safe to take the other drug?

3. Figure 4.66 on page 248 shows the concentration of nicotine in the blood during and after smoking a cigarette. Figure 4.77 shows the concentration of nicotine in the blood during and after using chewing tobacco or nicotine gum. (The chewing occurred during the first 30 minutes, and the experimental data shown represent the average values for 10 patients.)[17] Compare the three nicotine concentration curves (for cigarettes, chewing tobacco and nicotine gum) in terms of peak concentration, the time until peak concentration, and the rate at which the nicotine is eliminated from the bloodstream.

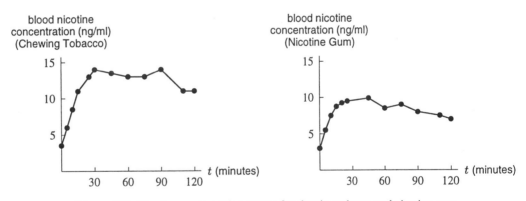

Figure 4.77: Nicotine concentration curves for chewing tobacco and nicotine gum

4. Hydrocodone bitartrate is a cough suppressant usually administered in a 10 mg oral dose. The peak concentration of the drug in the blood occurs 1.3 hours after consumption, and the peak concentration is 23.6 ng/ml. After 1.3 hours, the half-life for the elimination of the drug is 3.8 hours. Draw the drug concentration curve for hydrocodone bitartrate.

---

[17]Benowitz, Porchet, Skeiner, Jacob, "Nicotine Absorption and Cardiovascular Effects with Smokeless Tobacco Use: Comparison with Cigarettes and Nicotine Gum." *Clinical Pharmacology and Therapeutics* 44 (1988):24.

5. Figure 4.78 shows the drug concentration curves after oral administration of 0.5 mg of each of four digoxin products. All the tablets met current USP standards of potency, disintegration time and dissolution rate.[18]

    (a) Discuss differences and similarities in the peak concentration and the time to reach peak concentration.

    (b) Give possible values for minimum effective concentration and maximum safe concentration that would make Product C or Product B_2 the preferred drug.

    (c) Give possible values for minimum effective concentration and maximum safe concentration that would make Product A the preferred drug.

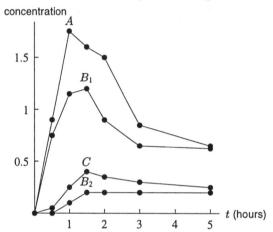

Figure 4.78: Drug concentration curves for four digoxin products

6. Figure 4.79 shows the concentration of bemetizide (a diuretic) in the blood after single oral doses of 25 mg alone, or 25 mg bemetizide and 50 mg triamterene in combination.[19] If the minimum effective concentration in a patient is 40 ng/ml, compare the effect of combining bemetizide with triamterene on peak concentration, time to reach peak concentration, time until onset of effectiveness, and duration of effectiveness. Under what circumstances might it be wise to use the triamterene with the bemetizide?

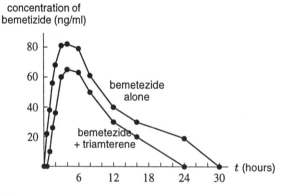

Figure 4.79: What effect does the drug triamterene have on bemetizide concentration?

[18]Graeme S. Avery, ed. *Drug Treatment: Principles and Practice of Clinical Pharmacology and Therapeutics*, (LOCATION: Adis Press, 1976).

[19]Wellin & Tse, *Pharmacokinetics of Cardiovascular, Central Nervous System, and Antimicrobial Drugs*, 1985, The Royal Society of Chemistry.

7. Let $C = ate^{-bt}$ represent a drug concentration curve.

   (a) Discuss the effect on peak concentration and time to reach peak concentration of varying the parameter a while keeping b fixed.

   (b) Discuss the effect on peak concentration and time to reach peak concentration of varying the parameter b while keeping a fixed.

   (c) Suppose $a = b$, so $C = ate^{-at}$. Discuss the effect on peak concentration and time to reach peak concentration of varying the parameter a.

8. Figure 4.80 shows the plasma levels of canrenone in a healthy volunteer after a single oral dose of spironolactone given on a fasting stomach and together with a standardized breakfast. (Spironolactone is a diuretic agent that is partially converted into canrenone in the body.)[20] Discuss the effect of food on peak concentration and time to reach peak concentration. Is the effect of the food strongest during the first 8 hours, or after 8 hours?

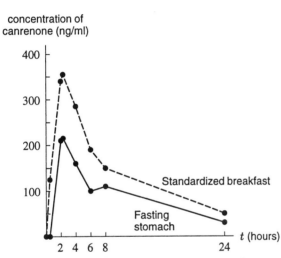

Figure 4.80: What effect does food have on the concentration of this drug?

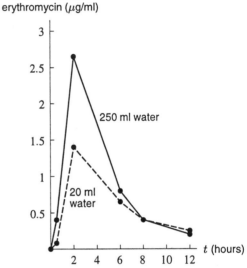

Figure 4.81: What effect does water have on the concentration of erythromycin?

9. Absorption of different forms of the antibiotic erythromycin may be increased, decreased, delayed or not affected by food. Figure 4.81 shows the drug concentration levels of erythromycin in healthy, fasting human volunteers who received single oral doses of 500 mg erythromycin tablets, together with either large (250 ml) or small (20 ml) accompanying volumes of water.[21] Discuss the effect of the water on the concentration of erythromycin in the blood. How are the peak concentration and the time to reach peak concentration affected? When does the effect of the volume of water wear off?

10. The drug concentration curve for a certain drug is given by $C = 17.2te^{-0.4t}$ ng/ml, where t is in hours. The minimum effective concentration is 10 ng/ml. Sketch the graph of this curve.

    (a) If the second dose of the drug is to be administered when the first dose becomes ineffective, when should the second dose be given?

    (b) If you want the onset of effectiveness of the second dose to coincide with termination of effectiveness of the first dose, when should the second dose be given?

[20]Welling & Tse, *Pharmacokinetics of Cardiovascular, Central Nervous System, and Antimicrobial Drugs*, (LOCATION: The Royal Society of Chemistry, 1985).

[21]J.W. Bridges and L.F. Chasseaud, *Progress in Drug Metabolism*, (New York: John Wiley and Sons, 1980).

11. Figure 4.82 shows a graph of the percentage of drug dissolved against time, for four tetracycline products A, B, C and D. Figure 4.83 shows the drug concentration curves for the same four tetracycline products.[22] Discuss the effect of dissolution rate on peak concentration and time to reach peak concentration.

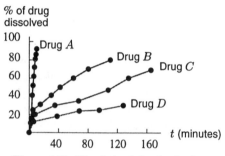

Figure 4.82: Dissolution behavior for four tetracycline products

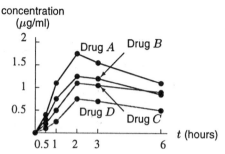

Figure 4.83: Drug concentration curves of four tetracycline products

12. The method of administering a drug can have a strong influence on the drug concentration curve. Figure 4.84 shows drug concentration curves for penicillin following various routes of administration. Three milligrams per kilogram of body weight were dissolved in water and administered intravenously (IV), intramuscularly (IM), subcutaneously (SC), and orally (PO). The same quantity of penicillin dissolved in oil was administered intramuscularly (P-IM). The minimum effective concentration (MEC) is labeled on the graph.[23]

(a) Which method reaches peak concentration the fastest? The slowest?

(b) Which method has the largest peak concentration? The smallest?

(c) Which method wears off the fastest? The slowest?

(d) Which method has the longest effective duration? The shortest?

(e) When penicillin is administered orally, for approximately what time interval is it effective?

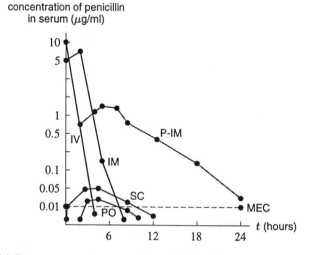

Figure 4.84: Drug concentration curves for penicillin administered five different ways

---

[22] J.W. Bridges and L.F. Chasseaud, *Progress in Drug Metabolism*, (New York: John Wiley and Sons, 1980).

[23] J.W. Bridges and L.F. Chasseaud, *Progress in Drug Metabolism*, 1980, John Wiley and Sons.

REVIEW PROBLEMS FOR CHAPTER FOUR

1. Each of the graphs in Figure 4.85 belongs to one of the following families of functions. In each case, identify which of the following the function is most likely to be:

   an exponential function
   a logarithmic function
   a polynomial (What is the degree? Is the leading coefficient positive or negative?)
   a periodic function
   a logistic function
   a surge function

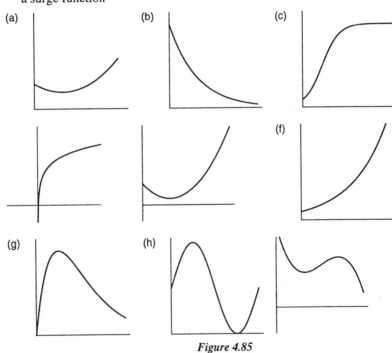

Figure 4.85

2. The graph of $f(x)$ is shown in Figure 4.86. Draw each of the following graphs:

   (a) $y = f(x) - 2$ (b) $y = 3f(x)$ (c) $y = -f(x)$
   (d) $y = f(x + 1)$ (e) $y = 4 - f(x)$ (f) $y = 1 + f(x - 2)$

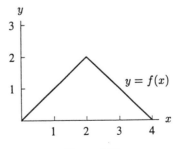

Figure 4.86

3. Table 4.20 gives values for $g(t)$, a periodic function.

   (a) Estimate the period and amplitude for this function. Explain your reasoning.
   (b) Estimate $g(34)$ and $g(60)$. Explain your reasoning.

TABLE 4.20

t	0	2	4	6	8	10	12	14	16	18	20	22	24	26	28
$g(t)$	14	19	17	15	13	11	14	19	17	15	13	11	14	19	17

4. A bank pays 8% interest, compounded continuously. If $10,000 is deposited now, how much will be in the account in 5 years? In 10 years?

5. If a bank pays 6% interest compounded continuously, how long will it take for the balance in an account to double?

6. A continuous exponential decay rate for a quantity is given as -0.02, accurate to 2 decimal places, with time measured in years.

   (a) If the actual decay rate is -0.015, find the half-life of the substance.
   (b) If the actual decay rate is -0.024, find the half-life of the substance.
   (c) Discuss the effect of round-off error in exponential growth and decay problems.

7. Figure 4.87 shows a graph of quarterly beer production (in millions of barrels) during the period 1990 to 1993. Quarter 1 reflects production during the first three months of the year, etc.

   (a) Explain clearly why a periodic function should be used to model this data.
   (b) Approximately when does the maximum usually occur? The minimum? Why might this make sense?
   (c) What is the period for this data? What is the amplitude?

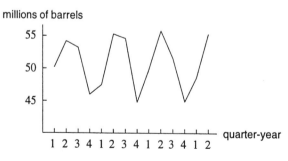

Figure 4.87: Quarterly beer production, 1990–1993

8. During the summer of 1988, one of the hottest on record in the midwest, a graduate student in environmental science studied the temperature fluctuations of a local river. Figure 4.88 shows a graph of the temperature of the river (in °C) taken every hour, with 0 being midnight of the first day.

   (a) Explain clearly why a periodic function should be used to model this data.
   (b) Approximately when does the maximum usually occur? The minimum? Why might this make sense?
   (c) What is the period for this data? What is the amplitude?

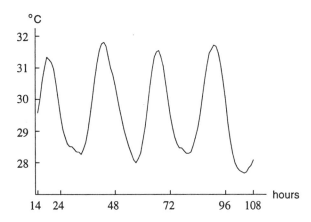

Figure 4.88: River temperature, 2 p.m. Day 1 — 1 p.m. Day 5

9. The concentration (in ng/ml) of a certain drug in a patient's bloodstream has been modeled by

$$C(t) = 20te^{-0.03t},$$

where t is in minutes since the drug was administered.

   (a) How long does it take for the drug to reach peak concentration, and what is the concentration at that time?
   (b) What is the concentration of the drug in the body after 15 minutes? After an hour?
   (c) If the minimum effective concentration is 10 ng/ml, when should the next dose be administered?

10. Suppose y_1, y_2, and y_3 are functions of x such that one of them is proportional to x, one is proportional to $1/x$, and one is proportional to x^2. Using the table below, write y_1, y_2, and y_3 as functions of x and find the constants of proportionality.

x	y_1	y_2	y_3
5	600	50	1.25
10	300	200	2.50
15	200	450	3.75
20	150	800	5.00
25	120	1250	6.25

11. Table 4.21 gives the world's population for three different years. If the world's population increased exponentially from 1950 through 1980 and continued to increase according to this pattern between 1980 and 1991, what would the world's population have been in 1991? How does this compare to the actual data and what conclusions, if any, can you draw?

TABLE 4.21   *Population of the world*

Year	Population (in billions)
1950	2.564
1980	4.478
1991	5.423

12. The profit function for a skateboard company is given by $\pi(p) = -p^2 + 70p - 125$, where p is the price in dollars charged by the company for a skateboard.

    (a) Sketch a graph of this function and find the price that will maximize profits.
    (b) For what prices will the company make a profit?

13. A company produces and sells customized shirts. The fixed costs for the company are $7000 and the variable costs are $5 per shirt.

    (a) Assume the company sells the shirts for $12 each. Find the cost and revenue functions, as functions of the quantity of shirts, q.
    (b) The company is considering changing the selling price of the shirts. Suppose the demand equation is $q = 2000 - 40p$, where p is the price in dollars charged by the company for a shirt and q is the number of shirts sold at that price. What quantity is sold at the current price of $12? What profit is realized at this price?
    (c) Use the demand equation to write cost and revenue as functions of the price, p. Then use the fact that profit = revenue − cost to write profit as a function of price.
    (d) Graph the profit function against price, and use the graph to find the price that will maximize profits. What is this price?

Convert the functions in Problems 14–15 into the form $P = P_0 a^t$.

14. $P = 2.91e^{0.55t}$

15. $P = (5 \cdot 10^{-3})e^{-1.9 \cdot 10^{-2}t}$

16. (a) Use the data from Table 4.22 to determine a formula of the form

$$Q = Q_0 e^{rt}$$

   which would give the number of rabbits, Q, at time t (in months).
   (b) What is the approximate doubling time for this population of rabbits?
   (c) Use your equation to predict when the rabbit population will reach 1000.

TABLE 4.22

t	0	1	2	3	4	5
Q	25	43	75	130	226	391

17. If you need $20,000 in your bank account in 6 years, how much must be deposited now? Assume an annual interest rate of 10%, compounded continuously.

18. What nominal annual interest rate has an effective annual yield of 5% under continuous compounding?

19. What is the effective annual yield, under continuous compounding, for a nominal annual interest rate of 8%?

20. Different kinds of the same element (called different *isotopes*) can have very different half-lives. The decay of plutonium-240 is described by the formula

$$Q = Q_0 e^{-0.00011t}$$

whereas the decay of plutonium-242 is described by

$$Q = Q_0 e^{-0.0000018t}.$$

Find the half-lives of plutonium-240 and plutonium-242.

21. An animal skull still has 20% of the carbon-14 that was present when the animal died. The half-life of carbon-14 is 5730 years. Find the approximate age of the skull.

22. Suppose prices are increasing by 0.1% a day.

    (a) By what percent do prices increase a year?
    (b) Looking at your answer to part (a), guess the approximate doubling time of prices increasing at this rate. Check your guess.

23. (a) Consider the functions graphed in Figure 4.89(a). Find the coordinates of C.
    (b) Consider the functions in Figure 4.89(b). Find the coordinates of C in terms of b.

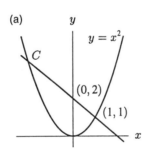
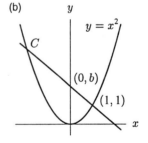

Figure 4.89

24. (a) What effect does the transformation

$$y = p(x) \quad \text{to} \quad y = p(1 + x)$$

have on the graph of $p(x)$?

    (b) If p is a polynomial of degree ≤ 2 such that for all x

$$p(x) = p(1 + x),$$

what can you say about p?

25.   Match the following formulas with the graphs in Figure 4.90:

(a) $y = 1 - 2^{-x}$ (d) $y = 1 - x^2$ (g) $y = -2\ln x$

(b) $y = x^2 + 4x + 5$ (e) $y = 2 + e^x$

(c) $y = 2\cos x$ (f) $y = x^3 - x^2 - x + 1$ (h) $y = 1 + \cos x$

(i) $y = \frac{1}{x}$

(I)
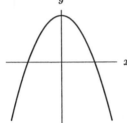

(II)
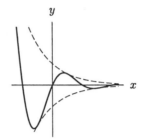

(III)
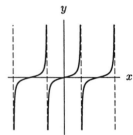

(IV)
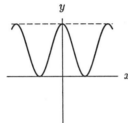

(V)
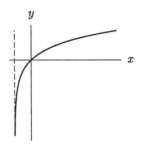

(VI)
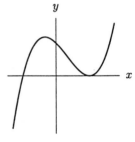

(VII)
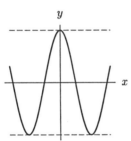

(VIII)
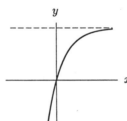

(IX)
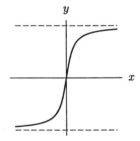

Figure 4.90

26.   For each of the following two conditions, find all polynomials, p, of degree ≤ 2 which satisfy the condition for all x.

(a) $p(x) = p(-x)$
(b) $p(2x) = 2p(x)$

27.   A *catalyst* in a chemical reaction is a substance which speeds up the reaction but which does not itself change. If the product of a reaction is itself a catalyst, the reaction is said to be *autocatalytic*. Suppose the rate, r, of a particular autocatalytic reaction is proportional to the quantity of the original material remaining times the quantity of product, p, produced. If the initial quantity of the original material is A and the amount remaining is $A - p$:

(a)   Express r as a function of p.
(b)   What is the value of p when the reaction is proceeding fastest?

28. Glucose is fed by intravenous injection at a constant rate, k, into a patient's bloodstream. Once there, the glucose is removed at a rate proportional to the amount of glucose present. If R is the net rate at which the quantity, G, of glucose in the blood is increasing:
    (a) Write a formula giving R as a function of G.
    (b) Sketch a graph of R against G.

29. In the early 1920s, Germany had tremendously high inflation, called hyperinflation. Photographs of the time show people going to the store with wheelbarrows full of money. If a loaf of bread cost 1/4 RM in 1919 and 2,400,000 RM in 1922, what was the average yearly inflation rate between 1919 and 1922?

30. A fish population is reproducing at an annual rate equal to 5% of the current population, P. Meanwhile, fish are being caught by fishermen at a constant rate, Y (measured in fish per year).
    (a) Write a formula for the rate, R, at which the fish population is increasing as a function of P.
    (b) Sketch a graph of R against P.

31. Find a formula for the function in the graph that follows:

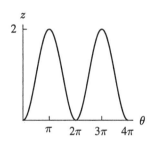

32. The depth of water in a tank oscillates once every 6 hours around an average depth of 7 feet. If the smallest depth is 5.5 feet and the largest depth is 8.5 feet, find a formula for the depth in terms of time, measured in hours. (There are many possible answers.)

33. Use a graphing calculator or computer to find the period of $2 \sin 4x + 3 \cos 2x$.

34. A software firm is trying to determine how much money to invest each year in research and development. An extensive study indicates that if it invests x million dollars each year in research and development then the company can expect to make a profit, $\pi(x)$, given in thousands of dollars, by

$$\pi(x) = -0.01x^4 + 0.33x^3 + 2.69x^2 + 4.35x - 500.$$

    (a) Use a graphing calculator or computer to sketch the graph of $\pi(x)$.
    (b) For what values of x is the company going to make a profit?
    (c) For what values of x is $\pi(x)$ increasing and for what values of x is $\pi(x)$ decreasing?
    (d) For what values of x is $\pi'(x)$ increasing and for what values of x is $\pi'(x)$ decreasing?
    (e) Using the fact that $\pi'(x) = -0.04x^3 + 0.99x^2 + 5.38x + 4.35$, graph $\pi'(x)$ and check your answer to part (d).
    (f) Which value of x maximizes the rate of change of $\pi(x)$ with respect to x?

35. The warehouse for a mail-order company receives shipment for a certain item on June 1, and the stock of the item is steadily depleted over the next few months. Suppose the inventory function for this item is given by

$$I(t) = 5000(0.9^t),$$

where t is measured in days since June 1.

(a) Find the average inventory in the warehouse during the 90 days after June 1.

(b) Graph the function $I(t)$ and illustrate the average inventory graphically.

36. Let $f(x) = \ln x$.

(a) Use small intervals to estimate $f'(1)$, $f'(2)$, $f'(3)$, $f'(4)$, and $f'(5)$.

(b) Use your answers to part (a) to guess a formula for the derivative of $\ln x$.

Evaluate the integrals in Problems 37–39.

37. $\displaystyle\int_1^3 \ln x\, dx$

38. $\displaystyle\int_{1.1}^{1.7} e^t \ln t\, dt$

39. $\displaystyle\int_{-3}^3 e^{-t^2}\, dt$

40. Suppose the cost function $C(q)$ represents the total cost to produce a quantity q units of a certain product. The fixed costs for the production are \$20,000. The marginal cost function is given by

$$C'(q) = 0.005q^2 - q + 56.$$

(a) On a graph of $C'(q)$, illustrate graphically the total variable cost of producing 150 units.

(b) Estimate $C(150)$, the total cost to produce 150 units.

(c) Find the value of $C'(150)$ and interpret your answer in terms of costs of production.

(d) Use your answers to parts (b) and (c) to estimate $C(151)$.

41. The number of hours, H, of daylight in Madrid as a function of date is approximated by the formula

$$H = 12 + 2.4\sin[0.0172(t - 80)]$$

where t is the number of days since the start of the year. Figure 4.91 shows a one-month portion of the graph of H.

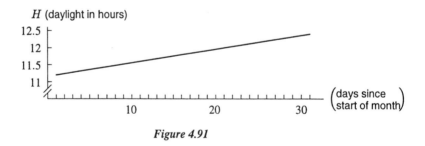

Figure 4.91

(a) Comment on the shape of the graph. Why does it look like a straight line?

(b) What month does this graph show? How do you know?

(c) What is the approximate slope of this line? What does the slope represent in practical terms?

CHAPTER FIVE

SHORT-CUTS TO DIFFERENTIATION

In this chapter we make a systematic study of the derivatives of functions given by formulas. These functions include powers, polynomials, exponential, logarithmic, and periodic functions. The chapter also contains general rules, such as the product rule and chain rule, which allow us to differentiate combinations of functions.

5.1 DERIVATIVE FORMULAS FOR POWERS AND POLYNOMIALS

In Chapter 2, we saw how the derivative represented a slope and a rate of change. We learned how to estimate the derivative of a function given by a graph (by estimating the slope of the tangent line at each point) and how to estimate the derivative of a function given by a table (by finding the average rate of change of the function). Now, we will learn how to find a formula for the derivative function when the function is given by a formula.

Derivative of a Constant Function

The graph of a constant function $f(x) = c$ is a horizontal line, with a slope of 0 everywhere. Therefore, its derivative is 0 everywhere. (See Figure 5.1.)

$$\boxed{\text{If } f(x) = c, \text{ then } f'(x) = 0.}$$

For example, $\dfrac{d}{dx}(5) = 0$, $\dfrac{d}{dx}(\pi) = 0$.

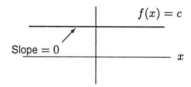

Figure 5.1: A constant function

Derivative of a Linear Function

We already know that the slope of a straight line is constant. This tells us that the derivative of a linear function is constant.

$$\boxed{\text{If } f(x) = b + mx, \text{ then } f'(x) = \text{Slope} = m.}$$

For example, $\dfrac{d}{dx}(5 - \dfrac{3}{2}x) = -\dfrac{3}{2}$.

To find the second derivative of a linear function, we must differentiate m, which is a constant. Thus, if f is linear,

$$f''(x) = 0.$$

So, as we would expect for a straight line, the graph of f is neither concave up nor concave down.

Derivative of a Constant Times a Function

Consider $y = f(x) = x$ and $y = g(x) = 5x$. We know that $f'(x) = 1$, the slope, and $g'(x) = 5$, the slope. In this case, g is 5 times f and the derivative of g is 5 times the derivative of f. Is this a coincidence? In Figure 5.2, you see the graph of $y = f(x)$ and of three multiples: $y = 3f(x)$, $y = \frac{1}{2}f(x)$, and $y = -2f(x)$. What is the relationship between the derivatives of these functions? In other words, for a particular x-value, how are the slopes of these graphs related?

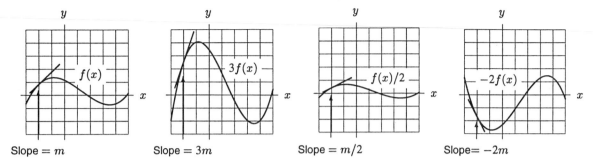

Slope $= m$ Slope $= 3m$ Slope $= m/2$ Slope$= -2m$

Figure 5.2: A function and its multiples: Derivative of multiple is multiple of derivative

Multiplying by a constant stretches or shrinks the graph vertically (and flips it over the x-axis if the constant is negative). The zeros of the function remain the same and the peaks and valleys occur at the same x-values. What does change is the slope of the curve at each point. If the graph has been stretched, the "rises" have all been increased by the same factor, whereas the "runs" remain the same. Thus, the slopes are all steeper by the same factor. If the graph has been shrunk, the slopes are all smaller by the same factor. If the graph has been flipped over the x-axis, the slopes will all have their signs reversed. In other words, if a function is multiplied by a constant, c, so is its derivative:

Derivative of a Constant Multiple: $\dfrac{d}{dx}\left[cf(x)\right] = cf'(x).$

Derivatives of Sums and Differences

Suppose we have two functions, $f(x)$ and $g(x)$, with the values listed in Table 5.1. Values of the sum $f(x) + g(x)$ are in the same table.

TABLE 5.1   *Sum of Functions*

x	$f(x)$	$g(x)$	$f(x) + g(x)$
0	100	0	100
1	110	0.2	110.2
2	130	0.4	130.4
3	160	0.6	160.6
4	200	0.8	200.8
5	250	1.0	251.0
6	310	1.2	311.2
7	380	1.4	381.4

You can see that when the increments of $f(x)$ and the increments of $g(x)$ are added together, they give the increments of $f(x) + g(x)$. For example, as x increases from 0 to 1, $f(x)$ increases by 10 and $g(x)$ increases by 0.2, while $f(x) + g(x)$ increases by $110.2 - 100 = 10.2$. Similarly, as x increases from 5 to 6, $f(x)$ increases by 60 and $g(x)$ by 0.2, while $f(x) + g(x)$ increases by $311.2 - 251.0 = 60.2$.

From this example, you can see that the rate at which $f(x) + g(x)$ is increasing is the sum of the rates at which $f(x)$ and $g(x)$ are increasing. Stating this with derivatives:

Derivative of Sum

$$\frac{d}{dx}\left[f(x) + g(x)\right] = f'(x) + g'(x).$$

Similarly for the difference:

Derivative of Difference

$$\frac{d}{dx}\left[f(x) - g(x)\right] = f'(x) - g'(x).$$

Derivatives of Powers of x

First, let's investigate the derivative of $f(x) = x^n$, with n a positive integer. We'll start by looking at $f(x) = x^2$ and $g(x) = x^3$.

Example 1   Estimate the derivative of each of the following at $x = 1$, $x = 2$, and $x = 3$. Use your answers to guess at a formula for the derivative function in each case.

(a) $f(x) = x^2$ (b) $g(x) = x^3$

Solution   (a)   We use a small interval of width $\Delta x = 0.001$ to estimate the three derivatives. We see

$$f'(1) \approx \frac{f(1.001) - f(1)}{1.001 - 1} = \frac{(1.001)^2 - 1^2}{0.001} \approx \frac{1.002 - 1}{0.001} = \frac{0.002}{0.001} = 2$$

$$f'(2) \approx \frac{f(2.001) - f(2)}{2.001 - 2} = \frac{(2.001)^2 - 2^2}{0.001} \approx \frac{4.004 - 4}{0.001} = \frac{0.004}{0.001} = 4$$

$$f'(3) \approx \frac{f(3.001) - f(3)}{3.001 - 3} = \frac{(3.001)^2 - 3^2}{0.001} \approx \frac{9.006 - 9}{0.001} = \frac{0.006}{0.001} = 6$$

A good guess at a formula for the derivative of x^2 is

$$f'(x) = 2x.$$

(b)   Similarly, if $g(x) = x^3$, we can use small intervals of width $\Delta x = 0.001$ to estimate the three derivatives for this function. We see that

$$g'(1) \approx \frac{g(1.001) - g(1)}{1.001 - 1} = \frac{(1.001)^3 - 1^3}{0.001} \approx \frac{1.003 - 1}{0.001} = \frac{0.003}{0.001} = 3$$

$$g'(2) \approx \frac{g(2.001) - g(2)}{2.001 - 2} = \frac{(2.001)^3 - 2^3}{0.001} \approx \frac{8.012 - 8}{0.001} = \frac{0.012}{0.001} = 12$$

$$g'(3) \approx \frac{g(3.001) - g(3)}{3.001 - 3} = \frac{(3.001)^3 - 3^3}{0.001} \approx \frac{27.027 - 27}{0.001} = \frac{0.027}{0.001} = 27$$

If you can't guess the formula yet, find a few more derivatives! A good guess at the formula for the derivative of x^3 is

$$g'(x) = 3x^2.$$

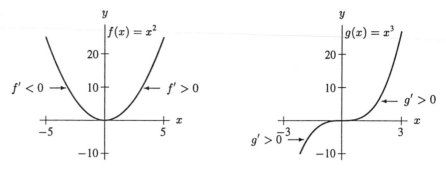

Figure 5.3: Graphs of $f(x) = x^2$ and $g(x) = x^3$

Let's see if our guesses for the derivatives of $f(x) = x^2$ and $g(x) = x^3$ make sense graphically. Graphs of these two functions are in Figure 5.3. The shape of the graph of $f(x) = x^2$ shows that its derivative is negative when $x < 0$, zero when $x = 0$, and positive for $x > 0$. Notice that $f'(x) = 2x$ has the behavior we expected: it is negative for $x < 0$, zero when $x = 0$, and positive for $x > 0$. The graph of f' is in Figure 5.4, along with the graph of f.

What about the derivative of $g(x) = x^3$? The graph suggests that its derivative is positive when $x < 0$, zero when $x = 0$, and becomes positive again when $x > 0$. Our guess at the derivative, $g'(x) = 3x^2$, has this behavior; it is zero when $x = 0$, but positive everywhere else. See Figure 5.5.

We will show in Section 5.6 that the formulas we have arrived at for the derivatives of x^2 and x^3 are accurate. We have

$$\frac{d}{dx}(x^2) = 2x \qquad \text{and} \qquad \frac{d}{dx}(x^3) = 3x^2.$$

Similar calculations will show you that

$$\frac{d}{dx}(x^4) = 4x^3, \quad \frac{d}{dx}(x^5) = 5x^4,$$

and so on. These results might well lead you to conjecture that, for any positive integer n, the following general relation holds:

$$\boxed{\frac{d}{dx}(x^n) = nx^{n-1},}$$

and indeed this turns out to be true.

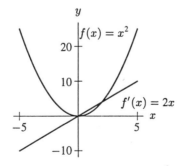

Figure 5.4: Graphs of $f(x) = x^2$ and its derivative

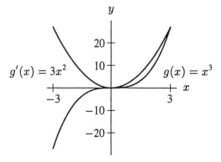

Figure 5.5: Graphs of $g(x) = x^3$ and its derivative

Example 2 Find the derivative of (a) x^8 (b) t^7

Solution (a) $\dfrac{d}{dx}(x^8) = 8x^7$.

(b) $\dfrac{d}{dt}(t^7) = 7t^6$.

Derivatives of Polynomials

We already know how to differentiate powers, constant multiples, and sums. For example,

$$\frac{d}{dx}(3x^5) = 3\frac{d}{dx}(x^5) = 3 \cdot 5x^4 = 15x^4$$

and $\quad \dfrac{d}{dx}(x^5 + x^3) = \dfrac{d}{dx}(x^5) + \dfrac{d}{dx}(x^3) = 5x^4 + 3x^2.$

Using these rules together, we can differentiate any polynomial.

Example 3 Differentiate $5x^2 - 7x^3$.

Solution

$$\frac{d}{dx}(5x^2 - 7x^3) = \frac{d}{dx}(5x^2) - \frac{d}{dx}(7x^3) \quad \text{Derivative of difference}$$

$$= 5\frac{d}{dx}(x^2) - 7\frac{d}{dx}(x^3) \quad \text{Derivative of multiple}$$

$$= 5(2x) - 7(3x^2) = 10x - 21x^2.$$

Example 4 Find the derivatives of

(a) $5x^2 + 3x + 2$, (b) $3x^7 - \dfrac{x^5}{5} + 2x^2 - 13$, (c) $x^3 - 0.2x^2 + 1.3x - 2.51$.

Solution (a)

$$\frac{d}{dx}(5x^2 + 3x + 2) = 5\frac{d}{dx}(x^2) + 3\frac{d}{dx}(x) + \frac{d}{dx}(2)$$

$$= 5 \cdot 2x + 3 \cdot 1 + 0 \quad \text{Since the derivative of a constant, } d(2)/dx, \text{ is zero}$$

$$= 10x + 3.$$

(b)

$$\frac{d}{dx}\left(3x^7 - \frac{x^5}{5} + 2x^2 - 13\right) = 3\frac{d}{dx}(x^7) - \frac{1}{5}\frac{d}{dx}(x^5) + 2\frac{d}{dx}(x^2) - \frac{d(13)}{dx}$$

$$= 3 \cdot 7x^6 - \frac{1}{5} \cdot 5x^4 + 2 \cdot 2x - 0 \quad \text{Since 13 is a constant, } d(13)/dx = 0$$

$$= 21x^6 - x^4 + 4x.$$

(c)

$$\frac{d}{dx}\left(x^3 - 0.2x^2 + 1.3x - 2.51\right) = \frac{d}{dx}(x^3) - 0.2\frac{d}{dx}(x^2) + 1.3\frac{d}{dx}(x) - \frac{d}{dx}(2.51)$$

$$= 3x^2 - 0.2 \cdot 2x + 1.3 \cdot 1 - 0$$

$$= 3x^2 - 0.4x + 1.3$$

We can also use these rules to differentiate expressions which are not polynomials.

Derivatives of Negative and Fractional Powers

The derivative formula for positive integer powers of x works for all powers. The rule

$$\frac{d}{dx}(x^n) = nx^{n-1}$$

holds for any constant real number n.

Example 5   Differentiate (a) $\dfrac{1}{x^3}$,   (b) $\sqrt{x}$,   (c) $2t^{4.5}$.

Solution   (a)   For $n = -3$: $\dfrac{d}{dx}\left(\dfrac{1}{x^3}\right) = \dfrac{d}{dx}(x^{-3}) = -3x^{-3-1} = -3x^{-4} = -\dfrac{3}{x^4}$, provided $x \neq 0$.

(b)   For $n = 1/2$: $\dfrac{d}{dx}(\sqrt{x}) = \dfrac{d}{dx}\left(x^{1/2}\right) = \dfrac{1}{2}x^{(1/2)-1} = \dfrac{1}{2}x^{-1/2} = \dfrac{1}{2\sqrt{x}}$.

(c)   For $n = 4.5$: $\dfrac{d}{dt}\left(2t^{4.5}\right) = 2\left(4.5t^{4.5-1}\right) = 9t^{3.5}$.

Using the Derivative Formulas

Example 6   Let $f(x) = x^2 + 1$. Compute the derivatives $f'(0)$, $f'(1)$, $f'(2)$, and $f'(-1)$. Check your answers graphically.

Solution   The graph of $f(x) = x^2 + 1$ is given in Figure 5.6. We will compute the derivatives analytically, and then check the answers by looking at the slope of the tangent line to this graph at the different values of x.

Since

$$f(x) = x^2 + 1$$

we see

$$f'(x) = 2x.$$

Thus, $f'(0) = 2(0) = 0$. At the point on the graph where $x = 0$, the slope of the tangent line is 0. We confirm this by seeing in Figure 5.6 that the tangent line at $x = 0$ is horizontal.

We have $f'(1) = 2(1) = 2$ and $f'(2) = 2(2) = 4$. The slope is 2 at $x = 1$ and the slope is 4 at $x = 2$. The slope is positive at both points and the graph is steeper at $x = 2$ than it is at $x = 1$. Again, this matches what we see in the graph in Figure 5.6.

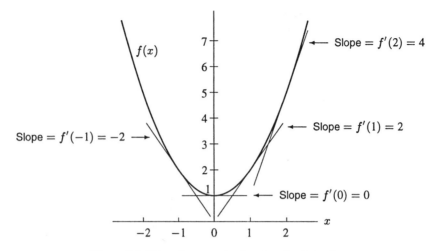

Figure 5.6: Using slopes to check values for derivatives

We have $f'(-1) = 2(-1) = -2$. This tells us that where $x = -1$, the slope is -2. Again, this agrees with the graph in Figure 5.6.

Example 7   Find the equation of the tangent line to the graph of

$$y = x^3 + 2x^2 - 5x + 7$$

at the point where $x = 1$. Sketch the graph of the curve and its tangent line on the same axes.

Solution   The slope of a tangent line is given by the derivative, so we begin by finding dy/dx:

$$\frac{dy}{dx} = 3x^2 + 2(2x) - 5(1) + 0 = 3x^2 + 4x - 5.$$

Therefore, the slope of the tangent line at $x = 1$ is given by

$$\left.\frac{dy}{dx}\right|_{x=1} = 3(1)^2 + 4(1) - 5 = 2.$$

Thus, the equation of the tangent line is of the form

$$y = b + 2x.$$

When $x = 1$, we have

$$y = 1^3 + 2(1^2) - 5(1) + 7 = 5,$$

so the point $x = 1$, $y = 5$ lies on the tangent line. Substituting gives

$$5 = b + 2(1), \quad \text{so} \quad b = 3.$$

Thus, the equation of the tangent line is

$$y = 3 + 2x.$$

The graphs of the curve and the tangent line are given in Figure 5.7.

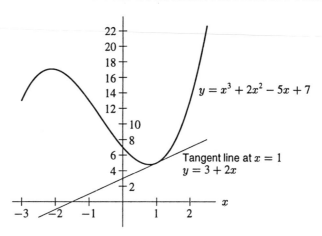

Figure 5.7: Find the equation for this tangent line

Example 8   Find and interpret the second derivatives of
(a) $f(x) = x^2$,   (b) $g(x) = x^3$.

Solution   (a)   For $f(x) = x^2$, $f'(x) = 2x$, so $f''(x) = \dfrac{d}{dx}(2x) = 2$. Since f'' is always positive, the graph of f is concave up, as expected for a parabola opening upwards. (See Figure 5.8.)

(b)   For $g(x) = x^3$, $g'(x) = 3x^2$, so $g''(x) = \dfrac{d}{dx}(3x^2) = 3\dfrac{d}{dx}(x^2) = 3 \cdot 2x = 6x$. This is positive for $x > 0$ and negative for $x < 0$, which means that the graph of $g(x) = x^3$ is concave up for $x > 0$ and concave down for $x < 0$. (See Figure 5.9.)

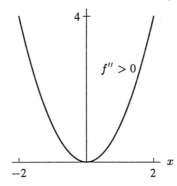

Figure 5.8: $f(x) = x^2$ and $f''(x) = 2$

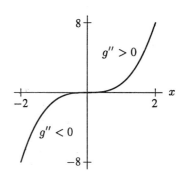

Figure 5.9: $g(x) = x^3$ and $g''(x) = 6x$

Example 9   If the position of a body, in meters, is given as a function of time t, in seconds, by

$$s = -4.9t^2 + 5t + 6,$$

find the velocity of the body at time t.

Solution   The velocity, v, is the derivative of the position:

$$v = \frac{ds}{dt} = \frac{d}{dt}(-4.9t^2 + 5t + 6) = -9.8t + 5 \text{ meters/sec.}$$

Example 10   Figure 5.10 shows the graph of a cubic polynomial. Both graphically and algebraically, describe the behavior of the derivative of this cubic.

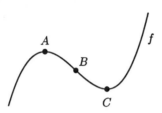

Figure 5.10: The cubic of Example 10

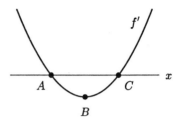

Figure 5.11: Derivative of the cubic of Example 10

Solution   Graphical approach: Suppose you move along the curve from left to right. To the left of A, the slope is positive; it starts very positive and decreases until the curve reaches A, where the slope is 0. Between A and C the slope is negative. Between A and B the slope is decreasing (getting more negative); it is most negative at B. Between B and C the slope is negative but increasing; at C the slope is zero. From C to the right, the slope is positive and increasing. The graph of the derivative function is shown in Figure 5.11.

Algebraic approach: f is a cubic that goes to $+\infty$ as $x \to +\infty$ so

$$f(x) = ax^3 + bx^2 + cx + d$$

with $a > 0$. Hence,

$$f'(x) = 3ax^2 + 2bx + c,$$

whose graph is a parabola opening upward, as in Figure 5.11.

Problems for Section 5.1

For Problems 1–21, find the derivative of the given function.

1. $y = 5$

2. $y = 3x$

3. $y = 5x + 13$

4. $y = x^{12}$

5. $y = x^{-12}$

6. $y = x^{4/3}$

7. $y = 8t^3$

8. $y = 3t^4 - 2t^2$

9. $f(x) = \dfrac{1}{x^4}$

10. $f(q) = q^3 + 10$

11. $f(x) = Cx^2$, C a constant.

12. $y = x^2 + 5x + 9$

13. $y = 6x^3 + 4x^2 - 2x$

14. $y = -3x^4 - 4x^3 - 6x + 2$

15. $y = 3x^2 + 7x - 9$

16. $y = 8t^3 - 4t^2 + 12t - 3$

17. $y = 4.2q^2 - 0.5q + 11.27$

18. $y = 3t^5 - 5\sqrt{t} + \dfrac{7}{t}$

19. $y = 3t^2 + \dfrac{12}{\sqrt{t}} - \dfrac{1}{t^2}$

20. $y = z^2 + \dfrac{1}{2z}$

21. $y = ax^2 + bx + c$
     (a, b, c are constants)

22. Let $f(t) = t^2 - 4t + 5$.
     (a) Find $f'(t)$.
     (b) Find $f'(1)$ and $f'(2)$.
     (c) Use a graph of $f(t)$ to check that your answers to part (b) are reasonable. Explain.

23. Let $f(x) = x^3 - 4x^2 + 7x - 11$. Find $f'(0)$, $f'(2)$, and $f'(-1)$.

24. Let $f(x) = x^2 + 3x - 5$. Find $f'(0)$, $f'(3)$, and $f'(-2)$.

25. (a) Use a graph of $P(q) = 6q - q^2$ to determine whether each of the following derivatives is positive, negative, or zero: $P'(1)$, $P'(3)$, $P'(4)$. Explain.
    (b) Find $P'(q)$ and then find the three derivatives in part (a).

26. A population has size $P(t) = t^3 + 4t + 1$ at time t. Find the rate of change of the population at time $t = 2$.

27. Let $f(x) = x^3 - 2x^2 + 3x + 2$.
    (a) Find $f'(x)$.
    (b) Find $f'(-1)$, $f'(0)$, $f'(1)$, and $f'(2)$.
    (c) Sketch a graph of $f(x)$. Consider the slope of a tangent line to the graph of this function at each of the points $x = -1, x = 0, x = 1$, and $x = 2$. Do these slopes appear to match your answers to part (b)? Explain.

28. If $f(t) = 2t^3 - 4t^2 + 3t - 1$, find $f'(t)$ and $f''(t)$.

29. If $f(t) = t^4 - 3t^2 + 5t$, find $f'(t)$ and $f''(t)$.

30. (a) Find the *eighth* derivative of $f(x) = x^7 + 5x^5 - 4x^3 + 6x - 7$. Think ahead! (The nth derivative is the result of differentiating n times.)
    (b) Find the seventh derivative of $f(x)$.

31. Find the equation of the line tangent to the graph of f at $(1, 1)$, where f is given by $f(x) = 2x^3 - 2x^2 + 1$.

32. Find the equation of the line tangent to the graph of $f(x) = 2x^3 - 5x^2 + 3x - 5$ at the point where $x = 1$.

33. Find the equation of the line tangent to the graph of $f(t) = 6t - t^2$ at the point where $t = 4$. Sketch the graph of $f(t)$ and the tangent line on the same axes.

34. A manufacturer has cost function $C(q) = 1000 + 2q^2$ where q is the quantity produced. Find the marginal cost of producing the 25th item, and interpret your answer in terms of costs of production.

35. The demand function for a certain product is given by $q = 300 - 3p$, where p is the price of the product and q is the quantity consumers will buy at that price. Recall that revenue equals price times quantity sold.
    (a) Write the revenue function as a function of price.
    (b) Find the marginal revenue when the price of the product is $10, and interpret your answer in terms of revenues.
    (c) For what prices is the marginal revenue positive? For what prices is it negative?

36. Zebra mussels are freshwater shellfish that attach themselves to anything they can find. They first appeared in the St. Lawrence River in the early 1980s. They are moving upriver, and may spread throughout the Great Lakes. Suppose that in one small bay, the number of zebra mussels at time t is given by $Z(t) = 300t^2$, where t is measured in months since the zebra mussel first appeared at that location. How many zebra mussels are in the bay after four months, and at what rate is the population growing? Give units with your answers.

37. The yield, Y, of an apple orchard (measured in bushels of apples per acre) is a function of the amount x of fertilizer in pounds used per acre. Suppose

$$Y = f(x) = 320 + 140x - 10x^2$$

   (a)   What is the yield if 5 pounds of fertilizer is used per acre?
   (b)   Find $f'(5)$. Give units with your answer and interpret it in terms of apples and fertilizer.
   (c)   Given your answer to part (b), should more or less fertilizer be used? Explain.

38.   The demand equation for a product is given by

$$p = f(q) = 50 - 0.03q^2,$$

where p is the price per unit (in dollars) and q is measured in thousands of units sold at that price.

   (a)   Find the p- and q-intercepts for this function and interpret them in terms of demand for this product.
   (b)   Find $f(20)$ and give units with your answer. Explain what it tells you in terms of demand for this product.
   (c)   Find $f'(20)$ and give units with your answer. Explain what it tells you in terms of demand for this product.

39.   A cost function (in dollars) is given by $C(q) = 0.08q^3 + 75q + 1000$, where q is the number of items produced.

   (a)   Find the marginal cost function.
   (b)   Find $C(50)$ and $C'(50)$. Give units with your answers and explain what they are telling you in terms of costs of production.

40.   The graph of the equation $y = x^3 - 9x^2 - 16x + 1$ has a slope equal to 5 at exactly two points. Find the coordinates of the points.

41.   Let $f(x) = x^3 - 12x$.

   (a)   Use $f'(x)$ to determine the intervals on which $f(x)$ is decreasing.
   (b)   Use $f''(x)$ to determine the intervals on which the graph of $f(x)$ is concave down.
   (c)   On what intervals is $f(x)$ both decreasing and its graph concave down?

42.   If the demand equation is linear, we can write $p = b + mq$, where p is the price of the product, q is the quantity sold at that price, and b and m are constants.

   (a)   Write the revenue function as a function of quantity sold.
   (b)   Find the marginal revenue function.

43.   A ball is dropped from the top of the Empire State building to the ground below. The height, y, of the ball above the ground (in feet) is given as a function of time, t, (in seconds) by

$$y = 1250 - 16t^2.$$

   (a)   Find the velocity of the ball at time t. What is the sign of the velocity? Why is this to be expected?
   (b)   When does the ball hit the ground, and how fast is it going at that time? Give your answer in feet per second and in miles per hour (1 ft/sec = 15/22 mph).

44.   (a)   Use the formula for the area of a circle of radius r, $A = \pi r^2$, to find $\dfrac{dA}{dr}$.
   (b)   The result from part (a) should look familiar. What does $\dfrac{dA}{dr}$ represent geometrically? Draw a picture.
   (c)   Use the difference quotient to explain the observation you made in part (b).

45. What is the formula for $V(r)$, the volume of a sphere of radius r? Find $\dfrac{dV}{dr}$. What is the geometrical meaning of $\dfrac{dV}{dr}$?

5.2 EXPONENTIAL AND LOGARITHMIC FUNCTIONS

The Exponential Function

What would you expect the graph of the derivative of the exponential function $f(x) = a^x$ to look like? The graph of the exponential function is shown in Figure 5.12. The function increases slowly when $x < 0$ and more rapidly for $x > 0$, so the values of f' are small but positive for $x < 0$ and larger for $x > 0$. Since the function is increasing for all real values of x, the graph of the derivative must lie above the x-axis. After some reflection, you may decide that the graph of f' must resemble the graph of f itself. We will see how this observation holds for $f(x) = 2^x$ and $g(x) = 3^x$.

The Derivatives of *2ˣ* and *3ˣ*

Let $f(x) = 2^x$. We suspect that the graph of the derivative of this function will be similar to the graph of f. The graphs of $f(x) = 2^x$ and and its derivative f' are shown together in Figure 5.13. The graph of the derivative f' lies below the graph of f, and it appears to be a vertical shrinking of the graph of f, which suggests that the derivative f' is proportional to f.

In Chapter 2, we saw that the derivative of $f(x) = 2^x$ at $x = 0$ is given by

$$f'(0) \approx \frac{f(0.001) - f(0)}{0.001 - 0} = \frac{2^{0.001} - 2^0}{0.001} = \frac{1.000693 - 1}{0.001} = 0.693.$$

The constant of proportionality appears to be 0.693. Indeed, we will show in Section 5.6 that

$$\frac{d}{dx}(2^x) = f'(x) \approx (0.693)2^x.$$

Graphs of $g(x) = 3^x$ and its derivative g' are shown in Figure 5.14. Notice that the graph of g' lies just above the graph of g, and it appears to be a vertical stretching of the graph of g, which suggests that the derivative g' is proportional to g. With a difference quotient, we can show that $g'(0) \approx 1.0986$, so the constant of proportionality is about 1.0986. We have the following:

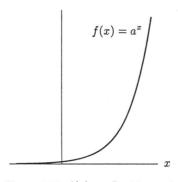

Figure 5.12: $f(x) = a^x$, with $a > 1$

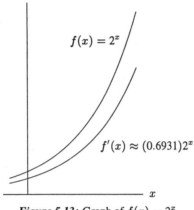

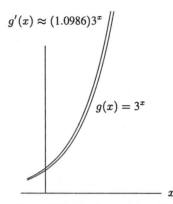

Figure 5.13: Graph of $f(x) = 2^x$
and its derivative

Figure 5.14: Graph of $g(x) = 3^x$
and its derivative

$$\frac{d}{dx}(3^x) = g'(x) \approx (1.0986)3^x.$$

The Derivative of e^x

We have seen that the derivatives of 2^x and 3^x are multiples of the original functions 2^x and 3^x. Is there an exponential function which equals its own derivative exactly? In Figure 5.13, we see that the graph of the derivative of $f(x) = 2^x$ lies below the graph of $f(x)$. In Figure 5.14, we see that the graph of the derivative of $f(x) = 3^x$ lies above the graph of $f(x)$. Therefore, if there is an exponential function for which the derivative exactly matches the original function, we would expect the base to be between 2 and 3. If you experiment with different bases, you will find that a base, a, of approximately 2.7 gives an exponential function a^x whose derivative almost exactly matches the original function, a^x. In fact, the base $e = 2.718\ldots$ has the remarkable property that its exponential function is its own derivative.

$$\frac{d}{dx}(e^x) = e^x$$

Figure 5.15 shows the graph of the derivative of 2^x below the graph of the function, and the graph of the derivative of 3^x above the graph of the function. With $e \approx 2.718$, the function e^x and its derivative are identical.

The Derivative of a^x, where a is a constant

It turns out that the constants involved in the derivatives of 2^x and 3^x are natural logarithms. In fact, since $0.6931 \approx \ln 2$ and $1.0986 \approx \ln 3$, this suggests that

$$\frac{d}{dx}(2^x) = (\ln 2)2^x \qquad \text{and} \qquad \frac{d}{dx}(3^x) = (\ln 3)3^x.$$

In Section 5.6, we will show that, in general,

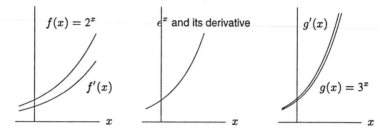

Figure 5.15: Graphs of the functions 2^x, e^x, and 3^x and their derivatives

For $a > 0$,
$$\frac{d}{dx}(a^x) = (\ln a)a^x.$$

Example 1   Differentiate $2 \cdot 4^x + 5e^x$.

Solution

$$\frac{d}{dx}(2 \cdot 4^x + 5e^x) = \frac{d}{dx}(2 \cdot 4^x) + \frac{d}{dx}(5e^x)$$
$$= 2\frac{d}{dx}(4^x) + 5\frac{d}{dx}(e^x)$$
$$= 2(\ln 4)4^x + 5e^x$$
$$\approx (2.7726)4^x + 5e^x.$$

The Derivative of $\ln x$

What would you expect the graph of the derivative of the logarithmic function $f(x) = \ln x$ to look like? A graph of the logarithmic function is shown in Figure 5.16. (Recall that the domain of the natural logarithm function is $x > 0$.) The function is increasing everywhere, so we expect the derivative to be positive for all $x > 0$. The graph of the logarithmic function is concave down everywhere, and so we expect the derivative to be decreasing for all $x > 0$. What else can we learn? The slope of the logarithmic function is very large near $x = 0$, and is very small for large x, so we expect the graph of the derivative to tend to $+\infty$ for x near 0 and to tend to 0 for very large x. A graph of the derivative of $f(x) = \ln x$ is shown in Figure 5.17.

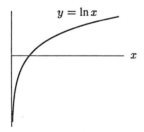

Figure 5.16: A graph of $y = \ln x$.

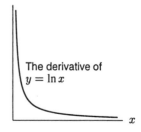

Figure 5.17: The derivative of $y = \ln x$.

The graph in Figure 5.17 should remind you of the function $y = 1/x$, since this is exactly what $f'(x)$ turns out to be. We will see an algebraic justification for this rule in Section 5.6.

$$\frac{d}{dx}(\ln x) = \frac{1}{x}.$$

Example 2   Differentiate $y = 5\ln t + 7e^t - 4t^2 + 12$.

Solution

$$\frac{d}{dt}(5\ln t + 7e^t - 4t^2 + 12) = 5\frac{d}{dt}(\ln t) + 7\frac{d}{dt}(e^t) - 4\frac{d}{dt}(t^2) + \frac{d}{dt}(12)$$

$$= 5\left(\frac{1}{t}\right) + 7(e^t) - 4(2t) + 0$$

$$= \frac{5}{t} + 7e^t - 8t$$

Using the Derivative Formulas

Example 3   In Chapter 1, we saw that the population of Mexico, P, can be modeled by

$$P = 67.39(1.026)^t,$$

where P is millions of people, and t is years since the start of 1980. At what rate is the population growing at the beginning of 1997? Give units with your answer.

Solution   The instantaneous rate of growth is the derivative, so we are being asked to find $\frac{dP}{dt}$ when $t = 17$. We first find the derivative function:

$$\frac{dP}{dt} = \frac{d}{dt}(67.39(1.026)^t) = 67.39(\ln 1.026)(1.026)^t$$

$$\approx 67.39(0.02567)(1.026)^t$$

$$\approx 1.730(1.026)^t.$$

At $t = 17$, we have

$$1.730(1.026)^{17} \approx 1.730(1.547) \approx 2.676.$$

The population of Mexico is growing at an instantaneous rate of about 2.676 million people per year at the start of 1997. (This means that the population is increasing at a rate of about 7300 people per day.)

Example 4   Find the equation of the tangent line to the graph of $f(x) = \ln x$ at the point where $x = 2$. Draw a graph with $f(x)$ and the tangent line on the same axes.

Solution   The graph of $f(x) = \ln x$ is shown in figure 5.16 on page 279, and we see that the derivative is positive at $x = 2$. To find the equation of the tangent line, we need to know a point on the line and the slope. The point with x-coordinate 2 has y-coordinate equal to $\ln 2 = 0.693$, so the point is $(2, 0.693)$.

Since $f'(x) = 1/x$, the slope at $x = 2$ equals $f'(2) = 1/2 = 0.5$.

Figure 5.18: Graph of $f(x) = \ln x$ and a tangent line

The slope is 0.5, and a point on the line is $(2, 0.693)$. Substituting into the equation for a line $y = b + mx$, we solve for b:

$$y = b + mx$$
$$0.693 = b + (0.5)(2)$$
$$0.693 = b + 1$$
$$-0.307 = b.$$

The equation of the tangent line is $y = -0.307 + 0.5x$. The graph of $f(x) = \ln x$ and this tangent line are shown in Figure 5.18.

Problems for Section 5.2

Differentiate the functions in Problems 1–20. You may assume that A, B, and C, where they appear, are constants.

1. $y = 5t^2 + 4e^t$

2. $f(x) = 2e^x + x^2$

3. $f(x) = 2^x + 2 \cdot 3^x$

4. $y = 4 \cdot 10^x - x^3$

5. $y = 3x - 2 \cdot 4^x$

6. $y = \dfrac{3^x}{3} + \dfrac{33}{\sqrt{x}}$

7. $f(x) = x^3 + 3^x$

8. $y = 5 \cdot 5^t + 6 \cdot 6^t$

9. $P(t) = Ce^t.$

10. $D = 10 - \ln p$

11. $R = 3 \ln q$

12. $y = t^2 + 5 \ln t$

13. $y = B + Ae^t$

14. $f(x) = Ae^x - Bx^2 + C$

15. $P = 3t^3 + 2e^t$

16. $P(t) = 3000(1.02)^t$

17. $P(t) = 12.41(0.94)^t$

18. $y = 5(2^x) - 5x + 4$

19. $R(q) = q^2 - 2 \ln q$

20. $y = x^2 + 4x - 3 \ln x$

21. Let $f(t) = 4 - 2e^t$. Find $f'(-1)$, $f'(0)$, and $f'(1)$. Sketch a graph of $f(t)$, and draw tangent line segments at $t = -1$, $t = 0$, and $t = 1$. Do the slopes of the line segments appear to match the derivatives you found? Explain.

22. Let $y = 3^x$. Find the equation of the tangent line to the graph of this function at the point where $x = 1$. Check your work by sketching a graph of the function and the tangent line on the same coordinate system.

23. Suppose the cost function for a certain product is given by $C = 1000 + 300 \ln q$, where q is the quantity produced. Find the cost and the marginal cost at a production level of 500. Interpret your answers in economic terms.

24. (a) Find the slope of the graph of $f(x) = 1 - e^x$ at the point where it crosses the x-axis.
    (b) Find the equation of the tangent line to the curve at this point.

25. With an inflation rate of 5%, prices are described by

$$P = P_0(1.05)^t$$

where P_0 is the price in dollars when $t = 0$ and t is time in years. Suppose $P_0 = 1$. How fast (in cents/year) is the price of the good rising when $t = 10$?

26. Since January 1, 1960, the population of Slim Chance has been described by the formula

$$P = 35{,}000(0.98)^t$$

where P is the population of the city t years after 1960. At what rate was the population of the city changing on January 1, 1983?

27. Certain pieces of antique furniture increased very rapidly in price in the 1970s and 1980s. For example, the value of a particular rocking chair is well approximated by

$$V = 75(1.35)^t,$$

where V is in dollars and t is the number of years since 1975. Find the rate, in dollars per year, that the price is increasing as a function of time.

28. The *Global 2000 Report* gave the world's population, P, as 4.1 billion in 1975 and growing at 2% annually.
    (a) Give a formula for P in terms of time, t, measured in years since 1975.
    (b) Find $\dfrac{dP}{dt}, \dfrac{dP}{dt}\Big|_{t=0}$, and $\dfrac{dP}{dt}\Big|_{t=15}$. What do each of these represent in practical terms?

29. Hungary is one of the few countries in the world where the population is decreasing, currently at about 0.2% a year. Thus, if t is time in years since 1990, the population P, in millions, of Hungary can be approximated by

$$P = 10.8(0.998)^t.$$

    (a) What does this model predict the population of Hungary will be in the year 2000?
    (b) How fast (in people/year) does this model predict Hungary's population will be decreasing in the year 2000?

30. Using the equation of the tangent line to the graph of e^x at $x = 0$, show that

$$e^x \geq 1 + x$$

for all values of x. A sketch will be helpful.

31. (a) Find the equation of the tangent line to $y = \ln x$ at $x = 1$.
    (b) Use it to calculate approximate values for $\ln(1.1)$ and $\ln(2)$.
    (c) Using a graph, explain whether the approximate values you have calculated are smaller or larger than the true values. Would the same result have held if you had used the tangent line to estimate $\ln(0.9)$ and $\ln(0.5)$? Why?

32. In this section, we stated that, for $a > 0$

$$\frac{d}{dx}(a^x) = (\ln a)a^x.$$

Use this expression for the derivative to explain for which values of a the function a^x is increasing and for which values it is decreasing.

33. Find all solutions of the equation

$$2^x = 2x.$$

How do you know that you found all solutions?

34. Find the value of c in Figure 5.19, where the line l tangent to the graph of $y = 2^x$ at $(0, 1)$ intersects the x-axis.

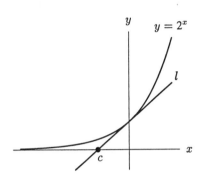

Figure 5.19

35. Find the quadratic polynomial $g(x) = ax^2 + bx + c$ which best fits the function $f(x) = e^x$ at $x = 0$, in the sense that

$$g(0) = f(0), \quad \text{and} \quad g'(0) = f'(0), \quad \text{and} \quad g''(0) = f''(0).$$

Using a computer or calculator, sketch graphs of f and g on the same axes. What do you notice?

5.3 THE CHAIN RULE

Composite Functions

A *composite function* is a "function of a function." One function is inside another function. There is an *inside function*, representing the calculation which is done first and an *outside function*, representing the calculation done second. Many functions are composite functions. One example is

$$f(t) = (t + 1)^4.$$

What is the inside function and what is the outside function? To find $f(2)$, we first add one ($2+1 = 3$) and then raise the sum to the 4th power ($3^4 = 81$). The inside function is $t + 1$ and the outside function is raising the argument to the fourth power. We have

$$f(2) = (2 + 1)^4 \quad = \underset{\substack{\text{first} \\ \text{calculation}}}{\nwarrow} 3^4 \quad = \underset{\substack{\text{second} \\ \text{calculation}}}{\nwarrow} 81.$$

We can separate the inside function and the outside function by using a new variable such as u to represent the inside function:

$$y = (t+1)^4 \quad \text{is the same as} \quad y = u^4 \quad \text{with} \quad u = t+1.$$

Composite functions such as this occur frequently in practice. Be sure you are comfortable in recognizing the inside function.

Example 1   Use a new variable u for the inside function to express each of the following as a composite function:
    (a) $y = \ln(3t)$ (b) $P = e^{-0.03t}$ (c) $w = 5(2r+3)^2$

Solution   (a)   The inside function is $3t$, so we have $y = \ln u$ with $u = 3t$.
    (b)   The inside function is $-0.03t$, so we have $P = e^u$ with $u = -0.03t$.
    (c)   The inside function is $2r+3$, so we have $w = 5u^2$ with $u = 2r+3$.

The Derivative of Composite Functions

Suppose that gas costs $1.20 per gallon and that we are using gas at a rate of 3 gallons per minute. What is the rate of change of cost with respect to time, in dollars per minute? Since we are using 3 gallons every minute and since each gallon costs $1.20, we are spending 3 times $1.20, or $3.60 every minute. We have

$$\text{Rate of change of cost} = \left(1.20\frac{\text{dollars}}{\text{gallon}}\right) \times \left(3\frac{\text{gallons}}{\text{minute}}\right) = 3.60 \text{ dollars per minute.}$$

Notice that the rate of change (in dollars per minute) is a product of two rates of change (dollars per gallon times gallons per minute). Since a rate of change is a derivative, we can write this in terms of derivatives.
    Since gas costs $1.20 per gallon, then total cost C (in dollars) is given by

$$C = 1.20g,$$

where g is the number of gallons of gas we use. Since we are using gas at a rate of 3 gallons per minute, then the amount of gas used is given by

$$g = 3t,$$

where t is the number of minutes. If we want to find the rate of change of cost with respect to time (dC/dt), we multiply the rate of change of cost with respect to gallons (dC/dg) by the rate of change of gallons with respect to time (dg/dt). We have

rate of change (dollars/minute) = rate of change (dollars/gallon) $\cdot$ rate of change (gallons/minute)

$$\frac{dC}{dt} = \frac{dC}{dg} \cdot \frac{dg}{dt}$$
$$= (1.20) \cdot (3)$$
$$= 3.60 \text{ dollars/minute.}$$

When can we expect one derivative to be computed as the product of two other derivatives?

The Chain Rule

Suppose y is a composite function of t. Then we can write $y = f(u)$ with $u = g(t)$ for some inside function g and outside function f. A small change in t, called Δt, generates a small change in u, called Δu. In turn, Δu generates a small change in y, called Δy. Provided Δt and Δu are not zero, we can say

$$\frac{\Delta y}{\Delta t} = \frac{\Delta y}{\Delta u} \cdot \frac{\Delta u}{\Delta t}.$$

Since the derivative $\dfrac{dy}{dt}$ is the limit of the quotient $\dfrac{\Delta y}{\Delta t}$ as Δt goes to zero, we have

The Chain Rule

$$\frac{dy}{dt} = \frac{dy}{du} \cdot \frac{du}{dt}.$$

In using the chain rule to find derivatives, it is necessary to be able to identify the inside function and the outside function in a composite function.

Using the Chain Rule to Find Derivatives

We learned the following derivative rules in earlier sections:

$$\frac{d}{dt}(t^n) = nt^{n-1} \qquad \frac{d}{dt}(e^t) = e^t \qquad \frac{d}{dt}(\ln t) = \frac{1}{t}.$$

These rules (and all others we will learn) can be extended to composite functions using the chain rule. We have the following:

If u is a function of t, then

$$\frac{d}{dt}(u^n) = nu^{n-1}\frac{du}{dt}$$

$$\frac{d}{dt}(e^u) = e^u\frac{du}{dt}$$

$$\frac{d}{dt}(\ln u) = \frac{1}{u}\frac{du}{dt}$$

Example 2   Find the derivative of the following functions: (a) $(4t^2 + 1)^7$,   (b) e^{3t}.

Solution   (a)   Here $u = 4t^2 + 1$ is the inside function; $f(u) = u^7$ is the outside function. Then

$$\frac{d}{dt}((4t^2 + 1)^7) = \frac{d}{dt}(u^7) = 7u^6\frac{du}{dt} = 7(4t^2 + 1)^6(8t) = 56t(4t^2 + 1)^6.$$

$$\text{since } \tfrac{du}{dt} = 8t$$

(b)   Let $u = 3t$ and $f(u) = e^u$. Then

$$\frac{d}{dt}(e^{3t}) = \frac{d}{dt}(e^u) = e^u\frac{du}{dt} = e^{3t} \cdot 3 = 3e^{3t}.$$

$$\text{since } \tfrac{du}{dt} = 3$$

Example 3   Differentiate (a) $(3x^3 - x)^5$,   (b) $\ln(q^2 + 1)$,   (c) e^{-x^2}

Solution   (a)   Here, $u = 3x^3 - x$ and $f(u) = u^5$, and so the derivative is

$$\frac{d}{dx}(3x^3 - x)^5 = 5(3x^3 - x)^4(9x^2 - 1).$$

(b)   We have $u = q^2 + 1$ and $f(u) = \ln u$, and so the derivative is

$$\frac{d}{dq}(\ln(q^2 + 1)) = \frac{1}{q^2 + 1}(2q).$$

(c)   Since $u = -x^2$ and $f(u) = e^u$, the derivative is

$$\frac{d}{dx}(e^{-x^2}) = e^{-x^2}(-2x) = -2xe^{-x^2}.$$

Notice in Example 2 (b) that the derivative of e^{3t} is $3e^{3t}$. As we have seen, functions of the form e^{kt} where k is a constant appear frequently in applications. What is the derivative of e^{kt}? We have $u = kt$, and so $du/dt = k$. Thus

---

If k is a constant,

$$\frac{d}{dt}(e^{kt}) = ke^{kt}.$$

Example 4   Find the derivative of $P = 5 + 3x^2 - 7e^{-0.2x}$.

Solution   The derivative is

$$\frac{dP}{dx} = 0 + 3(2x) - 7(-0.2e^{-0.2x}) = 6x + 1.4e^{-0.2x}.$$

Example 5   A payment of $1000 is made into an account which pays 8% annual interest, compounded continuously.
(a)   Find a formula $f(t)$ for the amount in the account at time t, where t is measured in years since the money was deposited.
(b)   Find $f(10)$ and $f'(10)$ and interpret your answers.

Solution   (a)   The amount of money in the account is $P = f(t) = 1000e^{0.08t}$.
(b)   We have

$$f(10) = 1000e^{(0.08)(10)} = 2225.54.$$

The account contains $2225.54 after 10 years.
To find $f'(10)$, we first note that $f'(t) = 1000(0.08e^{0.08t}) = 80e^{0.08t}$. Therefore,

$$f'(10) = 80e^{(0.08)(10)} = 178.04.$$

After 10 years, the account is growing at the rate of about 178 dollars per year.

Problems for Section 5.3

1. Use a new variable u for the inside function to express each of the following as a composite function: (a) $y = (5t^2 - 2)^6$ (b) $P = 12e^{-0.6t}$ (c) $C = 12\ln(q^3 + 1)$

2. Use a new variable u for the inside function to express each of the following as a composite function: (a) $y = 2^{3x-1}$ (b) $P = \sqrt{5t^2 + 10}$ (c) $w = 2\ln(3r + 4)$

Find the derivative of the functions in Problems 3–32.

3. $f(x) = (x + 1)^{99}$

4. $R = (q^2 + 1)^4$

5. $w = (t^2 + 1)^{100}$

6. $w = (t^3 + 1)^{100}$

7. $w = (5r - 6)^3$

8. $f(t) = e^{3t}$

9. $y = e^{0.7t}$

10. $y = e^{-4t}$

11. $y = \sqrt{s^3 + 1}$

12. $w = e^{\sqrt{s}}$

13. $P = e^{-0.2t}$

14. $w = e^{-3t^2}$

15. $y = \ln(5t + 1)$

16. $P = 50e^{-0.6t}$

17. $P = 200e^{0.12t}$

18. $y = 12 - 3x^2 + 2e^{3x}$

19. $C = 12(3q^2 - 5)^3$

20. $f(x) = 6e^{5x} + e^{-x^2}$

21. $y = 5e^{5t+1}$

22. $f(x) = \ln(1 - x)$

23. $f(t) = \ln(t^2 + 1)$

24. $f(x) = \ln(1 - e^{-x})$

25. $f(x) = \ln(e^x + 1)$

26. $f(t) = 5\ln(5t + 1)$

27. $g(t) = \ln(4t + 9)$

28. $y = 5 + \ln(3t + 2)$

29. $Q = 100(t^2 + 5)^{0.5}$

30. $y = 5x + \ln(x + 2)$

31. $y = (5 + e^x)^2$

32. $P = (1 + \ln x)^{0.5}$

33. Find the equation of the tangent line to the graph of

$$y = e^{-2t}$$

at $t = 0$. Check your work by sketching the graphs of $y = e^{-2t}$ and the tangent line on the same coordinate system.

34. Find the equation of the tangent line at $x = 1$ to $y = f(x)$ where $f(x)$ is the function in Problem 20.

35. Assume that the demand function for a certain product is given by

$$q = f(p) = 10,000e^{-0.25p},$$

where q is the quantity sold and p is the price of the product, in dollars. Find $f(2)$ and $f'(2)$. Explain in economic terms what information each of these answers gives you.

36. Assume the cost function for a certain product is

$$C(q) = 1000 + 30e^{0.05q}$$

where q is the quantity produced. Find the cost and the marginal cost when $q = 50$. Explain in economic terms what information each of these answers gives you.

37. One gram of radioactive carbon-14 decays according to the formula

$$Q = e^{-0.000121t}$$

where Q is the number of grams of carbon-14 remaining after t years.

   (a) Find the rate at which carbon-14 is decaying (in grams/year).
   (b) Sketch the rate you found in part (a) against time.

38. The temperature, H, in degrees Fahrenheit (°F), of a can of soda that is put into a refrigerator to cool is given as a function of time, t, in hours, by

$$H = 40 + 30e^{-2t}.$$

   (a) Find the rate at which the temperature of the soda is changing (in °F/hour).
   (b) What is the sign of $\dfrac{dH}{dt}$? Why?
   (c) When, for $t \geq 0$, is the magnitude of $\dfrac{dH}{dt}$ largest? In terms of the can of soda, why is this?

39. If you invest P dollars in a bank account at an annual interest rate of $r\%$, after t years you will have B dollars, where

$$B = P\left(1 + \frac{r}{100}\right)^t.$$

   (a) Find $\dfrac{dB}{dt}$, assuming P and r are constant. In terms of money, what does $\dfrac{dB}{dt}$ represent?

   (b) Find $\dfrac{dB}{dr}$, assuming P and t are constant. In terms of money, what does $\dfrac{dB}{dr}$ represent?

5.4 THE PRODUCT RULE

You now know how to find derivatives of powers and exponentials, of sums and constant multiples of functions, and of compositions of functions. This section will show you how to find the derivatives of products of functions.

Suppose we know the derivatives of $f(x)$ and $g(x)$ and want to calculate the derivative of the product, $f(x)g(x)$. We start by looking at an example. Suppose $f(x) = 3x + 1$ and $g(x) = 5x - 4$. Then $f'(x) = 3$ and $g'(x) = 5$. We first find the product

$$f(x)g(x) = (3x + 1)(5x - 4)$$
$$= 15x^2 - 7x - 4.$$

Thus, the derivative of the product is $30x - 7$. Notice that the derivative of the product is *not* equal to the product of the derivatives, since $f'(x)g'(x) = (3)(5) = 15$. In general, we have the following rule, which will be justified in Section 5.6.

The Product Rule

$$(fg)' = f'g + fg'.$$

In words:

The derivative of a product is the derivative of the first factor multiplied by the second, plus the first factor multiplied by the derivative of the second.

In the alternative notation, if $u = f(x)$ and $v = g(x)$, we have the following:

The Product Rule

$$\frac{d(uv)}{dx} = \frac{du}{dx} \cdot v + u \cdot \frac{dv}{dx}$$

We verify this rule for the example above with $f(x) = 3x + 1$ and $g(x) = 5x - 4$. Using the product rule, we see that the derivative of $f(x)g(x)$ is

$$\begin{aligned} f'(x)g(x) + f(x)g'(x) &= (3)(5x - 4) + (3x + 1)(5) \\ &= (15x - 12) + (15x + 5) \\ &= 30x - 7. \end{aligned}$$

We see that the formula for the product rule gives us the same answer as first multiplying out the functions and then finding the derivative.

Example 1   Differentiate (a) $x^2 e^{2x}$ (b) $t^3 \ln(t + 1)$ (c) $(3x^2 + 5x)e^x$

Solution   (a)

$$\begin{aligned} \frac{d}{dx}(x^2 e^{2x}) &= \frac{d}{dx}(x^2) \cdot e^{2x} + x^2 \frac{d}{dx}(e^{2x}) \\ &= (2x)e^{2x} + x^2(2e^{2x}) \\ &= 2xe^{2x} + 2x^2 e^{2x}. \end{aligned}$$

(b)

$$\begin{aligned} \frac{d}{dt}(t^3 \ln(t+1)) &= \frac{d}{dt}(t^3) \cdot \ln(t+1) + t^3 \frac{d}{dt}(\ln(t+1)) \\ &= (3t^2)\ln(t+1) + t^3\left(\frac{1}{t+1}\right) \\ &= 3t^2 \ln(t+1) + \frac{t^3}{t+1}. \end{aligned}$$

(c)

$$\begin{aligned} \frac{d}{dx}((3x^2 + 5x)e^x) &= \left(\frac{d}{dx}(3x^2 + 5x)\right)e^x + (3x^2 + 5x)\frac{d}{dx}(e^x) \\ &= (6x + 5)e^x + (3x^2 + 5x)e^x \\ &= (3x^2 + 11x + 5)e^x. \end{aligned}$$

Example 2   Differentiate

(a) $5xe^{x^2}$

(b) $t^2(3t+1)^3$

(c) $x\ln x$

Solution   (a)

$$\frac{d}{dx}(5xe^{x^2}) = \left(\frac{d}{dx}(5x)\right)e^{x^2} + 5x\frac{d}{dx}(e^{x^2}) = (5)e^{x^2} + 5x(e^{x^2}\cdot 2x)$$

$$= 5e^{x^2} + 10x^2e^{x^2}.$$

(b)

$$\frac{d}{dt}(t^2(3t+1)^3) = \frac{d}{dt}(t^2)\cdot(3t+1)^3 + t^2\frac{d}{dt}((3t+1)^3)$$

$$= (2t)(3t+1)^3 + t^2(3(3t+1)^2\cdot 3)$$

$$= 2t(3t+1)^3 + 9t^2(3t+1)^2$$

(c)

$$\frac{d}{dx}(x\ln x) = \left(\frac{d}{dx}(x)\right)\ln x + x\frac{d}{dx}(\ln x) = 1\cdot\ln x + x\cdot\frac{1}{x}$$

$$= \ln x + 1.$$

Example 3   Find the derivative of $C = \dfrac{e^{2t}}{t}$.

Solution   We write $C = e^{2t}t^{-1}$ and use the product rule:

$$\frac{d}{dt}(e^{2t}t^{-1}) = \frac{d}{dt}(e^{2t})\cdot t^{-1} + e^{2t}\frac{d}{dt}(t^{-1})$$

$$= (2e^{2t})\cdot t^{-1} + e^{2t}(-1t^{-2})$$

$$= \frac{2e^{2t}}{t} - \frac{e^{2t}}{t^2}.$$

Example 4   A demand curve for a product is modeled by the equation $p = 80e^{-0.003q}$, where p is the price of the product in dollars and q is the quantity sold at that price.

(a) How many items are sold when the price is $75? How many items are sold when the price is $10? What is the revenue at each of these prices?

(b) Find the revenue function, as a function of quantity sold.

(c) Find the marginal revenue function.

Solution   (a) We can use the graph of the demand curve in Figure 5.20 to find q for various values of p.
We see that when $p = 75$, the quantity sold is about 20. When $p = 10$, the quantity sold is about 700. As we would expect, when the price is lower, more products are sold.
Since revenue = price times quantity sold, when the price is $75, we have

$$R = p\cdot q = (75)(20) = \$1500.$$

When the price is $10, the revenue is

$$R = p\cdot q = (10)(700) = \$7000.$$

We see that revenue is much higher at the lower price, since the demand is so much higher.

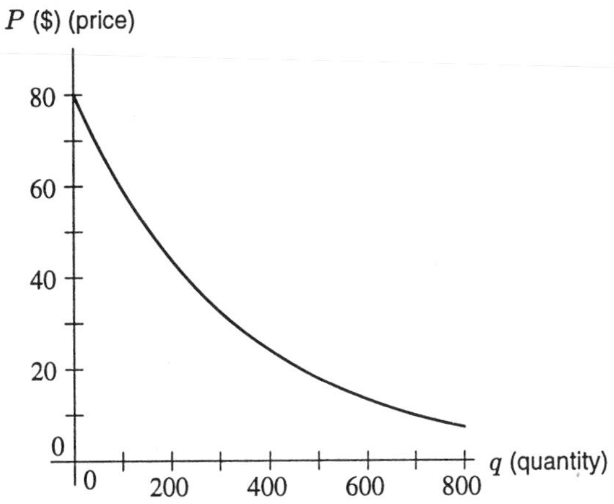

Figure 5.20: Demand curve: $p = 80e^{-0.003q}$

(b)   We have $R = pq = (80e^{-0.003q})q = 80qe^{-0.003q}$.

(c)   The marginal revenue function is the derivative of the revenue function, so we use the product rule. We have

$$\text{Marginal Revenue} = \frac{d}{dq}(80qe^{-0.003q})$$

$$= \left(\frac{d}{dq}(80q)\right)e^{-0.003q} + 80q\left(\frac{d}{dq}(e^{-0.003q})\right)$$

$$= (80)e^{-0.003q} + 80q(-0.003e^{-0.003q})$$

$$= 80e^{-0.003q} - 0.24qe^{-0.003q}.$$

Problems for Section 5.4

1.   If $f(x) = x^2(x^3 + 5)$, find $f'(x)$ two ways: by using the product rule and by multiplying out before taking the derivative. Do you get the same result? Should you?

2.   If $f(x) = (2x+1)(3x-2)$, find $f'(x)$ two ways: by using the product rule and by multiplying out. Do you get the same result?

For Problems 3–17, find the derivative. In some cases, it may be to your advantage to simplify first.

3. $f(x) = xe^x$ 4. $f(t) = te^{-2t}$ 5. $y = x \cdot 2^x$

6. $w = (t^3+5t)(t^2-7t+2)$ 7. $y = (t^2 + 3)e^t$ 8. $z = (3t + 1)(5t + 2)$

9. $y = (t^3 - 7t^2 + 1)e^t$ 10. $P = t^2 \ln t$ 11. $f(x) = \dfrac{x^2 + 3}{x}$

12. $R = 3qe^{-q}$ 13. $y = te^{-t^2}$ 14. $f(z) = \sqrt{z}e^{-z}$

15. $g(p) = p\ln(2p + 1)$ 16. $f(t) = te^{5-2t}$ 17. $f(w) = (5w^2 + 3)e^{w^2}$

18.   Find the equation of the tangent line to the graph of $f(x) = x^2e^{-x}$ at $x = 0$. Check your work by graphing this function and the tangent line on the same coordinate system.

19. Assume the demand function for a certain product is given by

$$q = 1000e^{-0.02p}$$

where p is the price of the product in dollars and q is the quantity sold at that price.
   (a) Write revenue, R, as a function of price.
   (b) Find the rate of change of revenue with respect to price.
   (c) Find the revenue and rate of change of revenue with respect to price when the price is $10, and interpret your answers in economic terms.

20. A drug concentration curve is given by $C = f(t) = 20te^{-0.04t}$, with C in mg/ml and t in minutes.

   (a) Sketch a graph of C against t. Is $f'(15)$ positive or negative? Is $f'(45)$ positive or negative? Explain.
   (b) Find analytically $f(30)$ and $f'(30)$, and interpret them in terms of the concentration of the drug in the body.

21. Find the equation of the tangent line to $P = t \ln t$ at $t = 2$. Graph the function P and the tangent line on the same coordinate system.

22. The quantity, q, of a certain skateboard sold depends on the selling price, p, in dollars, so we write $q = f(p)$. You are given that $f(140) = 15,000$ and $f'(140) = -100$.

   (a) What does $f(140) = 15,000$ and $f'(140) = -100$ tell you about the sale of the skateboards?
   (b) The total revenue, R, earned by the sale of skateboards is given by $R = pq$. Find
   $$\left.\frac{dR}{dp}\right|_{p=140}.$$
   (c) What is the sign of $\left.\dfrac{dR}{dp}\right|_{p=140}$? If the skateboards are currently selling for $140, how should the price be changed to increase revenue?

23. Let $f(v)$ be the gas consumption (in liters/km) of a car going at velocity v (in km/hr). In other words, $f(v)$ tells you how many liters of gas the car uses to go one kilometer, if it is going at velocity v. You are told that

$$f(80) = 0.05 \quad \text{and} \quad f'(80) = 0.0005.$$

Explain what this is telling you in terms of gas consumption.

5.5 DERIVATIVES OF PERIODIC FUNCTIONS

Since the sine and cosine functions are periodic, their derivatives must be periodic also. (Why?) Let's look at the graph of $f(x) = \sin x$ in Figure 5.21 and estimate the derivative function graphically.

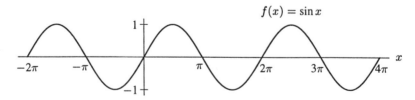

Figure 5.21: The sine function

First you might ask yourself where the derivative is zero. (At $x = \pm\pi/2$, $\pm 3\pi/2$, $\pm 5\pi/2$, etc.) Then ask yourself where the derivative is positive and where it is negative. (Positive for $-\pi/2 < x < \pi/2$; negative for $\pi/2 < x < 3\pi/2$, etc.) Since the largest positive slopes are at $x = 0, 2\pi$, and so on, and the largest negative slopes are at $x = \pi, 3\pi$, and so on, you get something like the graph in Figure 5.22.

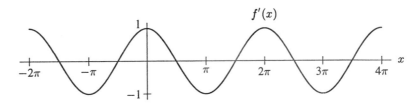

Figure 5.22: Derivative of $f(x) = \sin x$

The graph of the derivative looks suspiciously like the graph of the cosine function and this might lead you to conjecture, quite correctly, that the derivative of the sine is the cosine.

Of course, we cannot be sure, just from the graphs, that the derivative of the sine really is the cosine but it turns out that this is true. One thing we can do now is to give an informal check that the derivative function in Figure 5.22 has amplitude 1 (as it must if it is the cosine). That means we have to show that the derivative of $f(x) = \sin x$ is 1 when $x = 0$. The next example shows this is true when x is in radians.

Example 1   Using a calculator, estimate the derivative of $f(x) = \sin x$ at $x = 0$. Make sure your calculator is set in radians.

Solution   We use a small interval (we choose $\Delta x = 0.01$) and $f(x) = \sin x$ with x in radians to compute

$$f'(0) = \frac{\sin(0.01) - \sin(0)}{0.01 - 0} = \frac{0.0099998 - 0}{0.01} = 0.99998 \approx 1.0.$$

The derivative of $f(x) = \sin x$ at $x = 0$ is approximately 1.0. This matches our graphical hypothesis that the derivative of $\sin x$ is $\cos x$.

Warning: It is important to notice that in the previous example x was in *radians*; any conclusions we have drawn about the derivative of $\sin x$ are valid *only* when x is in radians.

Example 2   Starting with the graph of the cosine function, sketch a graph of its derivative.

Solution   The graph of $g(x) = \cos x$ is in Figure 5.23(a). Its derivative is 0 at $x = 0, \pm\pi, \pm 2\pi$, and so on; it is positive for $-\pi < x < 0$, $\pi < x < 2\pi$, and so on, and it is negative for $0 < x < \pi$, $2\pi < x < 3\pi$, and so on. The derivative is in Figure 5.23(b).

(a)

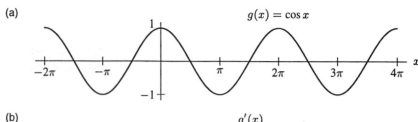

(b)

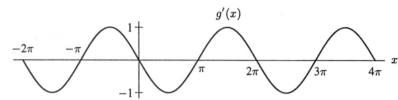

Figure 5.23: $g(x) = \cos x$ and its derivative, $g'(x)$

As we did with the sine, we'll use the graphs to make a conjecture. The derivative of the cosine in Figure 5.23(b) looks exactly like the graph of sine, except reflected about the x-axis. It turns out that the derivative of $\cos x$ is $-\sin x$.

For x in radians,

$$\frac{d}{dx}(\sin x) = \cos x \quad \text{and} \quad \frac{d}{dx}(\cos x) = -\sin x.$$

Example 3   Differentiate each of the following:

(a) $5\sin t - 8\cos t$
(b) $5 - 3\sin x + x^3$

Solution   (a)

$$\frac{d}{dt}(5\sin t - 8\cos t) = 5\frac{d}{dt}(\sin t) - 8\frac{d}{dt}(\cos t)$$
$$= 5(\cos t) - 8(-\sin t)$$
$$= 5\cos t + 8\sin t$$

(b)

$$\frac{d}{dx}(5 - 3\sin x + x^3) = \frac{d}{dx}(5) - 3\frac{d}{dx}(\sin x) + \frac{d}{dx}(x^3)$$
$$= 0 - 3(\cos x) + 3x^2$$
$$= -3\cos x + 3x^2.$$

The chain rule tells us how to differentiate all composite functions, including those involving the sine and cosine, such as $y = \sin(3t)$. Here, we have $y = \sin u$ with $u = 3t$, and

$$\frac{dy}{dt} = \frac{dy}{du} \cdot \frac{du}{dt}$$
$$= \cos u \frac{du}{dt}$$
$$= \cos(3t) \cdot 3$$
$$= 3\cos(3t)$$

In general, we have

If u is a function of t, then

$$\frac{d}{dt}(\sin u) = \cos u \frac{du}{dt} \quad \text{and} \quad \frac{d}{dt}(\cos u) = -\sin u \frac{du}{dt}$$

Example 4   Differentiate: (a) $\sin(t^2)$ (b) $5\cos(2t)$ (c) $t\sin t$

Solution   (a) We have $y = \sin u$ with $u = t^2$, so

$$\frac{d}{dt}(\sin(t^2)) = \frac{d}{dt}(\sin u) = \cos u \frac{du}{dt} = \cos(t^2) \cdot 2t = 2t\cos(t^2).$$

(b) We have $y = 5\cos u$ with $u = 2t$, so

$$\frac{d}{dt}(5\cos(2t)) = \frac{d}{dt}(5\cos u) = -5\sin u \frac{du}{dt} = -5\sin(2t) \cdot 2 = -10\sin(2t).$$

(c) We use the product rule:

$$\frac{d}{dt}(t\sin t) = \frac{d}{dt}(t) \cdot \sin t + t\frac{d}{dt}(\sin t)$$
$$= (1) \cdot \sin t + t(\cos t)$$
$$= \sin t + t\cos t.$$

Problems for Section 5.5

Differentiate the functions in Problems 1–16. You may assume that A, B, and C, where they appear, are constants.

1. $y = 5\sin x$

2. $P = 3 + \cos t$

3. $y = t^2 + 5\cos t$

4. $y = B + A\sin t$

5. $y = 5\sin x - 5x + 4$

6. $R(q) = q^2 - 2\cos q$

7. $R = \sin(5t)$

8. $W = 4\cos(t^2)$

9. $y = A\sin(Bt)$

10. $y = \sin(x^2)$

11. $y = 2\cos(5t)$

12. $y = 6\sin(2t) + 3\cos(4t)$

13. $f(x) = \sin(3x)$

14. $z = \cos(4\theta)$

15. $f(x) = x^2 \cos x$

16. $f(x) = 2x \sin(3x)$

17. Let $y = \sin x$. Find the equation of the tangent line to the graph of this function at the point where $x = \pi$. Check your work by sketching a graph of the function and the tangent line on the same coordinate system.

18. A company's sales are seasonal and can be modeled by

$$S = 2000 + 600 \sin(\frac{\pi}{6}t),$$

where S is sales per month and t is in months.

(a) Sketch a graph of S against t for $t = 0$ to $t = 36$. What is the maximum monthly sales? What is the minimum monthly sales? If $t = 0$ is January 1, when during the year are sales highest?

(b) Find $f(2)$ and $f'(2)$, and interpret each in terms of sales.

19. A boat at anchor is bobbing up and down in the sea. The vertical distance, y, in feet, between the sea floor and the boat is given as a function of time, t, in minutes, by

$$y = 15 + \sin 2\pi t.$$

(a) Find the vertical velocity, v, of the boat at time t.

(b) Make a rough sketch of y and v against t.

20. On page 231 of Section 4.4 the depth, y, of water in Boston harbor was given by

$$y = 5 + 4.9 \cos\left(\frac{\pi}{6}t\right),$$

where t is the number of hours since midnight.

(a) Find $\frac{dy}{dt}$. What does $\frac{dy}{dt}$ represent, in terms of water level?

(b) For $0 \leq t \leq 24$, when is $\frac{dy}{dt}$ zero? (Figure 4.43 on page 231 may help.) Explain what it means (in terms of water level) for $\frac{dy}{dt}$ to be zero.

5.6 VERIFYING THE DERIVATIVE FORMULAS

The Definition Of The Derivative

The derivative is the instantaneous rate of change of a function. We can estimate a derivative by computing average rates of change over smaller and smaller intervals. It is helpful to let h represent the size of the interval.

The average rate of change between x and $x + h$ is given by

$$\frac{f(x+h) - f(x)}{(x+h) - x}$$

which simplifies to

$$\frac{f(x+h) - f(x)}{h}.$$

Be sure that you recognize this difference quotient as the same rate of change formula that we have used so many times. For example, if $x = 2$ and the size of the interval, h, is 0.001, then the average rate of change between 2 and 2.001 is

$$\frac{f(2.001) - f(2)}{2.001 - 2} = \frac{f(2.001) - f(2)}{0.001} = \frac{f(x + h) - f(x)}{h}.$$

This new difference quotient is just another way to write the formula for the average rate of change. To find the derivative, or instantaneous rate of change, we use smaller and smaller intervals. To find the derivative exactly, we take the limit as h, the size of the interval, shrinks down to zero. The derivative is

$$\text{The limit, as } h \text{ approaches zero, of } \frac{f(x + h) - f(x)}{h}.$$

Finally, instead of writing the phrase "the limit, as h approaches 0", we use the shorthand notation

$$\lim_{h \to 0}.$$

We can now exhibit the formal definition of the derivative in all its finery:

For any function f, we define the **derivative function**, f', by

$$f'(x) = \lim_{h \to 0} \frac{f(x + h) - f(x)}{h}.$$

Verifying Derivative Formulas

By looking at the graph of $f(x) = x^2$ and then estimating the derivative of this function at several points, we guessed in Section 5.1 that the derivative of x^2 is $f'(x) = 2x$. In order to prove that this derivative is correct, we have to use the definition of the derivative.

In evaluating the expression

$$\lim_{h \to 0} \frac{f(x + h) - f(x)}{h},$$

we simplify the difference quotient first, and then take the limit as h approaches zero.

Example 1   Prove that the derivative of $f(x) = x^2$ is $f'(x) = 2x$.

Solution   We'll calculate the derivative of $f(x) = x^2$ using the definition:

$$f'(x) = \lim_{h \to 0} \frac{f(x + h) - f(x)}{h} = \lim_{h \to 0} \frac{(x + h)^2 - x^2}{h}$$

$$= \lim_{h \to 0} \frac{x^2 + 2xh + h^2 - x^2}{h} = \lim_{h \to 0} \frac{2xh + h^2}{h}$$

$$= \lim_{h \to 0} \frac{h(2x + h)}{h}.$$

To take the limit, we look at what happens when h is close to 0, but we do not let $h = 0$. Therefore, we can divide by h and say

$$f'(x) = \lim_{h \to 0} \frac{h(2x + h)}{h} = \lim_{h \to 0} (2x + h) = 2x,$$

because as h gets close to zero, $2x + h$ gets close to $2x$. So

$$f'(x) = \frac{d}{dx}(x^2) = 2x.$$

Example 2   Prove that the derivative of $g(x) = x^3$ is $g'(x) = 3x^2$.

Solution

$$g'(x) = \lim_{h \to 0} \frac{g(x+h) - g(x)}{h} = \lim_{h \to 0} \frac{(x+h)^3 - x^3}{h}$$

Multiplying out $\longrightarrow = \lim_{h \to 0} \dfrac{x^3 + 3x^2h + 3xh^2 + h^3 - x^3}{h}$

$$= \lim_{h \to 0} \frac{3x^2h + 3xh^2 + h^3}{h}$$

Dividing by $h \longrightarrow = \lim_{h \to 0} (3x^2 + 3xh + h^2) = 3x^2,$

Looking at what happens as $h \to 0$

So $g'(x) = \dfrac{d}{dx}(x^3) = 3x^2$.

Example 3   Give an informal justification that the derivative of $f(x) = e^x$ is $f'(x) = e^x$.

Solution   We have

$$f'(x) = \lim_{h \to 0} \frac{f(x+h) - f(x)}{h}$$

$$= \lim_{h \to 0} \frac{e^{x+h} - e^x}{h} \qquad \left(\text{since } f(x) = e^x\right)$$

$$= \lim_{h \to 0} \frac{e^x e^h - e^x}{h}$$

$$= \lim_{h \to 0} e^x \left(\frac{e^h - 1}{h}\right).$$

What is the limit of $\dfrac{e^h - 1}{h}$ as h approaches zero? The graph of $\dfrac{e^h - 1}{h}$ is shown in Figure 5.24 and it seems that as $h \to 0$, $\dfrac{e^h - 1}{h}$ approaches 1, and so the limit is 1. In fact, it can be proved that the limit equals 1. We have

$$f'(x) = \lim_{h \to 0} e^x \left(\frac{e^h - 1}{h}\right) = e^x \cdot 1 = e^x.$$

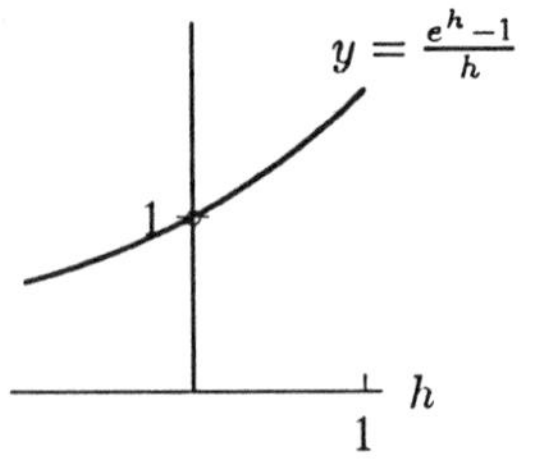

Figure 5.24: What is $\lim\limits_{h \to 0} \dfrac{e^h - 1}{h}$?

Using the Chain Rule to Establish Derivative Formulas

We will use the chain rule to justify the formulas for derivatives of logarithms and derivatives of exponentials with arbitrary base a.

Derivative of ln x

We'll differentiate an identity which involves $\ln x$. Since $e^{\ln x} = x$,

$$\frac{d}{dx}(e^{\ln x}) = \frac{d}{dx}(x) = 1.$$

On the other hand, the chain rule gives

$$\frac{d}{dx}(e^{\ln x}) = e^{\ln x} \cdot \frac{d}{dx}(\ln x). \qquad \text{Since } e^x \text{ is outside function and } \ln x \text{ is inside function.}$$

Thus,

$$\frac{d}{dx}(\ln x) = \frac{1}{e^{\ln x}} = \frac{1}{x},$$

so

$$\boxed{\frac{d(\ln x)}{dx} = \frac{1}{x}.}$$

Derivative of a^x

Earlier we showed that the derivative of a^x is proportional to a^x. Now we show that the constant of proportionality is $\ln a$. We will use the identity

$$\ln(a^x) = x \ln a.$$

Since $\ln a$ is a constant, $\dfrac{d}{dx}(\ln a^x) = \dfrac{d}{dx}(x \ln a) = \ln a$. On the other hand, using $\dfrac{d}{dx}(\ln x) = \dfrac{1}{x}$, the chain rule gives

$$\frac{d}{dx}(\ln a^x) = \frac{1}{a^x} \cdot \frac{d}{dx}(a^x) = \ln a.$$

Thus

$$\frac{1}{a^x}\frac{d}{dx}(a^x) = \ln a.$$

Solving for $\dfrac{d(a^x)}{dx}$ gives the result we stated in Section 5.2:

$$\boxed{\begin{array}{l} \text{For } a > 0, \\[2mm] \qquad \dfrac{d}{dx}(a^x) = (\ln a)a^x. \end{array}}$$

The Product Rule

Suppose we know the derivatives of $f(x)$ and $g(x)$ and want to calculate the derivative of the product, $f(x)g(x)$. The derivative of the product is calculated by taking the limit, namely,

$$\frac{d[f(x)g(x)]}{dx} = \lim_{h \to 0} \frac{f(x+h)g(x+h) - f(x)g(x)}{h}.$$

To picture the quantity $f(x+h)g(x+h) - f(x)g(x)$, imagine the rectangle with sides $f(x+h)$ and $g(x+h)$ in Figure 5.25, where $\Delta f = f(x+h) - f(x)$ and $\Delta g = g(x+h) - g(x)$.

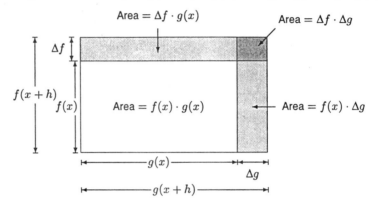

Figure 5.25: Illustration for the product rule (with Δf, Δg positive)

Then

$$f(x+h)g(x+h) - f(x)g(x) = \text{(Area of whole rectangle)} - \text{(Unshaded area)}$$
$$= \text{Area of the three shaded rectangles}$$
$$= \Delta f \cdot g(x) + f(x) \cdot \Delta g + \Delta f \cdot \Delta g$$

Now divide by h:

$$\frac{f(x+h)g(x+h) - f(x)g(x)}{h} = \frac{\Delta f}{h} \cdot g(x) + f(x) \cdot \frac{\Delta g}{h} + \frac{\Delta f \cdot \Delta g}{h}.$$

To evaluate the limit as $h \to 0$, let's examine the three terms on the right separately. Notice that

$$\lim_{h \to 0} \frac{\Delta f}{h} \cdot g(x) = f'(x)g(x) \quad \text{and} \quad \lim_{h \to 0} f(x) \cdot \frac{\Delta g}{h} = f(x)g'(x).$$

In the third term we multiply the top and bottom by h to get $\frac{\Delta f}{h} \cdot \frac{\Delta g}{h} \cdot h$. Then,

$$\lim_{h \to 0} \frac{\Delta f \cdot \Delta g}{h} = \lim_{h \to 0} \frac{\Delta f}{h} \cdot \frac{\Delta g}{h} \cdot h = \lim_{h \to 0} \frac{\Delta f}{h} \cdot \lim_{h \to 0} \frac{\Delta g}{h} \cdot \lim_{h \to 0} h = f'(x) \cdot g'(x) \cdot 0 = 0.$$

Therefore, we conclude that

$$\lim_{h \to 0} \frac{f(x+h)g(x+h) - f(x)g(x)}{h} = \lim_{h \to 0} \left(\frac{\Delta f}{h} \cdot g(x) + f(x) \cdot \frac{\Delta g}{h} + \frac{\Delta f \cdot \Delta g}{h} \right)$$
$$= \lim_{h \to 0} \frac{\Delta f}{h} \cdot g(x) + \lim_{h \to 0} f(x) \cdot \frac{\Delta g}{h} + \lim_{h \to 0} \frac{\Delta f \cdot \Delta g}{h}$$
$$= f'(x)g(x) + f(x)g'(x).$$

Thus we have the following rule:

The Product Rule

$$(fg)' = f'g + fg'.$$

In words:

The derivative of a product is the derivative of the first factor multiplied by the second, plus the first factor multiplied by the derivative of the second.

Problems for Section 5.6

1. Use the definition of the derivative to show that if $f(x) = 5x$, then $f'(x) = 5$. Be sure to justify all steps.

2. Use the definition of the derivative to show that if $f(x) = 2x + 1$, then $f'(x) = 2$. Be sure to justify all steps.

3. Use the definition of the derivative to show that if $f(x) = 3x^2$, then $f'(x) = 6x$. Be sure to justify all steps.

4. Use the definition of the derivative to show that if $f(x) = 5x^2$, then $f'(x) = 10x$. Be sure to justify all steps.

5. Use the definition of the derivative to show that if $f(x) = x^2 + x$, then $f'(x) = 2x + 1$. Be sure to justify all steps.

6. Use the definition of the derivative, and justify all steps, to show that if $f(x) = x^4$, then $f'(x) = 4x^3$. (Hint: You may use the fact that $(x+h)^4 = x^4 + 4x^3h + 6x^2h^2 + 4xh^3 + h^4$.)

7. Use the definition of the derivative, and justify all steps, to show that if $f(x) = x^5$, then $f'(x) = 5x^4$. (Hint: You may use the fact that $(x+h)^5 = x^5 + 5x^4h + 10x^3h^2 + 10x^2h^3 + 5xh^4 + h^5$.)

8. On a sketch of $y = f(x)$ similar to that in Figure 5.26, mark lengths that represent the quantities in parts (a) – (e). (Pick any convenient x, and assume $h > 0$.)
   (a) $f(x)$ (b) $x + h$ (c) $f(x + h)$ (d) $f(x + h) - f(x)$ (e) h
   (f) Using your answers to parts (a)–(e), show how the quantity $\dfrac{f(x + h) - f(x)}{h}$ can be represented as the slope of a line on the graph.

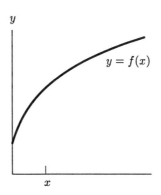

Figure 5.26

9. On a sketch of $y = f(x)$ similar to that in Figure 5.27, mark lengths that represent the quantities in parts (a) – (e). (Pick any convenient x, and assume $h > 0$.)

    (a) $f(x)$ (b) $x + h$ (c) $f(x + h)$ (d) $f(x + h) - f(x)$ (e) h

    (f) Using your answers to parts (a)–(e), show how the quantity $\dfrac{f(x + h) - f(x)}{h}$ can be represented as the slope of a line on the graph.

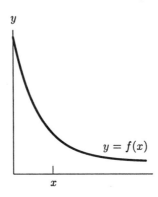

Figure 5.27

10. (a) Use a graph of $f(h) = \dfrac{2^h - 1}{h}$ to show that $\lim\limits_{h \to 0} \dfrac{2^h - 1}{h} \approx 0.6931$.

     (b) Use the definition of the derivative, and the result from part (a), to prove that if $f(x) = 2^x$, then $f'(x) = (0.6931)2^x$.

5.7 FINDING ANTIDERIVATIVES

What is an Antiderivative?

A function $F(x)$ is called an *antiderivative* of another function $f(x)$ if the derivative of $F(x)$ is $f(x)$. Can you think of an antiderivative of $f(x) = 2x$? Remember that the derivative of x^2 is $2x$, so we see that

$$x^2 \text{ is an antiderivative of } 2x.$$

Notice that we can always check our work when we find antiderivatives: the derivative of the antiderivative must be the original function.

We have seen that x^2 is an antiderivative of $2x$. Can you think of another function whose derivative is $2x$? How about $x^2 + 1$? Or $x^2 + 17$? Or $x^2 + C$ where C is any constant? Each of these has derivative $2x$, and so each of these is an antiderivative of $2x$. The good news is that the family of functions $x^2 + C$ includes all antiderivatives of $2x$. We say that

$$x^2 + C \text{ is the most general antiderivative of } 2x.$$

Once we have found one antiderivative $F(x)$ for a function, then all other antiderivatives will be in the form $F(x) + C$, and we call $F(x) + C$ the *most general antiderivative*.

Formulas for Antiderivatives

Finding antiderivatives of functions is like taking square roots of numbers: if you pick a number at random, such as 7 or 493, you will have trouble saying what its square root is without a calculator. But if you happen to pick a number such as 25 or 64, which you know is a perfect square, then you can find its square root exactly. Similarly, if you happen to pick a function which you recognize as a derivative, then you can find its antiderivative easily.

We saw this when we noticed that $2x$ was the derivative of x^2; this told us that x^2 was an antiderivative of $2x$. If we divide by 2, then we find that

$$\text{An antiderivative of } x \text{ is } \frac{x^2}{2}.$$

To check this statement, take the derivative of $x^2/2$:

$$\frac{d}{dx}\left(\frac{x^2}{2}\right) = \frac{1}{2} \cdot \frac{d}{dx} x^2 = \frac{1}{2} \cdot 2x = x.$$

What about an antiderivative of x^2? The derivative of x^3 is $3x^2$, so the derivative of $x^3/3$ is $3x^2/3 = x^2$. Thus,

$$\text{An antiderivative of } x^2 \text{ is } \frac{x^3}{3}.$$

Can you see the pattern?

$$\text{An antiderivative of } x^n \text{ is } \frac{x^{n+1}}{n+1}.$$

(We assume $n \neq -1$, or we would have $x^0/0$, which doesn't make sense.) It is easy to check this formula by differentiation:

$$\frac{d}{dx}\left(\frac{x^{n+1}}{n+1}\right) = \frac{(n+1)x^n}{n+1} = x^n.$$

For $n \neq 1$,
$$\frac{x^{n+1}}{n+1} \text{ is an antiderivative of } x^n.$$

Example 1   Find an antiderivative of each of the following:
(a) x^5 (b) t^8 (c) $12x^3$ (d) $q^3 - 6q^2$

Solution   (a)   An antiderivative of x^5 is $\dfrac{x^6}{6}$.

(b)   An antiderivative of t^8 is $\dfrac{t^9}{9}$.

(c)   An antiderivative of $12x^3$ is $12(\dfrac{x^4}{4}) = 3x^4$.

(d)   An antiderivative of $q^3 - 6q^2$ is $\dfrac{q^4}{4} - 6(\dfrac{q^3}{3}) = \dfrac{q^4}{4} - 2q^3$.

Notice that we can check all of our answers by differentiating the antiderivative.

The preceding example illustrates that the sum and constant multiplication rules of differentiation work in reverse.

Can you think of an antiderivative of the function $f(x) = 5$? We know that the derivative of $5x$ is 5, so $F(x) = 5x$ is an antiderivative of $f(x) = 5$. In general, what is the antiderivative of a constant, k? Since the derivative of kx is x, we have

$$kx \text{ is an antiderivative of } k.$$

Antiderivatives of $\dfrac{1}{x}$ and e^{kx}

Can you think of an antiderivative of $\dfrac{1}{x}$? Since the derivative of $\ln x$ is $\dfrac{1}{x}$, we have

An antiderivative of $\dfrac{1}{x}$ is $\ln x$.

What about e^{kx}? We know that the derivative of e^{kx} is ke^{kx}, so we have

An antiderivative of e^{kx} is $\dfrac{1}{k}e^{kx}$.

$$\ln x \text{ is an antiderivative of } \frac{1}{x},$$
$$\frac{1}{k}e^{kx} \text{ is an antiderivative of } e^{kx}.$$

Example 2 Find an antiderivative of (a) $8x^3 + \dfrac{1}{x}$ (b) $12e^{0.2t}$

Solution (a) An antiderivative of $8x^3 + \dfrac{1}{x}$ is $8(\dfrac{x^4}{4}) + \ln x = 2x^4 + \ln x$. (Differentiate this answer to check your work!)

    (b) An antiderivative of $12e^{0.2t}$ is $12(\dfrac{1}{0.2}e^{0.2t}) = 60e^{0.2t}$.

The Indefinite Integral

The Fundamental Theorem of Calculus tells us that if the derivative of $F(x)$ is $f(x)$, then

$$\int_a^b f(x)dx = F(b) - F(a).$$

If we want to compute $\int_1^3 2x\, dx$, we can either use Riemann sums or the Fundamental Theorem. Since

$$\frac{d}{dx}(x^2) = 2x,$$

the Fundamental Theorem with $F(x) = x^2$ tells us that

$$\int_1^3 2x\, dx = F(3) - F(1) = 3^2 - 1^2 = 8.$$

Which antiderivative should you use to compute a definite integral? It makes no difference because the constant cancels out when you subtract $F(a)$ from $F(b)$. For example, if we had used $x^2 + C$ in computing $\int_1^3 2x\,dx$, we would have arrived at the same answer as before:

$$\int_1^3 2x\,dx = F(3) - F(1) = (3^2 + C) - (1^2 + C) = 8.$$

The general antiderivative is written $F(x) + C$. Because of the connection with the definite integral, we introduce a notation for this general antiderivative that looks like the definite integral without the limits and is called the *indefinite integral*. If $F(x)$ is an antiderivative of $f(x)$, we write

$$\int f(x)\,dx = F(x) + C.$$

The **indefinite integral** $\int f(x)\,dx$ is the most general antiderivative of $f(x)$. If $F'(x) = f(x)$, we have

$$\int f(x)\,dx = F(x) + C.$$

It is important to understand the difference between

$$\int_a^b f(x)\,dx \qquad \text{and} \qquad \int f(x)\,dx.$$

The first is a number, and the second is actually a *family* of functions. They have such similar notations because the second is helpful in computing the first. Because the notation is similar, the word "integration" is frequently used for the process of finding the antiderivative as well as of finding the definite integral. The context usually makes clear which is intended.

We will also introduce a shorthand notation for $F(b) - F(a)$: we will write it as

$$\left. F(x) \right|_a^b$$

For example:

$$\int_1^3 2x\,dx = \left. x^2 \right|_1^3 = 3^2 - 1^2 = 8.$$

Example 3   Find: (a) $\displaystyle\int x^2\,dx,$ (b) $\displaystyle\int e^t\,dt.$

Solution   (a) $\int x^2\,dx$ means the general antiderivative of x^2, so

$$\int x^2\,dx = \frac{x^3}{3} + C.$$

(b) $\int e^t\,dt$ means the general antiderivative of e^t, so

$$\int e^t\,dt = e^t + C.$$

We have learned several formulas for computing antiderivatives. We summarize them below using the indefinite integral notation.

$$\int x^n \, dx = \frac{x^{n+1}}{n+1} + C, \ n \neq -1.$$

$$\int k \, dx = kx + C$$

$$\int e^{kx} \, dx = \frac{1}{k}e^{kx} + C$$

$$\int \frac{1}{x} \, dx = \ln x + C$$

Example 4 Find $\int (3x + x^2) \, dx$.

Solution We know that $x^2/2$ is an antiderivative of x and that $x^3/3$ is an antiderivative of x^2. Putting these together, we get

$$\int (3x + x^2) \, dx = 3\left(\frac{x^2}{2}\right) + \frac{x^3}{3} + C.$$

You should always check your antiderivatives by differentiation—it's easy to do. Here

$$\frac{d}{dx}\left(\frac{3}{2}x^2 + \frac{x^3}{3} + C\right) = \frac{3}{2} \cdot 2x + \frac{3x^2}{3} = 3x + x^2.$$

Using the Fundamental Theorem of Calculus

Remember what the Fundamental Theorem says: If $F(x)$ is an antiderivative of $f(x)$, then

$$\int_a^b f(x)dx = F(x)\Big|_a^b = F(b) - F(a).$$

Example 5 Use the Fundamental Theorem to compute each of the following definite integrals:

(a) $\int_0^2 6x^2 \, dx$ (b) $\int_1^2 (8x + 5)dx$ (c) $\int_0^1 8e^{2t} \, dt$

Solution (a) Since $6(\frac{x^3}{3})$ is an antiderivative of $6x^2$,

$$\int_0^2 6x^2 dx = 6(\frac{x^3}{3})\Big|_0^2 = 2x^3\Big|_0^2$$
$$= 2(2^3) - 2(0^3)$$
$$= 2 \cdot 8 - 0$$
$$= 16.$$

(b)   Since $4x^2 + 5x$ is an antiderivative of $8x + 5$,

$$\int_1^2 (8x + 5)dx = (4x^2 + 5x)\Big|_1^2 = (4(2^2) + 5(2)) - (4(1^2) + 5(1))$$
$$= (16 + 10) - (4 + 5)$$
$$= 26 - 9$$
$$= 17.$$

(c)

$$\int_0^1 8e^{2t}dt = 8(\frac{1}{2}e^{2t})\Big|_0^1 = 4e^{2t}\Big|_0^1$$
$$= 4e^2 - 4e^0$$
$$\approx 29.5562 - 4$$
$$= 25.5562.$$

Example 6   (a)   Write a definite integral to represent the area under the graph of $e^{0.5t}$ between $t = 0$ and $t = 4$.
(b)   Use a graph of $f(t) = e^{0.5t}$ to give a rough estimate of the area.
(c)   Use the Fundamental Theorem to find the exact area.

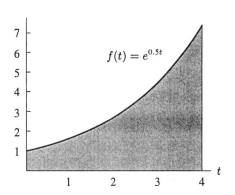

Figure 5.28: Estimate the shaded area

Solution   (a)   The area is equal to $\int_0^4 e^{0.5t}dt$.
(b)   The function is graphed in Figure 5.28. The shaded area appears to be about 4 times 3, or 12.
(c)   We have

$$\int_0^4 e^{0.5t}dt = 2e^{0.5t}\Big|_0^4 = 2e^{0.5(4)} - 2e^{0.5(0)} \approx 14.778 - 2 = 12.778.$$

A graph is an excellent way to check the analytical calculations. We see that the exact answer of 12.778 matches what we estimated from Figure 5.28.

Problems for Section 5.7

Find an antiderivative of each function in Problems 1–8.

1. x^4

2. $t^7 + t^3$

3. $5q^2$

4. $6x^3 + 4$

5. $3t^2 + 7t + 1$

6. $10 + 8x^3$

7. $x^2 - 6x + 17$

8. e^{-3t}

9. Find an antiderivative of $f(x) = 3x^2 + 5$. Differentiate your answer to check your work.

10. Find an antiderivative of $f(x) = 6x^2 - 8x + 3$. Differentiate your answer to check your work.

Find the indefinite integrals in Problems 11–18.

11. $\displaystyle \int t^{12}\, dt$

12. $\displaystyle \int 6x^2\, dx$

13. $\displaystyle \int (x^3 - x)\, dx$

14. $\displaystyle \int (x^2 + 1)\, dx$

15. $\displaystyle \int (x^3 + 4x + 8)\, dx$

16. $\displaystyle \int (8t + 3)\, dt$

17. $\displaystyle \int e^{2t}\, dt$

18. $\displaystyle \int \left(x^2 + \frac{1}{x}\right) dx$

Using the Fundamental Theorem, evaluate the definite integrals in Problems 19–26 exactly. Show your work.

19. $\displaystyle \int_0^3 t^3\, dt$

20. $\displaystyle \int_1^3 6x^2\, dx$

21. $\displaystyle \int_1^2 \frac{1}{x}\, dx$

22. $\displaystyle \int_0^2 (3t^2 + 4t + 3)\, dt$

23. $\displaystyle \int_0^1 (6q^2 + 4)\, dq$

24. $\displaystyle \int_0^1 e^{-0.2t}\, dt$

25. $\displaystyle \int_0^3 e^{0.05t}\, dt$

26. $\displaystyle \int_0^2 (x^2 + 1)\, dx$

27. Check your answer to Problem 19 by making a Riemann Sum estimate of the area under the graph of t^3 between $t = 0$ and $t = 3$.

28. Check your answer to Problem 24 by sketching a graph of $e^{-0.2t}$ on graph paper and counting boxes under the curve between $t = 0$ and $t = 1$.

29. Write the definite integral to represent the area under the graph of $6x^2 + 1$ between $x = 0$ and $x = 2$, and then use the Fundamental Theorem to evaluate the integral. Sketch a graph to check your work.

30. Use the Fundamental Theorem and the formula for average value given in Section 3.4 to find the average value of $f(x) = e^{0.5x}$ between $x = 0$ and $x = 3$. Show your work. Sketch a graph of $f(x)$ and indicate on it the graphical meaning of the average value.

REVIEW PROBLEMS FOR CHAPTER FIVE

Find the derivatives for the functions in Problems 1–22.

1. $f(t) = 6t^4$

2. $P(t) = e^{2t}$

3. $W = r^3 + 5r - 12$

4. $C = e^{0.08q}$

5. $f(x) = x^3 - 3x^2 + 5x - 12$

6. $y = 5e^{-0.2t}$

7. $s(t) = (t^2 + 4)(5t - 1)$

8. $g(t) = e^{(1+3t)^2}$

9. $f(x) = x^2 + 3\ln x$

10. $Q(t) = 5t + 3e^{1.2t}$

11. $g(z) = (z^2 + 5)^3$

12. $f(x) = 6(5x - 1)^3$

13. $f(z) = \ln(z^2 + 1)$

14. $y = xe^{3x}$

15. $q = 100e^{-0.05p}$

16. $y = x^2 \ln x$

17. $s(t) = t^2 + 2\ln t$

18. $P = 4t^2 + 7\sin t$

19. $R(t) = (\sin t)^5$

20. $h(t) = \ln\left(e^{-t} - t\right)$

21. $f(x) = \sin(2x)$

22. $y = x^2 \cos x$

23. Given $r(2) = 4$, $s(2) = 1$, $s(4) = 2$, $r'(2) = -1$, $s'(2) = 3$, and $s'(4) = 3$, compute the following derivatives, or state what additional information you would need to be able to compute the derivative.

   (a) $H'(2)$ if $H(x) = r(x) + s(x)$

   (b) $H'(2)$ if $H(x) = 5s(x)$

   (c) $H'(2)$ if $H(x) = r(x) \cdot s(x)$

   (d) $H'(2)$ if $H(x) = \sqrt{r(x)}$

24. Suppose the distance, s, of a moving body from a fixed point is given as a function of time by $s = 20e^{t/2}$. Find the velocity, v, of the body as a function of t.

25. Suppose that demand for a certain product is given by

$$q = 5000e^{-0.08p},$$

where p is the price in dollars of the product and q is the quantity sold at that price.

   (a) What quantity is sold at a price of $10?

   (b) Find the derivative of demand with respect to price when the price is $10 and interpret your answer in terms of demand for the product.

26. Suppose the demand equation for a product is as given in Problem 25. Find the revenue and the derivative of revenue with respect to price at a price of $10. Interpret your answers in economic terms.

27. Given a number $a > 1$, the equation

$$a^x = 1 + x$$

has the solution $x = 0$. Are there any other solutions? How does your answer depend on the value of a? [Hint: Graph the functions on both sides of the equation.]

28. Suppose the depth of the water, y, in meters, in the Bay of Fundy, Canada, is given as a function of time, t, in hours after midnight, by the function

$$y = 10 + 7.5\cos(0.507t).$$

Is the tide rising or falling, and how fast (in meters/hour), at each of the following times?
    (a) 6:00 am   (b) 9:00 am   (c) Noon   (d) 6:00 pm

29. The temperature Y in degrees Fahrenheit of a yam in a hot oven t minutes after it is placed there is given by
$$Y(t) = 350(1 - 0.7e^{-0.008t}).$$

    (a) What was the temperature of the yam when it was placed in the oven?
    (b) What is the temperature of the oven?
    (c) When will the yam reach $175°$ F in temperature?
    (d) Estimate the rate at which the temperature of the yam is increasing when $t = 20$.

Find each of the indefinite integrals in Problems 30–33.

30. $\displaystyle\int 9x^2\,dx$ 31. $\displaystyle\int (t^2 - 6t + 5)\,dt$

32. $\displaystyle\int e^{-0.05t}\,dt$ 33. $\displaystyle\int \left(8x^3 + \frac{1}{x}\right) dx$

34. Suppose the rate at which the world's oil is being consumed can be modeled by

$$r = 32e^{0.05t},$$

where r is in billions of barrels per year, t is in years, and $t = 0$ is January 1, 1990.

    (a) Write a definite integral which measures the total quantity of oil used between the start of 1990 and the start of 1995.
    (b) Use the Fundamental Theorem of Calculus to evaluate the integral. Show your work and give units with your answer.

35. Use the definition of the derivative to prove that the derivative of $f(x) = 3x^2$ is $f'(x) = 6x$. Be sure to justify all steps.

36. On what intervals is the function $f(x) = x^4 - 4x^3$ both decreasing and concave up? Use derivatives and show your work.

37. Given $p(x) = x^n - x$, find the intervals over which p is a decreasing function when: (a) $n = 2$ (b) $n = \frac{1}{2}$ (c) $n = -1$

38. Using a graph to help you, find the equations of all lines through the origin tangent to the parabola
$$y = x^2 - 2x + 4.$$

Sketch the lines on the graph.

39. Find the equations of the tangent lines to the graph of $f(x) = \sin x$ at $x = 0$ and at $x = \pi/3$. Use each tangent line to approximate $\sin \pi/6$. Would you expect these results to be equally accurate, since they are taken equally far away from $x = \pi/6$ but on opposite sides? If the accuracy is different, can you account for the difference?

40. A museum has decided to sell one of its paintings and to invest the proceeds. The price the painting will fetch changes with time, and is denoted by $P(t)$, where t is the number of years since 1990. If the picture is sold between 1990 and 2010 (so $0 \le t \le 20$), and the money from

the sale is invested in a bank account earning 5% annual interest compounded once a year, the balance, $B(t)$, in the account in the year 2010 is given by

$$B(t) = P(t)(1.05)^{20-t}.$$

(a) Explain why $B(t)$ is given by this formula.
(b) Show that the formula for $B(t)$ is equivalent to

$$B(t) = (1.05)^{20} \frac{P(t)}{(1.05)^t}.$$

(c) Find $B'(10)$, given that $P(10) = 150{,}000$ and $P'(10) = 5000$.

41. Imagine you are zooming in on the graph of each of the following functions near the origin:

$$y = x \qquad y = \sqrt{x} \qquad y = x^2 \qquad y = x^3 + \tfrac{1}{2}x^2$$

$$y = x^3 \qquad y = \ln(x+1) \qquad y = \tfrac{1}{2}\ln(x^2+1) \qquad y = \sqrt{2x - x^2}$$

Which of them look the same? Group together those functions which become indistinguishable near the origin, and give the equations of the lines they look like.

CHAPTER SIX

APPLICATIONS

In Chapter 2 we introduced the derivative and saw how to interpret it as slope and as rate of change. In Chapter 3, we introduced the definite integral and saw how to interpret it as area and as total change. In Chapter 5 we learned short-cut formulas for differentiating all of the standard functions. In this chapter, we learn more of the language of calculus (critical points and inflection points), and we see how to use calculus to solve some specific problems in economics, business, and life sciences.

6.1 CRITICAL POINTS AND INFLECTION POINTS

Why is it Useful to Know Where a Function is Increasing and Decreasing?

The following example shows how to use the derivative of a function to understand its graph. When we graph a function on a computer or graphing calculator, we see only part of the picture. The derivative can often direct our attention to important features on the graph.

Example 1   Sketch a helpful graph of the function $f(x) = x^3 - 9x^2 - 48x + 52$.

Solution   Since f is a cubic polynomial, we expect a graph that is roughly S-shaped. Graphing this function with $-10 \leq x \leq 10$, $-10 \leq y \leq 10$, gives the two nearly vertical lines in Figure 6.1. We know that there is more going on than this, but how do we know where to look?

We'll use the derivative to determine where the function is increasing and where it is decreasing. The derivative of f is

$$f'(x) = 3x^2 - 18x - 48.$$

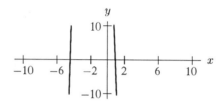

Figure 6.1: Unhelpful graph of $f(x) = x^3 - 9x^2 - 48x + 52$

To find where $f' > 0$ or $f' < 0$, we first find where $f' = 0$, that is, where $3x^2 - 18x - 48 = 0$. Solving this quadratic equation, we get $x = -2$ and $x = 8$. Since $f' = 0$ *only* at $x = -2$ and $x = 8$, and since f' is continuous, f' cannot change sign on any of the intervals $x < -2$, $-2 < x < 8$, or $x > 8$. How can we tell the sign of f' on each of these intervals? The easiest way is to pick a point and substitute into f'. For example, since $f'(-3) = 33 > 0$, we know f' is positive for $x < -2$, so f is increasing for $x < -2$. Similarly, since $f'(0) = -48$ and $f'(10) = 72$, we know that f decreases between $x = -2$ and $x = 8$ and increases for $x > 8$. We summarize the behavior of f on each interval:

f	Increasing ↗	$x = -2$	Decreasing ↘	$x = 8$	Increasing ↗	
f'	$+$	0	$-$	0	$+$	x

We find that $f(-2) = 104$ and $f(8) = -396$. Hence on the interval $-2 < x < 8$ the function decreases from a high of 104 to a low of -396. (Now we see why not much showed up in our first calculator graph.) One more point on the graph is easy to get: the y-intercept, $f(0) = 52$. With just these three points we can get a much more helpful graph. By setting the domain and range in our calculator to $-10 \leq x \leq 20$ and $-400 \leq y \leq 400$, we get Figure 6.2. Where does the graph in Figure 6.1 fit into this one?

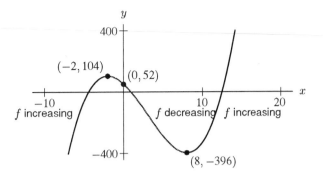

Figure 6.2: Helpful graph of $f(x) = x^3 - 9x^2 - 48x + 52$

Critical Points

In the preceding example, the points $x = -2$ and $x = 8$, where $f'(x) = 0$, played a key role. Now we will give a name to such points.

> For any function f, a point p in the domain of f where $f'(p) = 0$ or $f'(p)$ is undefined is called a **critical point** of the function. In addition, the point $(p, f(p))$ on the graph of f is also called a critical point. A **critical value** of f is the value, $f(p)$, of the function at a critical point, p.

Notice that "critical point of f" can refer either to special points in the domain of f or to special points on the graph of f. You will know which meaning is intended from the context.

What Do the Critical Points Tell Us?

Geometrically, at a critical point where $f'(p) = 0$, the line tangent to the graph of f at p is horizontal. At a critical point where $f'(p)$ is undefined, there is no horizontal tangent to the graph—there's either a vertical tangent or no tangent at all. (For example, $x = 0$ is a critical point for the absolute value function $f(x) = |x|$.) However, most of the functions we will work with will be differentiable everywhere, and therefore most of our critical points will be of the $f'(p) = 0$ variety.

The points where $f'(p) = 0$ (or $f'(p)$ is undefined) divide the domain of f into intervals on which the sign of the derivative stays the same, either positive or negative. Therefore, *between two successive critical points the graph of a function cannot change direction; it either goes up or down.*

A function may have any number of critical points or none at all. (See Figures 6.3–6.5.)

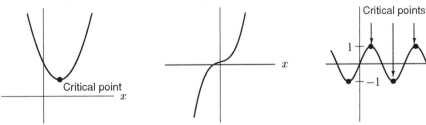

Figure 6.3: A quadratic:
One critical point

Figure 6.4: $f(x) = x^3 + x + 1$:
No critical points

Figure 6.5: Infinitely many
critical points

Local Maxima and Minima

What happens to a function at a critical point? Suppose that $f'(p) = 0$. We know that *at* p the graph has a horizontal tangent, but what happens *near* p? If f' has different signs on either side of p, then the graph changes direction at p, so the graph must look like one of those in Figure 6.6.

In the graph on the left in Figure 6.6 we say that f has a local minimum at p, and in the graph on the right we say that f has a local maximum at p. We use the adjective "local" because we are describing only what happens near p.

Suppose p is a critical point of a function f. Then

- f has a **local minimum** at p if, near p, the values of f get no smaller than $f(p)$.
- f has a **local maximum** at p if, near p, the values of f get no larger than $f(p)$.

How Do We Decide Which Critical Points are Local Maxima and Which are Local Minima?

Every local maximum and every local minimum must occur at a critical point. If we know where the critical points are, we know where all the potential "turning points" are. We can use a graph of the function to see which critical points are local maxima, which are local minima, and which are neither. As Figure 6.6 shows, we can also use information about the derivative of a function on either side of a critical point to determine whether it is a local maximum, a local minimum, or neither.

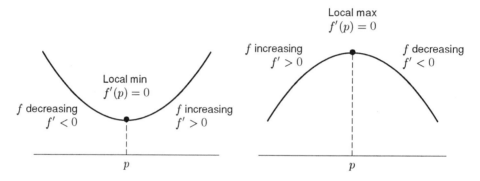

Figure 6.6: Changes in direction: Local maxima and minima

Example 2   (a) Sketch a graph of a function with the following three properties: (1) it has critical points at $x = 2$ and $x = 5$; (2) the derivative of the function is positive to the left of 2 and positive to the right of 5; (3) the derivative is negative between 2 and 5.

(b) Identify the critical points as local maxima, local minima, or neither.

Solution   (a) We know that $f(x)$ is increasing when $f'(x)$ is positive, and $f(x)$ is decreasing when $f'(x)$ is negative. The function is increasing to the left of 2 and increasing to the right of 5, and it is decreasing between 2 and 5. A possible sketch is given in Figure 6.7.

(b) We see that the function has a local maximum at $x = 2$ and a local minimum at $x = 5$.

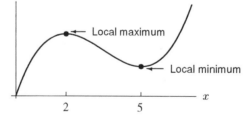

Figure 6.7: A function with two critical points

Warning!

The sign of f' doesn't *have* to change at a critical point. Consider $f(x) = x^3$, whose graph is in Figure 6.8. The derivative, $f'(x) = 3x^2$, is positive on both sides of $x = 0$, so f increases on both sides of $x = 0$, and there is neither a local maximum nor a local minimum at $x = 0$. In other words, *a function doesn't have to have a local maximum or local minimum at every critical point.*

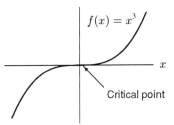

Figure 6.8: Critical point which is not a local maximum or minimum.

Example 3   The rate of growth of a tumor is given in Figure 6.9, as a function of time t. This is the derivative of a function, $S(t)$, giving the size of the tumor at time t.
(a)   What are the critical points of the function $S(t)$?
(b)   Identify each critical point as either a local maximum, a local minimum, or neither.

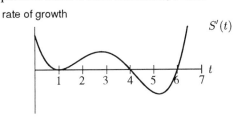

Figure 6.9: The rate of growth of a tumor

Solution   (a)   Figure 6.9 shows a graph of $S'(t)$. The critical points of S occur at times t such that $S'(t) = 0$. We see in Figure 6.9 that $S'(t) = 0$ at $t = 1, 4$, and 6, and so the critical points are at $t = 1$, $t = 4$, and $t = 6$.

(b)   In Figure 6.9, we see that $S'(t)$ is positive to the left of 1 and between 1 and 4, that it is negative between 4 and 6, and that it is positive to the right of 6. Therefore the function $S(t)$ is increasing to the left of 1 and between 1 and 4 (with a slope of zero at 1), decreasing between 4 and 6, and increasing again to the right of 6. A possible sketch of the function representing the size of the tumor at time t, $S(t)$, is given in Figure 6.10. We see that S has neither a local maximum nor a local minimum at the critical point $t = 1$, that it has a local maximum at $t = 4$, and that it has a local minimum at $t = 6$.

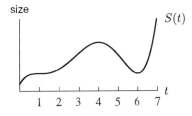

Figure 6.10: Possible graph of the function representing the size of the tumor at time t

Concavity and Inflection Points

A study of the points on the graph of a function where the slope changes sign led us to critical points. Now we will study the points on the graph where the concavity changes.

> A point at which the graph of a function f changes concavity is called an **inflection point** of f.

The words "inflection point of f" can refer either to a point in the domain of f or to a point on the graph of f. The context of the problem will tell you which is meant.

How Do You Locate an Inflection Point?

Since the concavity of the graph of f changes at an inflection point, the sign of f'' changes there. It is positive on one side of the inflection point, and negative on the other; so at the inflection point, f'' is zero or undefined. (See Figure 6.11.)

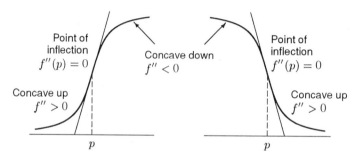

Figure 6.11: Change in concavity

Example 4   Find the inflection points of $f(x) = x^3 - 9x^2 - 48x + 52$.

Solution   From the graph of $f(x)$ in Figure 6.12 we see that part of the graph of f is concave up and part is concave down, and so the function must have an inflection point. To find the inflection point exactly, we find where the second derivative is zero: we calculate $f'(x) = 3x^2 - 18x - 48$. Thus,

$$f''(x) = 6x - 18 \qquad \text{and so} \qquad f''(x) = 0 \text{ when } x = 3$$

We see in the graph that $f(x)$ changes concavity at $x = 3$ and so $x = 3$ is an inflection point.

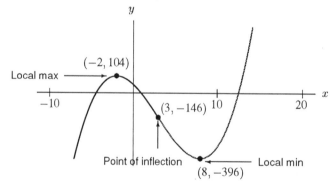

Figure 6.12: Graph of $f(x) = x^3 - 9x^2 - 48x + 52$, again

Example 5   (a)   How many critical points and how many inflection points does the surge function $f(x) = xe^{-x}$ have?

(b)   Use derivatives to find the critical points and inflection points exactly.

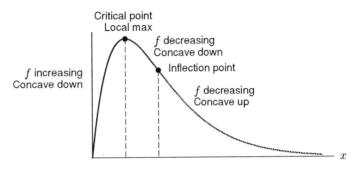

Figure 6.13: Graph of $f(x) = xe^{-x}$

Solution   (a)   The graph of $f(x) = xe^{-x}$ is given in Figure 6.13. We see that it has one critical point, which is also a local maximum. Are there any inflection points? Since the graph of the function is concave down at the critical point and concave up for large x, the graph of the function changes concavity and so there must be an inflection point. We see that the inflection point is to the right of the critical point.

(b)   To find the critical point, we find the point where the first derivative of f equals zero. Using the product rule, we have

$$f'(x) = x(-e^{-x}) + (1)(e^{-x}) = (1 - x)e^{-x}.$$

We have $f'(x) = 0$ when $x = 1$, so the critical point is at $x = 1$. To find the inflection point, we find where the second derivative of f equals zero. Using the product rule on the first derivative, we have

$$f''(x) = (1 - x)(-e^{-x}) + (-1)(e^{-x}) = (x - 2)e^{-x}.$$

We have $f''(x) = 0$ when $x = 2$, so the inflection point is at $x = 2$.

Warning!

Not every point x where $f''(x) = 0$ (or f'' is undefined) is an inflection point (just as not every point where $f' = 0$ is a local maximum or minimum). For instance $f(x) = x^4$ has $f''(x) = 12x^2$ so $f''(0) = 0$, but $f'' > 0$ when $x > 0$ and when $x < 0$, so there is *no* change in concavity at $x = 0$. See Figure 6.14.

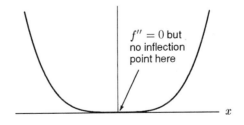

Figure 6.14: Graph of $f(x) = x^4$

Example 6   Suppose that water is being poured into the vase in Figure 6.15 at a constant rate measured in volume per unit time. Graph $y = f(t)$, the depth of the water against time, t. Explain the concavity, and indicate the inflection points.

Solution   At first the water level, y, rises quite slowly because the base of the vase is wide, and so it takes a lot of water to make the depth increase. However, as the vase narrows, the rate at which the water is rising increases. This means that initially y is increasing at an increasing rate, and the graph is concave up. The rate of increase in the water level is at a maximum when the water reaches the middle of the vase, where the diameter is smallest; this is an inflection point. After that, the rate at which y increases starts to decrease again, and so the graph is concave down. (See Figure 6.16.)

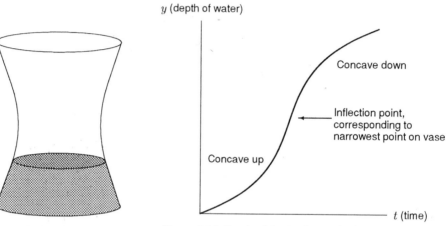

Figure 6.15: A vase

Figure 6.16: Graph of depth of water in the vase, y, against time, t

Problems for Section 6.1

In Problems 1–4, on a sketch similar to the graph given,

(a)   Indicate all critical points, and identify each as a local maximum, a local minimum, or neither. How many critical points are there?

(b)   Indicate the approximate locations of all inflection points. How many inflection points are there?

1.

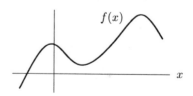

Figure 6.17

2.

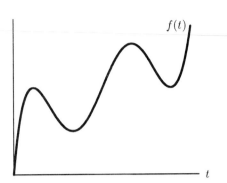

Figure 6.18

3.

Figure 6.19

4.

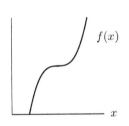

Figure 6.20

5. Sketch graphs of two continuous functions f and g, each of which has exactly five critical points, the points A–E in Figure 6.21, and which satisfy the following conditions:

(a) $\displaystyle \lim_{x \to -\infty} f(x) = \infty$ and
 $\displaystyle \lim_{x \to \infty} f(x) = \infty$

(b) $\displaystyle \lim_{x \to -\infty} g(x) = -\infty$ and
 $\displaystyle \lim_{x \to \infty} g(x) = 0$

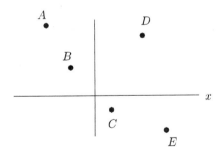

Figure 6.21

In Problems 6–7, the graph of the derivative function f' is given. Indicate on a sketch of the graph the x-values that are critical points of the function f itself. Identify each critical point as a local maximum, a local minimum, or neither.

6.

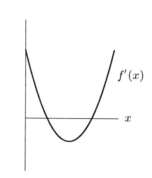

Figure 6.22

7.

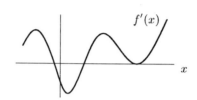

Figure 6.23

In each of Problems 8–13, use the first derivative to find all critical points and use the second derivative to find all inflection points. Show your work. Use a graph to identify each critical point as a local maximum, a local minimum, or neither.

8. $f(x) = x^2 - 5x + 3$

9. $f(x) = 2x^3 + 3x^2 - 36x + 5$

10. $f(x) = 3x^4 - 4x^3 + 6$

11. $f(x) = x^4 - 8x^2 + 5$

12. $y = x^4 - 4x^3 + 10$

13. $f(x) = 3x^5 - 5x^3$

14. How many real roots does the equation $x^5 + x + 7 = 0$ have? How do you know? [Hint: How many critical points does this function have?]

15. Assume the polynomial f has exactly two local maxima and one local minimum.

(a) Sketch a possible graph of f.
(b) What is the largest number of zeros f could have?
(c) What is the least number of zeros f could have?
(d) What is the least number of inflection points f could have?
(e) Is the degree of f even or odd? How can you tell?
(f) What is the smallest degree f could have?
(g) Find a possible formula for $f(x)$.

16. The rabbit population on a small Pacific island is approximated by

$$P(t) = \frac{2000}{1 + e^{(5.3 - 0.4t)}}$$

with t measured in years since 1774, when Captain James Cook left 10 rabbits on the island. Using a calculator or computer:

(a) Graph P. Does the population level off?
(b) Estimate when the rabbit population grew most rapidly. How large was the population at that time?

(c)   What natural causes could lead to the shape of the graph of P?

17.   Find constants a and b in the function $f(x) = x^2 + ax + b$ so that the minimum for this parabola is at the point $(3, 5)$. (Hint: Begin by finding the critical point in terms of a.)

18.   Choose the constants a and b in the function $f(x) = x^2 + ax + b$ so that the global minimum for this parabola is at the point $(-2, -3)$.

19.   Choose the constants a and b in the function

$$f(x) = axe^{bx}$$

such that $f(\frac{1}{3}) = 1$ and the function has a maximum at $x = \frac{1}{3}$.

20.   Sketch the graph of $f(x) = 2x^3 - 9x^2 + 12x + 1$. Use the graph to decide how many solutions the following equations have: (a) $f(x) = 10$ (b) $f(x) = 5$ (c) $f(x) = 0$
(d) $f(x) = 2e$
      You need not find these solutions.

21.   Suppose that consumer demand for a certain product is changing over time, and the rate of change of this demand, $r(t)$, is given in Table 6.1.

TABLE 6.1   *Rate of change of demand for a certain product*

Time, t (weeks)	0	1	2	3	4	5	6	7	8	9	10
Rate, $r(t)$ (units/week)	12	10	4	-2	-3	-1	3	7	11	15	10

(a)   When is the demand for this product increasing? When is it decreasing?
(b)   Estimate the times at which demand is at a local maximum, and the times at which demand is at a local minimum.

22.   Suppose f has a continuous derivative everywhere. From the values of $f'(\theta)$ in the table below, estimate the θ values with $1 < \theta < 2.1$ at which $f(\theta)$ has a local maximum or minimum, and identify which is which.

θ	$f'(\theta)$	θ	$f'(\theta)$
1.0	2.37	1.6	0.76
1.1	0.31	1.7	2.80
1.2	-2.00	1.8	3.61
1.3	-3.45	1.9	2.76
1.4	-3.34	2.0	0.69
1.5	-1.70	2.1	-1.62

23.   (a)   On a computer or calculator, graph $f(\theta) = \theta - \sin\theta$. Can you tell whether the function has any zeros in the interval $0 \le \theta \le 1$?
(b)   Find f'. What does the sign of f' tell you about the zeros of f in the interval $0 \le \theta \le 1$?

24.   Assume the function f is differentiable everywhere and has just one critical point, at $x = 3$. In parts (a) through (d), you are given additional conditions. In each case decide whether $x = 3$ is a local maximum, a local minimum, or neither. Explain your reasoning. Also sketch possible graphs for all four cases.
      (a) $f'(1) = 3$ and $f'(5) = -1$ (c) $f(1) = 1, f(2) = 2, f(4) = 4, f(5) = 5$
      (b) $\lim_{x \to \infty} f(x) = \infty$ and   (d) $f'(2) = -1, f(3) = 1, \lim_{x \to \infty} f(x) = 3$
            $\lim_{x \to -\infty} f(x) = \infty$

25. For what values of a and b will the function $f(x) = a(x - b \ln x)$ have a global minimum at the point $(2, 5)$? See Figure 6.24 for a graph of $f(x)$ when $a = 1$ and $b = 1$.

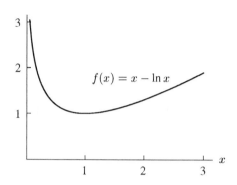

Figure 6.24: Graph of $f(x) = x - \ln x$

26. (a) Water is flowing at a constant rate into a cylindrical container standing vertically. Sketch a graph showing the depth of water against time.
    (b) Water is flowing at a constant rate into a cone-shaped container standing on its point. Sketch a graph showing the depth of the water against time.

27. A graph is given in Figure 6.25. Indicate on a sketch of the graph approximately where the inflection points of $f(x)$ are if

    (a) the graph shows the function $f(x)$,
    (b) the graph shows the derivative $f'(x)$,
    (c) the graph shows the second derivative $f''(x)$.

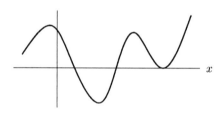

Figure 6.25

For Problems 28–31, use a calculator or computer to sketch graphs of f, f' and f''. Use these graphs to help you answer the questions.

28. For $f(x) = e^{x/2} - \ln(x^2 + 1)$, find the coordinates of all intercepts, maxima, minima, and inflection points to two decimal places.

29. For $f(x) = \sin(x^2)$ between $x = 0$ and $x = 3$, find the coordinates of all intercepts, critical points, and inflection points to two decimal points.

30. Sketch a graph of $f(x) = e^{1/x}$, with $x \neq 0$, on the interval $-5 \leq x \leq 5$.

    (a) For what values of x is the function increasing? Decreasing? Neither?
    (b) For what values of x is the graph of the function concave up? Concave down? Neither?

31. Let $f(x) = \ln(x^3 + 1)$, on the interval $-1 \leq x \leq 5$.

    (a) For what values of x is the function increasing? Decreasing? Neither?

    (b) For what values of x is the the graph of the function concave up? Concave down? Neither?

32. Suppose you know that a certain function f has $f(0) = 0$ and derivative given by

$$f'(x) = (\ln x)^2 - 2(\sin x)^4 \qquad \text{on the interval} \qquad 0 < x \leq 7.5.$$

    Use a calculator or computer to graph f' and its derivative. Clearly state where $f(x)$ is increasing and where it is decreasing, and where the graph of $f(x)$ is concave up and where it is concave down. Use this information to sketch a graph of $f(x)$.

33. If water is flowing at a constant rate (i.e., constant volume per unit time) into the Grecian urn in Figure 6.26, sketch a graph of the depth of the water against time. Mark on the graph the time at which the water reaches the widest point of the urn.

Figure 6.26

34. If water is flowing at a constant rate (i.e., constant volume per unit time) into the vase in Figure 6.27, sketch a graph of the depth of the water against time. Mark on the graph the time at which the water reaches the corner of the vase.

Figure 6.27

For Problems 35–38, sketch a possible graph of $y = f(x)$, using the given information about the derivatives $y' = f'(x)$ and $y'' = f''(x)$. Assume that the function is defined and continuous for all real x.

35.

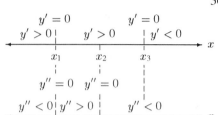

36.

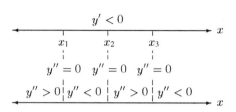

37.

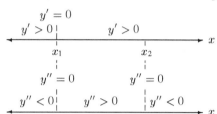

38.

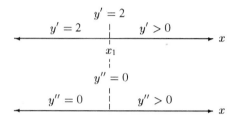

39. For the function, f, given in the graph in Figure 6.28:

    (a) Sketch $f'(x)$.

    (b) Where does $f'(x)$ change its sign?

    (c) Where does $f'(x)$ have local maxima or minima?

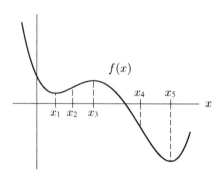

Figure 6.28

40. Using your answer to Problem 39 as a guide, write a short paragraph (using complete sentences) which describes the relationship between the following features of a function f:

    (a) The local maxima and minima of f.

    (b) The points at which the graph of f changes concavity.

    (c) The sign changes of f'.

    (d) The local maxima and minima of f'.

6.2 OPTIMIZATION

Global Maxima and Minima

We encounter numerous problems in the world where it is important to find the maximum or minimum value of some quantity. For example, a firm will try to realize the maximum possible profit; to do so it may try to keep its costs to a minimum. The techniques for finding maximum and minimum values of functions make up the field called *optimization*. The local maxima and minima tell us where a function is locally largest or smallest. However we are often more interested in where a function is absolutely largest or smallest in a given domain. We say

---

For any function f:

- f has a **global minimum** at p if $f(p)$ is less than or equal to all other values of f.
- f has a **global maximum** at p if $f(p)$ is greater than or equal to all other values of f.

How Do We Find Global Maxima and Minima?

If f is a continuous function defined on a closed interval $a \leq x \leq b$ (i.e., an interval containing its endpoints), Figure 6.29 illustrates that the global maximum or minimum of f occurs either at a local maximum or a local minimum respectively, or at one of the endpoints $x = a$ or $x = b$ of the interval.

To find the global maximum and minimum of a continuous function on a closed interval:
Compare values of the function at all the critical points in the interval and at the endpoints.

What if the function is defined on an open interval $a < x < b$ or on the entire real line? The function graphed in Figure 6.30 has no global maximum because the function has no largest value. The global minimum of this function coincides with one of the local minima and is marked. A function defined on the entire real line or on an open interval may or may not have a global maximum or a global minimum.

To find the global maximum and minimum of a continuous function on an open interval or on the entire real line: Find the value of the function at all the critical points and sketch a graph.

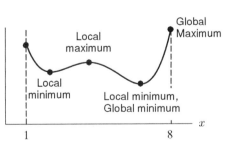

Figure 6.29: Global maximum and minimum on a closed interval $a \leq x \leq b$

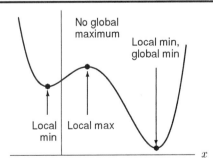

Figure 6.30: Global maximum and minimum when the domain is the real line.

Warning!

Notice that local maxima or minima of a function f occur at critical points, where $f'(p) = 0$ (or $f'(p)$ is undefined). Since global maxima or minima can occur at endpoints (where f' is not necessarily 0 or undefined), *not every global maximum or minimum is a local maximum or minimum.*

Example 1   Find the global maxima and minima of $f(x) = x^3 - 9x^2 - 48x + 52$ on the interval $-5 \le x \le 14$.

Solution   We have calculated the critical points of this function previously using

$$f'(x) = 3x^2 - 18x - 48 = 3(x + 2)(x - 8),$$

so $x = -2$ and $x = 8$ are critical points. We know that the global maxima and minima must occur at a critical point or at an endpoint of the interval, so we evaluate f at these four points:

$$f(-5) = -58, \qquad f(-2) = 104, \qquad f(8) = -396, \qquad f(14) = 360.$$

Comparing these four values, we see that the global maximum is 360 and occurs at $x = 14$, and that the global minimum is -396 and occurs at $x = 8$. Figure 6.31 shows a graph of this function on the interval $-5 \le x \le 14$.

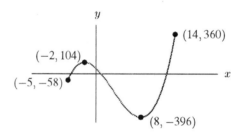

Figure 6.31: Global maximum and minimum on a closed interval

Maximizing Profit

The fundamental issue for a producer of goods is how to maximize profit. Recall that the revenue function, $R(q)$, gives total revenue from selling a quantity of goods q, and that the cost function, $C(q)$, gives the total cost of producing a quantity q. The profit $\pi(q)$ is the difference: $\pi(q) = R(q) - C(q)$.

We begin by revisiting Example 5 in Section 2.6, in which we were asked to find the maximum profit if the total revenue and total cost functions are given by the curves R and C, respectively, in Figure 6.32. Since profit = revenue − cost = $R(q) - C(q)$, we see that profit is maximized when $R(q) > C(q)$ and the vertical distance between the two curves is at a maximum. This occurs at about $q = 140$.

In Section 2.6, we determined that this is the point where the slopes of the two curves $R(q)$ and $C(q)$ are equal. Does this always have to be so? The slopes of the curves are given, respectively, by the marginal revenue, $MR = R'(q)$, and the marginal cost $MC = C'(q)$. If $R' > C'$, we

$ (thousands)

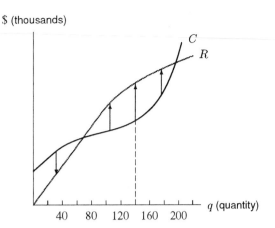

Figure 6.32: Maximum profit at $q = 140$

increase our profit by producing more items – the revenue from the additional items exceeds the cost of producing them. If $R' < C'$, we increase our profit by producing fewer items – the cost of producing the items outweighs the revenues and so we are better off not making the items. The profit, $R(q) - C(q)$, is largest when the slopes of the curves are equal, which is at the point where marginal revenue equals marginal cost.

Let's look at the general situation: To maximize or minimize profit over an interval, we optimize the profit function π where

$$\pi(q) = R(q) - C(q).$$

We have seen that global maxima and minima of a function can only occur at critical points of the function or at the endpoints, if any, of the interval. To find critical points of π, look for zeros of the derivative:

$$\pi'(q) = R'(q) - C'(q) = 0.$$

So

$$R'(q) = C'(q),$$

that is, the slopes of the graphs of $R'(q)$ and $C'(q)$ at q are equal. In economic language

---

The maximum (or minimum) profit can occur when

Marginal cost = Marginal revenue.

Of course, maximum or minimum profit does not *have* to occur where $MR = MC$; there are always the endpoints to consider. However, this relationship is quite powerful, because it is the condition that helps determine the maximum (or minimum) profit in general.

Example 2   Find the quantity q which will maximize profit if the total revenue and total cost (in dollars) are given by

$$R(q) = 5q - 0.003q^2$$
$$C(q) = 300 + 1.1q$$

where $0 \leq q \leq 1000$ units. What production level will give the minimum profit?

Solution   We can begin by looking for production levels that give Marginal revenue = Marginal cost:

$$MR = R'(q) = 5 - 0.006q$$
$$MC = C'(q) = 1.1.$$

So

$$5 - 0.006q = 1.1$$
$$q = \frac{3.9}{0.006} = 650 \text{ units.}$$

Does this represent a local maximum or minimum of the profit π? We can tell by looking at what is going on at production levels of 649 units and 651 units. When $q = 649$ we have $MR = \$1.106$, which is greater than the (constant) marginal cost of $1.10. This means that producing one more unit will bring in more revenue than its cost, so profit will increase. When $q = 651$, $MR = \$1.094$, which is *less* than MC, so it is not profitable to produce the 651st unit. We conclude that $q = 650$ is a local maximum for the profit function π. The profit earned by producing and selling this quantity is $\pi(650) = R(650) - C(650) = \$1982.50 - \$1015 = \$967.50$.

To check for global maxima we need to look at the endpoints. If $q = 0$, the only cost is $300 (the fixed costs) and there is no revenue, so $\pi(0) = -300$. At the upper limit of $q = 1000$, $R(1000) = \$2000$, while $C(1000) = \$1400$, and so $\pi(1000) = \$600$. Therefore, the maximum profit is obtained when $MR = MC$, which occurs at a production level of $q = 650$ units, and the minimum profit occurs when $q = 0$ and there is no production at all.

Example 3   The total revenue and total cost curves for a product are given in Figure 6.33.
   (a)   Sketch the curves for marginal revenue and marginal cost on the same axes, and indicate on this graph the quantities where marginal revenue equals marginal cost. What is the significance of these two quantities?
   (b)   Sketch the general shape of the profit function $\pi(q)$. Assume that the fixed costs of production are positive. At which quantity is profit maximized?

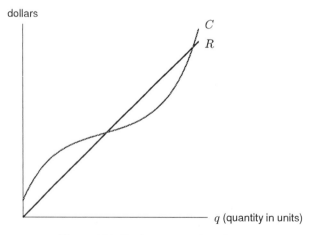

Figure 6.33: Total revenue and total cost

Solution   (a)   The graphs of total revenue $R(q)$ and total cost $C(q)$ are given in Figure 6.33. To graph marginal revenue and marginal cost, we sketch graphs of the derivatives of $R(q)$ and $C(q)$, respectively. Since $R(q)$ is a straight line with positive slope, the graph of marginal revenue, MR, is a horizontal line. (See Figure 6.34.) We see that $C(q)$ is always increasing, so marginal cost, MC, is always positive. As q increases, the cost curve changes from concave down to concave up, so the derivative of the cost function changes from decreasing to increasing. Therefore, the graph of marginal cost, MC, is decreasing and then increasing, as shown in Figure 6.34. The local minimum on the marginal cost curve corresponds to the inflection point of $C(q)$.

Where is profit maximized? We know that the maximum profit can occur when marginal revenue equals marginal cost. We see in Figure 6.34 that the marginal cost curve crosses the marginal revenue curve at two points, labeled q_1 and q_2, and so there are two points for which $MR = MC$. Which of these gives maximum profit?

We first consider a production size of q_1. To the left of the point q_1, we have $MC > MR$. Since profit $\pi = R - C$, we have $\pi' = R' - C' = MR - MC$. But $MR < MC$ to the left of q_1, so we see that π' is negative to the left of q_1 and the profit function is decreasing there. To the right of q_1, we have $MR > MC$, so π' is positive and the profit function is increasing. This behavior, decreasing first then increasing, indicates that the profit function has a local minimum at q_1. This is certainly not the production level we want.

What happens at q_2? To the left of q_2, we have $MR > MC$, so π' is positive and the profit function is increasing. To the right of q_2, we have $MC > MR$, so π' is negative and the profit function is decreasing. This behavior, increasing at first and then decreasing, shows that the profit function has a local maximum at q_2. The global maximum we seek will occur either at an endpoint (the largest and smallest possible production levels) or at the production level q_2.

(b)   We saw that the profit function is decreasing to the left of q_1, increasing between q_1 and q_2, and decreasing to the right of q_2. Thus, as production increases, the profit function decreases, then increases, and then decreases again. Since $\pi(q) = R(q) - C(q)$, the vertical intercept is $\pi(0) = R(0) - C(0)$. We know that $R(0) = 0$ and that $C(0)$ is equal to the fixed costs of production. Therefore the vertical intercept of the profit function is a negative number, equal in magnitude to the size of the fixed cost. Thus, the graph of the profit function has the general shape shown in Figure 6.35. The maximum profit occurs at a production level of $q = q_2$ units.

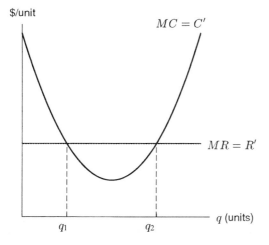

Figure 6.34: Marginal revenue and marginal cost

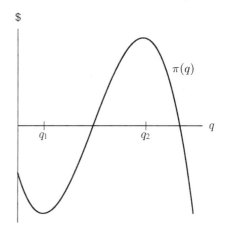

Figure 6.35: Profit function

Maximizing Revenue

Suppose you own a city bus company with a fixed schedule. Your costs will be the same no matter how many people ride the buses, so you will maximize profit by maximizing your revenue. What price should you charge for a ticket to maximize revenue?

In practice, the quantity of a good sold will often depend on the price charged. We know that revenue is equal to price times quantity. Recall that the demand curve introduced in Chapter 1 tells us how consumer demand depends on the price of the product. Using the demand equation, we can sometimes write the quantity explicitly as a function of price, $q = f(p)$. If we substitute $f(p)$ for q in the revenue function, we obtain revenue as a function of price. Similarly, if we write the demand equation with price as an explicit function of quantity and substitute this for price in the revenue function, we will have revenue as a function of quantity. In either case, it makes sense to talk about the price (or quantity, respectively) that will maximize revenue.

Example 4   Suppose the demand equation for all-day passes to an amusement park is given by $p = 70 - 0.02q$, where p is the price of a pass in dollars and q is the number of people attending at that price.
  (a) What price corresponds to an attendance of 3000 people? What is the total revenue at that price? What is the total revenue if the price is \$20?
  (b) Write the revenue function as a function of attendance at the amusement park, q.
  (c) What attendance will maximize revenue?
  (d) What price should be charged to maximize revenue?
  (e) What is the maximum revenue? What is the corresponding profit?

Solution   (a) If $q = 3000$, we see from the demand equation that $p = 70 - 0.02(3000) = 10$. A price of \$10 corresponds to an attendance of 3000 people. At this price, total revenue is (3000 people)(10 dollars/person), or \$30,000. In order to find total revenue at a price of \$20, we first find the attendance corresponding to this price. We have

$$p = 70 - 0.02q$$
$$20 = 70 - 0.02q$$
$$-50 = -0.02q$$
$$2500 = q.$$

At a price of \$20, attendance will be 2500 people, and so revenue will be $(2500)(20)$, or \$50,000. Notice that, although demand is reduced, the revenue is still higher at a price of \$20 than at a price of \$10.
  (b) Since Revenue = price × quantity, we have

$$R(q) = p \cdot q$$
$$= (70 - 0.02q)q$$
$$= 70q - 0.02q^2.$$

  (c) To maximize revenue, we find the critical points of the revenue function $R(q) = 70q - 0.02q^2$:

$$R'(q) = 70 - 0.02(2q)$$
$$0 = 70 - 0.04q$$
$$70 = 0.04q$$
$$1750 = q.$$

Maximum revenue is achieved when attendance at the amusement park is 1750 people.
  (d) We find the price corresponding to an attendance of 1750, using the demand equation:

$$p = 70 - 0.02(1750) = 70 - 35 = 35.$$

The optimal price for an all-day pass at the amusement park is \$35.

(e)   When the optimal price of $35 is charged, the attendance at the park is 1750 people. Thus, the maximum revenue is $R = pq = (35)(1750) = 61{,}250$. The maximum revenue is $61,250. The corresponding profit cannot be determined without knowledge of the costs.

Example 5   A white-water rafting company knows that at a price of $80 for a half-day trip, they will attract 300 customers. For every $5 decrease in price, they attract about an additional 30 customers. What price should the company charge in order to maximize revenue?

Solution   We first find the demand equation relating price to demand. Let's make a table of values first to be sure that we understand what is going on. If price, p, in dollars, is 80, the number of trips sold, q, is 300. If p is 75, then q is 330, and so on as in Table 6.2.

You should notice that this table of values corresponds to a linear function, so the demand q is a linear function of the price p. The slope is $30/(-5) = -6$, so the demand function is $q = -6p + b$, where b is the vertical intercept. Since $(80, 300)$ is a point on the line, we have

$$q = -6p + b$$
$$300 = -6(80) + b$$
$$300 = -480 + b$$
$$780 = b$$

The demand equation is $q = -6p + 780$. Since revenue $R = p \cdot q$, the revenue equation as a function of price is $R = p(-6p + 780) = -6p^2 + 780p$. The graph of this revenue function is given in Figure 6.36, and we see the maximum there. To find the maximum analytically, we differentiate the revenue function and find the critical points:

$$R' = -12p + 780 = 0$$
$$p = \frac{780}{12} = 65.$$

Maximum revenue is achieved when the price is $65.

TABLE 6.2   *Demand for rafting trips*

Price, p (dollars)	80	75	70	65	$\cdots$
Trips sold, q	300	330	360	390	$\cdots$

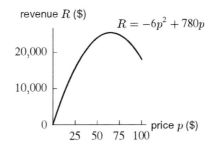

Figure 6.36: Revenue for a rafting company

Problems for Section 6.2

1. Plot the graph of $f(x) = x^3 - e^x$ using a graphing calculator or computer to find all local and global maxima and minima for: (a) $-1 \leq x \leq 4$ (b) $-3 \leq x \leq 2$

2. For $y = f(x) = x^{10} - 10x$, and $0 \leq x \leq 2$, find the value(s) of x for which:

   (a) $f(x)$ has a local maximum or local minimum. Indicate which ones are maxima and which are minima.

   (b) $f(x)$ has a global maximum or global minimum. Indicate which ones are maxima and which are minima.

3. For $f(x) = x - \ln x$, and $0.1 \leq x \leq 2$, find the value(s) of x for which:

   (a) $f(x)$ has a local maximum or local minimum. Indicate which ones are maxima and which are minima.

   (b) $f(x)$ has a global maximum or global minimum. Indicate which ones are maxima and which are minima.

4. For $f(x) = \sin^2 x - \cos x$, and $0 \leq x \leq \pi$, find, to two decimal places, the value(s) of x for which:

   (a) $f(x)$ has a local maximum or local minimum. Indicate which ones are maxima and which are minima.

   (b) $f(x)$ has a global maximum or global minimum. Indicate which ones are maxima and which are minima.

5. If $R(q) = 450q$ and $C(q) = 10,000 + 3q^2$, at what quantity is profit maximized? What is the total profit at this production level?

6. Assume the cost function is given by $C(q) = q^3 - 60q^2 + 1200q + 1000$ for $0 \leq q \leq 50$. If the product sells for \$588 each, what production level will maximize profit? Find the total cost, total revenue, and total profit at this production level. Sketch a graph of the cost and revenue functions on the same axes, and label the production level at which profit is maximized, and the corresponding cost, revenue, and profit on your graph. (As you sketch the graph, it may help to know that the costs can go as high as \$35,000.)

7. Cost and revenue functions for a certain product are shown in Figure 6.37, with quantity q in thousands of units and cost and revenue in thousands of dollars. For what production levels is the profit function positive? Negative? Estimate the point at which profit would be maximized.

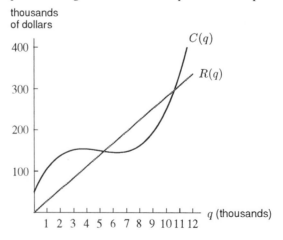

Figure 6.37: At what quantity is profit maximized?

8. Suppose that at a production level of 2000 for a given product, marginal revenue is $4 per unit and marginal cost is $3.25 per unit. Do you expect maximum profit to occur at a production level above or below 2000? Explain.

9. In Figure 6.34 on page 331, the points where marginal revenue equals marginal cost are labeled q_1 and q_2.

    (a) On the graph of the corresponding total cost and total revenue functions given in Figure 6.38, label the corresponding points q_1 and q_2. Explain in terms of slopes the significance of these points.

    (b) Explain in terms of profit why one is a local minimum and one is a local maximum.

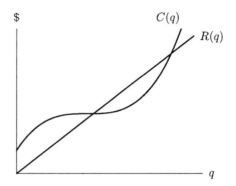

Figure 6.38: Total cost and total revenue

10. Graphs of the marginal revenue and marginal cost curves for a certain product are shown in Figure 6.39. At approximately what quantity is profit maximized? Explain.

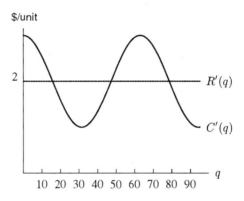

Figure 6.39: Marginal cost and marginal revenue

11. Suppose the total cost $C(q)$ of producing q goods is given by:

$$C(q) = 0.01q^3 - 0.6q^2 + 13q.$$

    (a) What is the fixed cost?

    (b) What is the maximum profit if each item is sold at a price of $7? (Assume you can sell everything you produce.)

    (c) At a fixed production level of 34 goods, for each $1 increase in price, 2 fewer goods are sold. Should you raise the price, and if so by how much?

12. Assume that the demand equation for a certain product is given by $p = 45 - 0.01q$. Write the revenue function as a function of q, and find the quantity that will maximize revenue. What price corresponds to this quantity, and what is the total revenue at this point?

13. At a price of $8 per ticket, a musical theater group can fill every seat in the theater, which has a maximum capacity of 1500. For every additional dollar charged, the number of people buying tickets goes down by 75. What ticket price will maximize revenue?

14. Suppose you run a small independent furniture business. Your assistant signs a deal with a customer to deliver up to 400 chairs, the exact number to be determined by the customer later. The price will be $90 per chair up to 300 chairs, and above 300, the price will be reduced by $0.25 per chair (on the whole order) for every additional chair over 300 ordered. What are the largest and smallest revenues your company can make under this deal?

15. Assume that the demand equation for a product is $p = b_1 - a_1 q$, and that the cost function is $C(q) = b_2 + a_2 q$, where p is the price of the product and q is the quantity sold. (Assume b_1, a_1, b_2, and a_2 are all positive.) Find the value of q, in terms of other variables, that will maximize profit.

16. Suppose a company manufactures only one product. The quantity produced, q, of this product depends on the amount of capital, K, invested (i.e., the number of machines the company owns, the size of its building, and so on) and the amount of labor, L, available. It is often assumed that q can be expressed as a function of K and L by a *Cobb-Douglas production function*:

$$q = cK^\alpha L^\beta$$

where c, α, β are positive constants, with $0 < \alpha < 1$ and $0 < \beta < 1$.

In this problem we will see how the Russian government could use a Cobb-Douglas function to estimate how many people a newly privatized industry might employ. A company in such an industry will have only a small amount of capital available to it, and will need to use all of it; K is therefore fixed. Suppose L is measured in man-hours, and that each man-hour costs the company w rubles (a ruble is the unit of Russian currency). Suppose that the company has no other costs besides labor, and that each unit of the good can be sold for a fixed price of p rubles. How many man-hours of labor should the company use in order to maximize its profit?

6.3 MORE OPTIMIZATION: AVERAGE COST

An individual company in a large industry always wants to maximize its profit. However it alone cannot affect demand for its product because other companies beyond its control also produce and sell the same product. The mechanism of the market will tend to force the supply of the product for the industry as a whole to the level at which price equals the minimum average cost of each firm, even though this does not maximize the profit for the industry as a whole. (This is why companies have an incentive to collude - by doing so they can increase their profit.) In this section we study the quantity of production that yields the minimum average cost.

What Is Average Cost?

We start with an example:

Example 1   A regional yogurt company has cost function $C(q) = 0.01q^3 - 0.6q^2 + 13q + 1000$ (in dollars), where q is the number of cases of yogurt produced. Find the average cost per case if 100 cases are produced.

Solution   The total cost of producing the 100 cases is given by

$$C(100) = 0.01(100^3) - 0.6(100^2) + 13(100) + 1000 = \$6300.$$

The total cost of producing the 100 cases is \$6300. We can find the average cost per case by dividing by 100, the number of cases produced.

$$\text{Average cost} = \frac{6300}{100} = 63 \text{ dollars/case.}$$

If 100 cases of yogurt are produced, the average cost per case is \$63.

The *average cost* is the cost per unit of producing a certain quantity; it is the total cost divided by the number produced.

The **average cost**, $a(q)$, of producing a quantity q is given by $a(q) = C(q)/q$.

Be careful not to confuse the average cost (the cost per unit of producing a certain quantity) with the marginal cost (the cost of producing the next one).

Example 2   If the cost function, in dollars, is $C(q) = 1000 + 20q$, where q is the number of units produced, find the marginal cost to produce the 100th unit, and find the average cost of producing 100 items.

Solution   This is a linear cost function with constant variable costs of \$20 per unit, so the marginal cost for each unit is \$20. (Another way to see this is that the graph of the cost function has a constant slope of 20, and so the derivative of the cost function is 20 at every point. Since marginal cost is equal to the derivative of the cost function, the marginal cost is \$20 per unit.) After 99 units have been produced, it costs \$20 to produce the next one.

The average cost of producing 100 units is given by

$$a(100) = \frac{C(100)}{100} = \frac{3000}{100} = 30 \text{ dollars/unit.}$$

The average cost of producing 100 units is \$30 per unit. Notice that the average cost includes the fixed costs spread over the entire production, whereas marginal cost does not, so average cost is greater than marginal cost for this example.

How Can We Visualize Average Cost?

How can we visualize the average cost on a graph? We know that average cost, $a(q)$, is total cost divided by quantity, so we have $a(q) = C(q)/q$. Since we can subtract zero from any number without changing it, we have

$$a(q) = \frac{C(q)}{q} = \frac{C(q) - 0}{q - 0}.$$

This is a difference quotient, equal to the slope of the line between the points $(0,0)$ and $(q, C(q))$. In other words, it is the slope of the line from the origin to the point $(q, C(q))$ on the cost curve. See Figure 6.40.

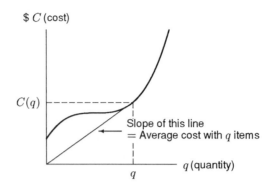

Figure 6.40: Visualizing average cost as a slope

Average cost $= \frac{C(q)}{q} =$ Slope of the line from the origin to the point $(q, C(q))$.

Minimizing Average Cost

Example 3   The graph of a cost function is given in Figure 6.41. Mark on the graph the quantity at which the average cost is minimized.

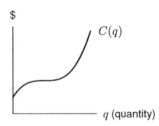

Figure 6.41: A cost function

Solution   The average cost of producing a quantity q is given by the slope of the line through the origin to the graph of $C(q)$ at q. In Figure 6.42, several of these lines have been drawn at points q_1, q_2, q_3, and q_4. The slopes of these lines are steep for small q, become less steep as q increases, and then get steeper again as q continues to increase. Thus, as q increases, the average costs decrease and then increase, and so there is a minimum value. This minimum occurs at the point q_0 in Figure 6.42.

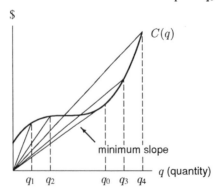

Figure 6.42: Finding the minimum average cost

Let's examine the production level q_0 in Figure 6.42 more closely. At this point, a line from the origin to the cost curve has minimum slope. We see in Figure 6.42 that this minimum occurs at the point where the line from the origin is tangent to the cost curve. At this point, the slope of the line from the origin to the cost curve is equal to the slope of the tangent line to the cost curve. Since the slope of the line from the origin to the cost curve is equal to average cost, and the slope of the tangent line to the cost curve is equal to marginal cost, we see that the minimum average cost occurs at the production level for which average cost equals marginal cost.

In the next example we consider what happens when marginal cost and average cost are not equal.

Example 4 Suppose 100 items are produced at an average cost of $2 per item. Find the average cost of producing 101 items:
(a) if the marginal cost to produce the 101st item is $1.
(b) if the marginal cost to produce the 101st item is $3.

Solution (a) If 100 items are produced at an average cost of $2 per item, the total cost of producing the items is $200. Since the marginal cost to produce the 101st item is $1, it costs $1 more to produce the additional item. Total costs for producing 101 items will therefore be $201. The average cost to produce these items is 201/101, or $1.99 per item. The average cost has gone down. This makes sense: if it costs less than the average to produce additional items, producing them will decrease the average cost.
(b) In this case, the marginal cost to produce the 101st item is $3. The total cost to produce 101 items will be $203 and the average cost will be 203/101, or $2.01 per item. Our average cost has gone up. Again, this makes sense: if it costs more than the average to produce additional items, average costs will increase when these items are produced.

Our graphical analysis of Figure 6.42 and the results of Example 4 reveal the relationship between marginal cost and average cost.

Average cost and marginal cost are related as follows:
1. Minimum average cost occurs when marginal cost equals average cost.
2. If marginal cost is less than average cost, average cost is reduced by increasing production.
3. If marginal cost is greater than average cost, average cost is increased by increasing production.

We can demonstrate (1) analytically. Recall that average cost is given by $a(q) = C(q)/q$. To minimize average cost, we find the critical points of $a(q)$.

Since $a(q) = C(q)/q = C(q)q^{-1}$, we use the product rule to find $a'(q)$:

$$a'(q) = C'(q)(q^{-1}) + C(q)(-q^{-2}) = \frac{C'(q)}{q} + \frac{-C(q)}{q^2} = \frac{qC'(q) - C(q)}{q^2}.$$

At critical points we have $a'(q) = 0$, so

$$\frac{q \cdot C'(q) - C(q)}{q^2} = 0$$

$$q \cdot C'(q) - C(q) = 0$$
$$q \cdot C'(q) = C(q)$$
$$C'(q) = \frac{C(q)}{q}$$

Marginal cost $=$ Average cost.

Example 5 Assume that the total cost function is given by $C(q) = q^3 - 6q^2 + 12q$, in thousands of dollars, where q is measured in thousands and $0 \le q \le 5$.

(a) Sketch a graph of $C(q)$ and estimate visually the quantity at which average cost is minimized.

(b) Determine analytically the exact value of q at which average cost is minimized.

(c) Sketch a graph of the average cost function, and use the graph to find the minimum average cost.

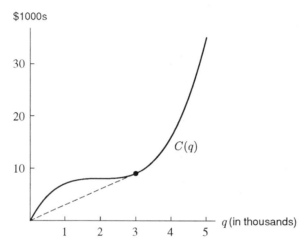

Figure 6.43: Minimizing average cost

Solution (a) A graph of $C(q)$ is given in Figure 6.43. Average cost is minimized at the point where a line between the origin and a point on the curve has minimum slope. This minimum slope occurs where the line through the origin is tangent to the curve. In Figure 6.43, this point is at approximately $q = 3$, corresponding to a production of 3000 units.

(b) We must find the quantity for which marginal cost equals average cost. Marginal cost equals the derivative $C'(q) = 3q^2 - 12q + 12$, so we have

$$\text{Marginal cost } = \text{ Average cost}$$
$$3q^2 - 12q + 12 = \frac{q^3 - 6q^2 + 12q}{q}$$
$$3q^2 - 12q + 12 = q^2 - 6q + 12$$
$$2q^2 - 6q = 0$$
$$2q(q - 3) = 0$$

There are two roots, at $q = 0$ and at $q = 3$. Since $q = 0$ is not a meaningful answer, we see that the average cost is minimized at $q = 3$. To minimize average costs, production should be set at 3000 units.

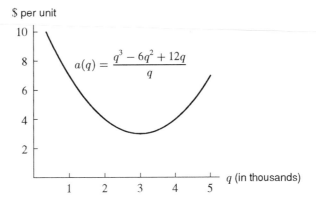

Figure 6.44: Graph of the average cost function

(c)   Since average cost is equal to total cost divided by quantity, we have

$$a(q) = \frac{C(q)}{q} = \frac{q^3 - 6q^2 + 12q}{q} = q^2 - 6q + 12.$$

A graph of this function is shown in Figure 6.44. We see graphically that the minimum average cost occurs at $q = 3$, which is a production of 3000 units. At this production level, the average cost is 3 dollars per unit, and the total cost is $9000.

Problems for Section 6.3

1.   The graph of a cost function is given in Figure 6.45.

(a)   Estimate the average cost at $q = 30$, and represent your answer graphically.

(b)   Estimate the marginal cost at $q = 30$, and represent your answer graphically.

(c)   At approximately what value of q is average cost minimized?

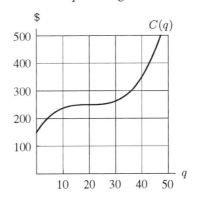

Figure 6.45: A cost function

2.   Suppose a cost function, in dollars, is given by $C(q) = 2500 + 12q$, where q is the number of items produced.

(a)   What is the marginal cost of producing the 100th item? the 1000th item?

(b)   What is the average cost of producing 100 items? 1000 items?

3. An agricultural worker in Uganda is interested in planting clover to increase the number of bees making their home in the region. There are 100 bees in the region naturally, and for every acre put under clover, 20 more bees are found in the region.

   (a) Draw a graph of the total number, $N(x)$, of bees as a function of x, the number of acres devoted to clover.

   (b) Explain, both geometrically and algebraically, the shape of the graph of:

      (i) The marginal rate of increase of the number of bees with acres of clover, $N'(x)$.

      (ii) The average number of bees per acre of clover, $N(x)/x$.

4. Two different possible cost functions are given in Figure 6.46.

   (a) The graph labeled (a) is concave down. For this cost function, is there a value of q at which average cost is minimized? If so, approximately where?

   (b) The graph labeled (b) is concave up. For this cost function, is there a value of q at which average cost is minimized? If so, approximately where?

   Explain your answers graphically.

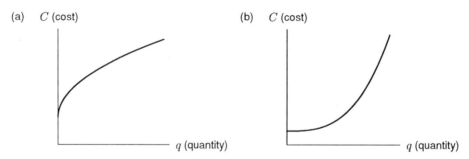

Figure 6.46: Cost functions with different concavity

5. Consider the cost function whose graph is given in Figure 6.47. Sketch a graph of the corresponding average cost function.

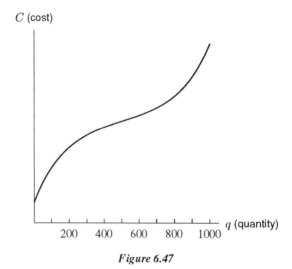

Figure 6.47

6. Let $C(q) = 0.04q^3 - 3q^2 + 75q + 96$ be the total cost of producing q items.

   (a) Find the average cost per item as a function of q.
   (b) Use a graphing calculator or computer to sketch a graph of average cost against q.
   (c) For what values of q is the average cost per item decreasing?
   (d) For what values of q is the average cost per item increasing?
   (e) For what value of q is the average cost per item smallest and what is the average cost per item at that point?

7. Assume that the total cost function for a product is given by $C(q) = q^3 - 12q^2 + 48q$, in thousands of dollars, where q is measured in thousands, and $0 \le q \le 12$.

   (a) Sketch a graph of $C(q)$ and visually estimate on your graph the quantity where average cost is minimized.
   (b) Determine analytically the exact value of q at which average cost is minimized.

8. A cost function is given in Figure 6.48.

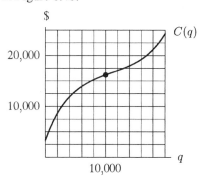

Figure 6.48

   (a) Find the average cost when the production level is 10,000 units and interpret it.
   (b) Represent your answer to part (a) graphically.
   (c) At approximately what production level is average cost minimized?

9. A reasonably realistic model of a firm's costs is given by the *short-run Cobb-Douglas cost curve*
   $$C(q) = Kq^{1/a} + F,$$
   where a is a positive constant, F is the fixed costs, and K measures the technology available to the firm.

   (a) Show that C is concave down if $a > 1$.
   (b) Assuming that average cost is minimized when average cost equals marginal cost, find what value of q minimizes the average cost.

10. Suppose a firm produces a quantity q of some good and that the average cost per item is given by:
    $$a(q) = 0.01q^2 - 0.6q + 13, \quad \text{for} \quad q > 0.$$

    (a) What is the total cost, $C(q)$, of producing q goods?
    (b) What is the minimum marginal cost? What is the practical interpretation of this result?
    (c) At what production level is the average cost a minimum? What is the lowest average cost?
    (d) Compute the marginal cost at $q = 30$. How does this relate to your answer to part (c)? Explain this relationship both analytically and qualitatively.

11. Suppose you are given the graph of the average cost $a(q)$ in Figure 6.49.

    (a) Show that if

    $$a(q) = b + mq$$

    then

    $$C'(q) = b + 2mq.$$

    (b) Sketch a graph of the marginal cost $C'(q)$.

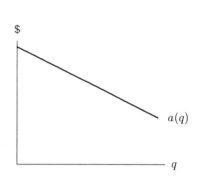

Figure 6.49: Average cost of a good   **Figure 6.50**

12. $C(q)$ is the total cost of producing a quantity q. The average cost $a(q)$ is given in Figure 6.50. The following rule is used by economists to determine the marginal cost $C'(q_0)$, for any q_0:

    • Construct the tangent t_1 to $a(q)$ at q_0.
    • Let t_2 be the line with the same vertical intercept as t_1 but with twice the slope of t_1.

    Then $C'(q_0)$ is as shown in Figure 6.50. Explain why this rule works.

6.4 APPLICATIONS OF THE DEFINITE INTEGRAL

The Definite Integral Revisited

In Chapter 3, we defined the definite integral as a limit of left or right Riemann sums. We begin by summarizing what we have learned about the definite integral.

Computing the Definite Integral

• If $f(x)$ is given by a table of values, then we can estimate $\int_a^b f(x)dx$ by using a left-hand sum or a right-hand sum, or (for better accuracy), the average of the left and right sums.

• If $f(x)$ is given by a graph, then we can estimate $\int_a^b f(x)dx$ by estimating the area between the graph of f and the x-axis (with areas below the x-axis counted negatively) between a and b.

• If $f(x)$ is given by a formula, then we can use technology to compute $\int_a^b f(x)dx$, or (if the function has a known antiderivative) we can use the Fundamental Theorem of Calculus.

Interpretations of the Definite Integral

Several interpretations of the definite integral were introduced in Chapter 3 , and we summarize them here.

- The integral $\int_a^b f(x)dx$ represents the **area** between the graph of f and the x-axis (with areas below the x-axis counted negatively) between a and b. (See Figure 6.51.)

- The **average value** of f over the interval $a \le x \le b$ is defined in terms of the integral:

$$\begin{array}{c} \text{Average value of } f \\ \text{from } a \text{ to } b \end{array} = \frac{1}{b-a} \int_a^b f(x)\,dx.$$

The average value of f is the height of the rectangle with base $(b - a)$ and whose area equals the area under the graph of $f(x)$ between $x = a$ and $x = b$. (See Figure 6.52.)

- If $f(x)$ gives the rate of change of a quantity, then the integral $\int_a^b f(x)dx$ gives the **total change** in the quantity between a and b.

- If $f(x)$ is the derivative of a function $F(x)$, then the integral $\int_a^b f(x)dx$ helps us understand the function $F(x)$, since we know from the **Fundamental Theorem of Calculus** that, if $F'(x) = f(x)$, then

$$\int_a^b f(x)dx = F(b) - F(a).$$

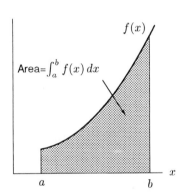

Figure 6.51: Area as an integral

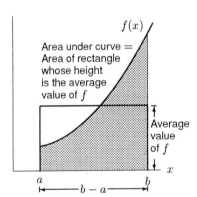

Figure 6.52: Average value

Example 1   The graph of a function $f(x)$ is given in Figure 6.53.

(a)   Evaluate $\int_0^5 f(x)\,dx$.

(b)   Find the average value of $f(x)$ on the interval $x = 0$ to $x = 5$. Check your answer graphically.

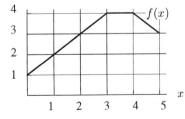

Figure 6.53: Estimate $\int_0^5 f(x)dx$

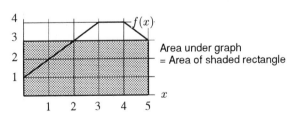

Figure 6.54: Average value of $f(x)$ is 3

Solution (a) Since $f(x) \geq 0$ on the interval from 0 to 5, the integral is equal to the area of the region under the graph of $f(x)$ between $x = 0$ and $x = 5$. From Figure 6.53 we see that this region consists of 13 full boxes and 4 half boxes, each box of area 1, for a total area of 15, and so we have

$$\int_0^5 f(x)\, dx = 15.$$

(b) The average value of $f(x)$ on the interval from 0 to 5 is given by

$$\text{Average value} = \frac{1}{5-0} \int_0^5 f(x)\, dx = \frac{1}{5}(15) = 3.$$

We can check our answer graphically by drawing a horizontal line at $y = 3$ on a graph of $f(x)$. (See Figure 6.54.) We see that, between $x = 0$ and $x = 5$, the area under the graph of $f(x)$ is equal to the area of the rectangle with height 3.

Graphing a Function Given a Graph of its Derivative

Suppose we have the graph of f' and we want to sketch the graph of f. We know that when f' is positive, f is increasing, and when f' is negative, f is decreasing. In other words, when the graph of f' lies above the x-axis, f is increasing, and when the graph of f' lies below the x-axis, f is decreasing. If we want to know exactly how much f increases or decreases, we compute the area between the x-axis and the graph of f'.

Example 2 The graph of the derivative $f'(x)$ of a function $f(x)$ is shown in Figure 6.55, and the values of some areas are given. If $f(0) = 10$, sketch a graph of the function $f(x)$ and give the coordinates of the local maxima and minima.

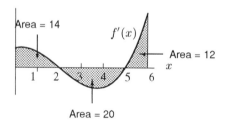

Figure 6.55: The graph of a derivative f'

Solution We see in Figure 6.55 that the derivative f' is positive between 0 and 2, negative between 2 and 5, and positive between 5 and 6. Therefore, the function f is increasing between 0 and 2, decreasing between 2 and 5, and increasing between 5 and 6. It must have the general shape shown in Figure 6.56. We see that there is a local maximum at $x = 2$ and a local minimum at $x = 5$. Notice that we can sketch the general shape of the graph of f without knowing any areas. We use the areas to sketch a more precise graph of this function.

We are told that $f(0) = 10$, so we begin by plotting the point $(0, 10)$. We see in Figure 6.55 that

$$\int_0^2 f'(x)\, dx = 14.$$

Therefore, the total change in f between 0 and 2 is 14, and the value of f increases 14 units between $x = 0$ and $x = 2$. Since $f(0) = 10$, we have $f(2) = 10 + 14 = 24$. The point $(2, 24)$ is a point on the graph of $f(x)$.

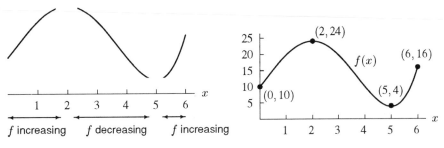

| *Figure 6.56*: The shape of f | *Figure 6.57*: The graph of f |

The total change in f between 2 and 5 is $\int_2^5 f'(x)\,dx$. We see in Figure 6.55 that the area between $x = 2$ and $x = 5$ is 20. Since this area lies entirely below the x-axis, we have

$$\int_2^5 f'(x)\,dx = -20.$$

The total change in f is -20. In other words, $f(x)$ decreases by 20 units between $x = 2$ and $x = 5$. Since $f(2) = 24$, we have $f(5) = 24 - 20 = 4$. Thus, the point $(5, 4)$ lies on the graph. Finally, between $x = 5$ and $x = 6$, the total change in $f(x)$ is an increase of 12 units, so $f(6) = 4 + 12 = 16$. We plot the point $(6, 16)$ on our graph. See Figure 6.57.

Applications of the Definite Integral

Example 3   After a forest fire, vegetation grows back quickly. For a certain species of tree, the number of new trees per square mile taking root each year is approximated for 30 years after the fire by

$$r(t) = 70 - 9\sqrt{t},$$

where t is the number of years since the forest fire. How many trees will grow during the first 10 years after a fire?

Solution   The function $r(t)$ gives the rate in trees per year at which new trees are appearing. The total number of new trees during the first 10 years is equal to the definite integral:

$$\text{Total number of new trees } = \int_0^{10} r(t)\,dt = \int_0^{10} (70 - 9\sqrt{t})\,dt.$$

We compute this integral using a calculator or by estimating the area under the graph of $r(t)$. (See Figure 6.58.) We have

$$\int_0^{10} (70 - 9\sqrt{t})\,dt = 510.3$$

About 510 new trees per square mile will appear during the first 10 years after the forest fire.

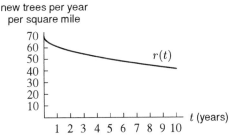

Figure 6.58: Rate of population growth of trees

Example 4   The marginal cost function to produce q units of a certain product is approximated by

$$C'(q) = q^2 - 40q + 500 \text{ dollars per unit.}$$

If the fixed costs of production are \$2000, estimate the total cost to produce 30 units.

Solution   The total cost to produce 30 units is equal to $C(30)$. Using the Fundamental Theorem of Calculus, we have

$$C(30) = C(0) + \int_0^{30} C'(q)\, dq.$$

Since the fixed costs of production are \$2000, we know $C(0) = 2000$. We can estimate $\int_0^{30} C'(q)\, dq$ using a calculator or by estimating the area under the graph of $C'(q)$. (See Figure 6.59.) We have

$$\int_0^{30} C'(q)\, dq = \int_0^{30} (q^2 - 40q + 500)\, dq = \$6000.$$

Therefore, $C(30) = \$2000 + \$6000 = \$8000$. The total cost to produce 30 items is approximately \$8000.

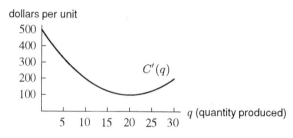

Figure 6.59: A marginal cost function

Example 5   Table 6.3[1] gives the emissions of nitrogen oxides in millions of metric tons per year in the United States from 1940 to 1990. Estimate the total emissions of nitrogen oxide during this 50 year period.

TABLE 6.3   *Emissions of nitrogen oxides*

Year	1940	1950	1960	1970	1980	1990
NO_x (millions of tons/year)	6.9	9.4	13.0	18.5	20.9	19.6

Solution   Table 6.3 gives the rate of emissions of nitrogen oxide in millions of metric tons per year. To find the total emissions, we use left-hand and right-hand Riemann sums. We have

Left-hand sum $= (6.9)(10) + (9.4)(10) + (13.0)(10) + (18.5)(10) + (20.9)(10) = 687.$

Right-hand sum $= (9.4)(10) + (13.0)(10) + (18.5)(10) + (20.9)(10) + (19.6)(10) = 814.$

Average of left- and right-hand sums $= \dfrac{687 + 814}{2} = 750.5.$

The total emissions of nitrogen oxide between 1940 and 1990 is about 750 million metric tons.

Population Growth Rates

Example 6   The rates of growth of the populations of two species of plants (measured in new plants per year) are shown in Figure 6.60. Assume that the populations of the two species are equal at time $t = 0$.

(a)   Which population is larger after one year?

(b)   Which population is larger after two years?

(c)   How much does the population of species 1 increase during the first two years?

---
[1] Statistical Abstract of the U.S., 1992

new plants per year

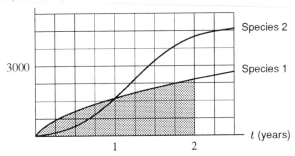

Figure 6.60: Population growth rates for two species

Solution   (a)   The rate of growth of the population of species 1 is higher than that of species 2 throughout the first year, and so the population of species 1 is larger after one year.

(b)   After two years, the situation is less clear, since the population of species 1 was increasing faster for the first year and that of species 2 for the second. However, we can use the fact that, if $r(t)$ is the rate of growth of the population

$$\text{Total change in population during first two years} = \int_0^2 r(t)dt.$$

This definite integral is the area under the graph of the function $r(t)$ between 0 and 2. Since the area in Figure 6.60 representing the total population change for species 2 is clearly larger than the area representing the total population change for species 1, the population of species 2 is larger after two years.

(c)   The population change for species 1 equals the area of the region under the graph of $r(t)$ between $t = 0$ and $t = 2$ in Figure 6.60. The region consists of approximately 16.5 boxes, each of area (750 plants/year)(0.25 year) = 187.5 plants, giving a total of (16.5)(187.5) = 3093.75 plants. The population of species 1 goes up by about 3100 plants during the two years.

Bioavailability in Drug Treatments

An important application of the definite integral in pharmacology and drug toxicity is to measure the overall presence of a drug in the bloodstream during the course of a treatment, known as the *bioavailability* of the drug. Unit bioavailability represents 1 unit concentration of the drug in the bloodstream for 1 hour. For example, a concentration of $3\mu g/cm^3$ in the blood for 2 hours would have bioavailability equal to $(3)(2) = 6(\mu g/cm^3)$-hours.

Ordinarily the concentration of a drug in the blood is not constant. If the drug is given orally or injected into the muscle, the concentration in the blood at first increases as the drug is absorbed into the bloodstream, and then the concentration gradually decreases back to zero as the drug is broken down and excreted. See Figure 6.61 for a typical graph of concentration as a function of time.[2]

Suppose that we want to calculate the bioavailability of a drug that is in the blood with concentration $C(t)\mu g/cm^3$ at time t for the time period $0 \leq t \leq T$. In order to use what we know about bioavailability for constant concentrations, we first divide up the time interval into many short intervals during each of which the concentration changes very little and so is approximately constant. During the interval from t to $t + \Delta t$

$$\text{bioavailability} \approx \text{Concentration} \times \text{Time}$$
$$= C(t)\Delta t$$

---

[2]*Drug Treatment*, Graeme S. Avery (Ed.), Adis Press, 1976.

concentration of drug in blood stream

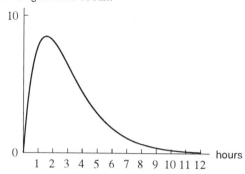

Figure 6.61: Drug concentration curve following oral dose

Summing over all subintervals gives

$$\text{Total bioavailability} \approx \sum C(t)\Delta t.$$

In the limit as $\Delta t \to 0$, we get the following integral:

$$\text{Bioavailability} = \int_0^T C(t)\,dt$$

It follows that the total bioavailability of a drug is equal to the area under the graph of the drug concentration function $C(t)$.

Example 7   Blood concentration profiles of two drugs given at time $t = 0$ are shown in Figure 6.62.[3] Discuss the differences and similarities between the two drugs in terms of peak concentration, speed of absorption into the bloodstream, and total bioavailability.

concentration of drug in plasma

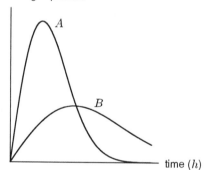

Figure 6.62: Comparing two drug products

Solution   We see that product A has a peak concentration more than twice as high as that of product B. Because product A achieves peak concentration sooner than product B, we can say that product A is absorbed more rapidly into the blood stream than product B. Finally, product A has greater total bioavailability, since we see that the area under the graph of the concentration function for product A is greater than the area under the graph for product B.

---

[3]*Drug Treatment*, Graeme S. Avery (Ed.), Adis Press, 1976.

Problems for Section 6.4

1.   If the graph of f is in Figure 6.63:
     (a)   What is $\int_0^6 f(x)\,dx$?
     (b)   What is the average value of f on the interval $x = 0$ to $x = 6$?

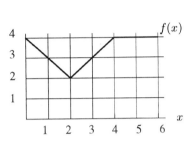

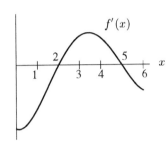

Figure 6.63   *Figure 6.64*

2.   The graph of the derivative f' of a function f is given in Figure 6.64.
     (a)   Where is f increasing and where is it decreasing? What are the x-coordinates of the local maxima and minima of f?
     (b)   Give a possible rough sketch of f. (You don't need a scale on the vertical axis.)

3.   The graph of the derivative g' of g is given in Figure 6.65. If $g(0) = 0$, sketch a graph of g. Give (x, y)-coordinates for all local maxima and minima.

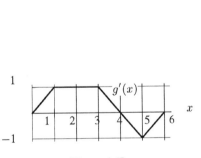

Figure 6.65   *Figure 6.66*

4.   The graph of the derivative F' of F is given in Figure 6.66, with some areas labeled. If $F(0) = 14$, sketch a graph of F. Give (x, y)-coordinates for all local maxima and minima.

5.   Consider the costs of drilling an oil well. The marginal costs depend on the depth at which you are drilling; drilling becomes more expensive, per meter, as you dig deeper into the earth. Suppose the fixed costs are 1,000,000 riyals (the riyal is the unit of currency of Saudi Arabia), and the marginal costs are

$$C'(x) = 4000 + 10x$$

in riyals/meter, where x is the depth in meters. Find the total cost of drilling a well 500 meters deep in Saudi Arabia.

6. Ice is forming on a pond at a rate given by

    $$\frac{dy}{dt} = \frac{\sqrt{t}}{2} \text{ inches per hour,}$$

    where y is the thickness of the ice in inches at time t measured in hours since the ice started forming.

    (a) Estimate the thickness of the ice after 8 hours.
    (b) At what rate is the thickness of the ice increasing after 8 hours?

7. In 1987 the average per capita income in the US was $26,000. Suppose that average per capita income is increasing at a rate in dollars per year given by

    $$r(t) = 480(1.024)^t,$$

    where t is the number of years since 1987. Estimate the average per capita income in 1995.

8. The marginal revenue function for selling a product is given by $R'(q) = 200 - 12\sqrt{q}$ dollars per unit, where q indicates the number of units sold.

    (a) Sketch the graph of $R'(q)$.
    (b) Estimate the total revenue in selling 100 units.
    (c) What is the marginal revenue at 100 units? Use this value and your answer to part (b) to estimate the total revenue in selling 101 units.

9. A manufacturer of mountain bikes has marginal cost function

    $$C'(q) = \frac{600}{0.3q + 5},$$

    where q is the quantity of bicycles produced.

    (a) If the fixed cost in producing the bicycles is $2000, find the total cost to produce 30 bicycles.
    (b) If the bikes are sold for $200 each, what is the profit (or loss) on the first 30 bicycles?
    (c) Find the marginal profit on the 31st bicycle.

10. The rate of sales (measured in sales per month) of a company during the year is given by

    $$r(t) = t^4 - 20t^3 + 118t^2 - 180t + 200,$$

    where t is measured in months since January 1.

    (a) Sketch a graph of the rate of sales per month during the first year ($t = 0$ to $t = 12$). Does it appear that more sales were made during the first half of the year, or during the second half?
    (b) Estimate the total sales of the company during the first 6 months of the year.
    (c) Estimate the total sales of the company during the last 6 months of the year.
    (d) What are the total sales for the entire year?
    (e) What are the average sales per month for the company during the year?

11. The rate at which the world's oil is being consumed is continuously increasing. The rate of consumption (in billions of barrels per year) can be approximated by $r(t) = 32e^{0.05t}$, where t is measured in years since the start of 1990.

    (a) Sketch the graph of $r(t)$.
    (b) Represent the total quantity of oil used between the start of 1990 and the start of 1995 as a definite integral.
    (c) Estimate the total quantity of oil used during this five year period.

12. A service station orders 100 cases of motor oil every 6 months, and the inventory of oil over a six-month period can be modeled by

$$f(t) = 100e^{-0.5t} \text{ cases of oil at time } t,$$

where t is measured in months since the order arrives.

   (a) How many cases are there at the start of the six-month period? How many cases are left at the end of the six-month period?
   (b) Find the average number of cases in inventory over the six-month period.

13. Annual coal production in the United States (in quadrillion BTU per year) is given in Table 6.4. Estimate the total amount of coal produced in the United States between 1960 and 1990. [4]

TABLE 6.4

Year	1960	1965	1970	1975	1980	1985	1990
Coal production (quadrillion BTU per year)	10.82	13.06	14.61	14.99	18.60	19.33	22.46

14. One of the earliest pollution problems brought to the attention of the Environmental Protection Agency (EPA) was the case of the Sioux Lake in eastern South Dakota. For years a small paper plant located nearby had been discharging waste containing carbon tetrachloride (CCl_4) into the waters of the lake. At the time the EPA learned of the situation, the chemical was entering at a rate of 16 cubic yards/year.

   The agency immediately ordered the installation of filters designed to slow (and eventually stop) the flow of CCl_4 from the mill. Implementation of this program took exactly three years, during which the flow of pollutant was steady at 16 cubic yards/year. Once the filters were installed, the flow declined. From the time the filters were installed until the time the flow stopped, the rate of flow was well approximated by

$$\text{Rate (in cubic yards/year)} = t^2 - 14t + 49,$$

where t is time measured in years since the EPA learned of the situation (thus $t \geq 3$).

   (a) Draw a graph showing the rate of CCl_4 flow into the lake as a function of time, beginning at the time the EPA first learned of the situation.
   (b) How many years elapsed between the time the EPA learned of the situation and the time the pollution flow stopped entirely?
   (c) How much CCl_4 entered the waters during the time shown in the graph in part (a)?

15. Two species of plants have been introduced into a region. Past history has shown that the rates of growth of the populations of the two plants follow the patterns shown in the graph in Figure 6.67. The populations of the two species are equal at time $t = 0$, and time t is measured in years.

   (a) Which species has a larger population at the end of 5 years? At the end of 10 years?
   (b) Which species do you think will have the larger population after 20 years? Explain.

---

[4]*World Almanac*, 1995

new plants per year

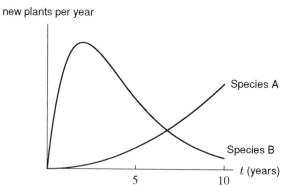

Figure 6.67: Growth rates for two species

16. Consider a bacteria population whose birth rate, B, is given by the curve in Figure 6.68 as a function of time in hours. The birth rate is in births per hour. The curve marked D gives the death rate (in deaths per hour) of the same population.

    (a) Explain what the shape of each of these graphs tells you about the population.
    (b) Use the graphs to find the time at which the net rate of increase of the population is at a maximum.
    (c) Suppose at time $t = 0$ the population has size N. Sketch the graph of the total number born by time t. Also sketch the graph of the number alive at time t. Use the given graphs to find the time at which the population size is a maximum.

bacteria/hour

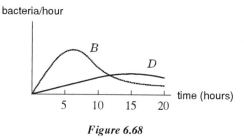

Figure 6.68

17. Figure 6.69 compares the concentration in blood plasma for two pain relievers. Compare the two products in terms of level of peak concentration, time until peak concentration, and overall bioavailability.

concentration of drug in plasma

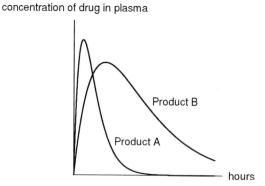

Figure 6.69: Plasma concentration curves for two pain relievers

18. Figure 6.70 shows plasma concentration curves for two drugs used to slow a rapid heart rate. Compare the two products in terms of level of peak concentration, time until peak concentration, and overall bioavailability.

concentration of drug in plasma

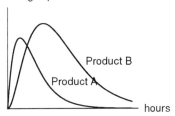

Figure 6.70: Plasma concentration curves for two drug products

19. Assume that, during the first four minutes after a foreign substance is introduced into the blood, the rate at which antibodies are made (in thousands of antibodies per minute) is given by

$$r(t) = \frac{t}{t^2 + 1},$$

where t is measured in minutes and $0 \leq t \leq 4$. Find the total quantity of antibodies in the blood at the end of the first four minutes.

20. The rate, r, at which people get sick during an epidemic of the flu can be approximated by $r = 1000te^{-0.5t}$, where r is measured in people/day and t is measured in days since the start of the epidemic.

   (a) Sketch a graph of r as a function of t.   (b) When are people getting sick fastest?
   (c) How many people get sick altogether?

21. The Quabbin Reservoir in the western part of Massachusetts provides most of Boston's water. The graph in Figure 6.71 represents the flow of water in and out of the Quabbin Reservoir throughout 1993.

   (a) Sketch a possible graph for the quantity of water in the reservoir, as a function of time.

rate of flow
(millions of gallons/day)

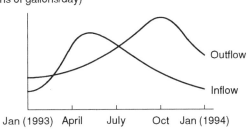

Figure 6.71

   (b) When, in the course of 1993, was the quantity of water in the reservoir largest? Smallest? Mark and label these points on the graph you drew in part (a).
   (c) When was the quantity of water decreasing most rapidly? Again, mark and label this time on both graphs.
   (d) By July 1994 the quantity of water in the reservoir was about the same as in January 1993. Draw plausible graphs for the flow into and the flow out of the reservoir for the first half of 1994. Explain your graph.

6.5 CONSUMER AND PRODUCER SURPLUS

Supply and Demand Curves

In a free market, the quantity of a certain good produced and sold can be described by the supply and demand curves of the good. As we saw in Chapter 1, the *supply curve* shows what quantity of the good the producers will supply at different price levels, and the *demand curve* shows what quantity of goods are bought at various prices. Typical supply and demand curves are shown in Figure 6.72. As we would expect, the supply curve is increasing (as price increases, the quantity supplied goes up) and the demand curve is decreasing (as price increases, the demand for the product goes down.) Recall that it is traditional to place price along the vertical axis and quantity along the horizontal axis, as we see in Figure 6.72.

It is assumed that the market will settle to the *equilibrium price* and *quantity*, p^* and q^*, where the graphs cross. This means that at the equilibrium point, a quantity q^* of a good will be produced and sold for a price of p^* per unit.

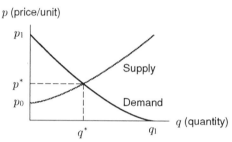

Figure 6.72: Supply and demand curves

Consumer and Producer Surplus

Notice that at equilibrium, a number of consumers have bought the item at a lower price than they would have been willing to pay. (For example, in Fig. 6.72 there are some consumers who would have been willing to pay prices up to p_1.) Similarly, there are some suppliers who would have been willing to produce the item at a lower price (down to p_0, in fact). We define the following terms:

> The **consumer surplus** represents the buyers' gain from trade. It equals the total amount gained by consumers by buying the good at the current price, rather than at the price they would have been willing to pay.
>
> The **producer surplus** represents the suppliers' gain from trade. It equals the total amount gained by producers by selling at the current price, rather than at the price they would have been willing to accept.
>
> In the absence of artificial price controls, the current price will equal the equilibrium price.

The consumers and producers both individually and collectively are richer merely for having traded. The consumer and producer surplus measure how much richer they are.

Suppose that all consumers who would buy at the equilibrium price actually buy the good at the maximum price they are willing to pay. To see how much they would pay, divide the interval from 0 to q^* into subintervals of length Δq. Figure 6.73 shows that a quantity Δq of items are sold at a price of about p_1, another Δq are sold for a slightly lower price of about p_2, the next Δq for a price of about p_3, and so on. Thus, the consumers' total expenditure is about

$$p_1\Delta q + p_2\Delta q + p_3\Delta q + \cdots = \sum p_i\Delta q.$$

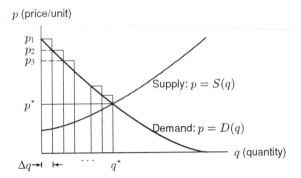

Figure 6.73: Calculation of consumer surplus

If the demand curve is given by the function $p = D(q)$, and if all consumers who were willing to pay more than p^* paid as much as they were willing, then as $\Delta q \to 0$, we would have

$$\begin{array}{c} \text{Consumer} \\ \text{expenditure} \end{array} = \int_0^{q^*} D(q)dq = \begin{array}{l} \text{Area under demand} \\ \text{curve from 0 to } q^*. \end{array}$$

Now if all goods are sold at the equilibrium price, the consumers' actual expenditure is only p^*q^*, the area of the rectangle between the axes and the lines $q = q^*$ and $p = p^*$. Thus the consumer surplus may be calculated as follows:

$$\begin{array}{c} \text{Consumer} \\ \text{surplus} \end{array} = \left(\int_0^{q^*} D(q)dq \right) - p^*q^* = \begin{array}{l} \text{Area between demand} \\ \text{curve and line } p = p^*. \end{array}$$

See Figure 6.74(a). Similarly, if the supply curve is given by the function $p = S(q)$, the producer surplus is represented in Figure 6.74(b), and defined as follows:

$$\begin{array}{c} \text{Producer} \\ \text{Surplus} \end{array} = p^*q^* - \left(\int_0^{q^*} S(q)dq \right) = \begin{array}{l} \text{Area between supply} \\ \text{curve and line } p = p^*. \end{array}$$

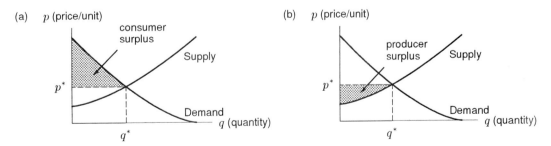

Figure 6.74: Consumer and producer surplus

Example 1   Suppose that the demand curve for a product is given by $p = 100e^{-0.008q}$ and the supply curve is given by $p = 4\sqrt{q} + 10$ for $0 \leq q \leq 500$, as in Figure 6.75. The price is in dollars.

(a) At a price of $50, what quantity will consumers be willing to buy and what quantity will producers be willing to supply? Do you think that the market will push prices up or down?

(b) Find the equilibrium price and equilibrium quantity. Since market forces tend to push prices closer to the equilibrium price, was your answer to the last question in part (a) right?

(c) If the price is equal to the equilibrium price, calculate the consumer surplus and interpret it.

(d) If the price is equal to the equilibrium price, calculate the producer surplus and interpret it.

Solution   (a) We compute the quantity demanded at a price of $50 using the demand equation $p = 100e^{-0.008q}$. Since $p = 50$, we set $50 = 100e^{-0.008q}$ and solve for q. Equivalently, we can work with the demand curve, locating the quantity q on the horizontal axis that corresponds to a price of $50 on the vertical axis. We see that $q \approx 86.6$. In other words, at a price of $50, consumer demand would be about 87 units.

We compute the quantity supplied at a price of $50 using the supply equation $p = 4\sqrt{q}+10$. Set $50 = 4\sqrt{q} + 10$, and solve for q, or use your graphing calculator or computer to find the corresponding point. We see that $q = 100$, so at a price of $50, producers supply about 100 units.

At a price of $50, the supply is larger than the demand, so some goods will remain unsold. We can expect prices to be pushed down.

(b) The supply and demand curves are shown in Figure 6.75. We see that the equilibrium price is about $p^* = \$48$, and the equilibrium quantity is about $q^* = 91$ units. The market will indeed push prices downward from $50, closer to the equilibrium price of $48.

(c) The consumer surplus is the area under the demand curve and above the line $p = 48$. (See Figure 6.76.) We have

$$\text{Consumer surplus} = \int_0^{q^*} D(q)dq - p^*q^*$$
$$= \int_0^{91} (100e^{-0.008q})dq - (48)(91)$$
$$\approx 6464 - 4368$$
$$= 2096.$$

The consumer surplus is $2096. Consumers gained $2096 in buying goods at the equilibrium price instead of the price they would have been willing to pay.

(d) The producer surplus is the area above the supply curve and below the line $p = 48$. (See

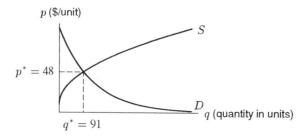

Figure 6.75: Demand and supply curves for a product

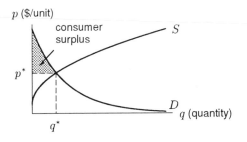

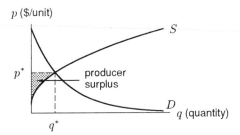

Figure 6.76: Consumer surplus   *Figure 6.77:* Producer surplus

Figure 6.77.) We have

$$\text{Producer surplus} = p^* q^* - \int_0^{q^*} S(q)\,dq$$

$$= (48)(91) - \int_0^{91} (4\sqrt{q} + 10)\,dq$$

$$\approx 4368 - 3225$$

$$= 1143.$$

The producer surplus is $1143. Producers gained $1143 in supplying goods at the equilibrium price instead of the price at which they would have been willing to provide the goods. Notice that the consumer gains from trade are larger than the producer gains from trade for this product.

Wage and Price Controls

In a free market, the price of a product will generally move to the equilibrium price. In some circumstances, however, outside forces will keep the price artificially high or artificially low. Rent control, for example, will keep prices below what the market would bear, whereas cartel pricing or the minimum wage law will raise prices above the market price. In these situations, the definitions of consumer and producer surplus are still valid and it makes sense to see what happens to these quantities in a situation where the price is artificially high or low.

Example 2   The dairy industry is a good example of cartel pricing: the government has set milk prices artificially high. What effect does forcing the price up from the equilibrium price have:

(a)   on the consumer surplus?

(b)   on the producer surplus?

(c)   on the total gains from trade (consumer surplus + producer surplus)?

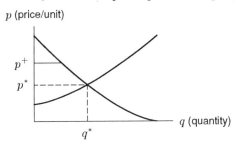

Figure 6.78: What effect does the artificially high price, p^+, have?

Solution   (a)   A graph of possible demand and supply curves for the milk industry is given in Figure 6.78, with the equilibrium price and quantity labeled p^* and q^* respectively. Suppose that the price is fixed at the artificially high price labeled p^+ in Figure 6.78. Recall that the consumer surplus is the difference between the amount the consumers did pay (p^+) and the amount they would have been willing to pay (given on the demand curve). This is the area shaded in Figure 6.79(a). Notice that this consumer surplus is clearly less than the consumer surplus at the equilibrium price, shown in Figure 6.79(b)

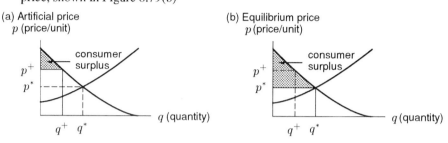

Figure 6.79: Consumer surplus for the milk industry

  (b)   At a price of p^+, the quantity sold, q^+, is less than it would have been at the equilibrium price. The producer surplus is the area between p^+ and the supply curve *at this reduced demand*. This area is shaded in Figure 6.80(a). Compare this producer surplus (at the artificially high price) to the producer surplus in Figure 6.80(b) (at the equilibrium price). It appears that in this case, producer surplus is greater at the artificial price than at the equilibrium price. (Different supply and demand curves might have led to a different answer.)

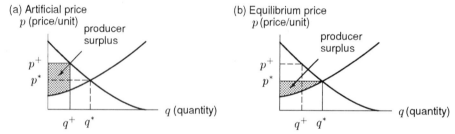

Figure 6.80: Producer surplus for the milk industry

  (c)   The total gains from trade (consumer surplus + producer surplus) at the artificially high price of p^+ is the area shaded in Figure 6.81(a). The total gains from trade at the equilibrium price of p^* is the area shaded in Figure 6.81(b). It is clear that, under artificial price conditions, total gains from trade go down. The total financial effect of the artificially high price on all producers and consumers combined is a negative one.

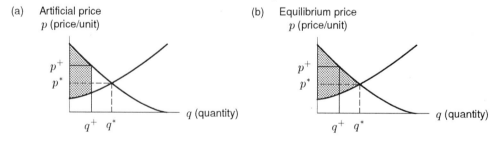

Figure 6.81: Total gains from trade

Problems for Section 6.5

1. A graph of the demand and supply curves for a certain product is shown in Figure 6.82.
   (a) What are the equilibrium price and equilibrium quantity?
   (b) Estimate the consumer surplus, and, on a sketch similar to Figure 6.82, shade the consumer surplus.
   (c) Estimate the producer surplus, and, on a sketch similar to Figure 6.82, shade the producer surplus.
   (d) What are the total gains from trade for this product?

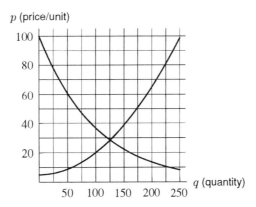

p (price/unit)

Figure 6.82: Demand and supply curves

2. Consider the demand and supply curves given in Figure 6.82. Suppose a price of $40 is artificially imposed.
   (a) At the price of $40, estimate the consumer surplus.
   (b) At the price of $40, estimate the producer surplus.
   (c) At the price of $40, estimate the total gains from trade.
   (d) Compare your answers in this problem to your answers in Problem 1, and discuss the effect of price controls on the consumer surplus, producer surplus, and gains from trade in this case.

3. Suppose that the demand for a product is given by $D(q) = 20e^{-0.002q}$ and the supply curve is given by $S(q) = 0.02q + 1$ for $0 \leq q \leq 1000$.
   (a) If the quantity is 300, will supply or demand be higher? Will this tend to push the quantity produced higher or lower? Closer to the equilibrium quantity or farther away from it?
   (b) Sketch graphs of these two functions and find the equilibrium price and equilibrium quantity.
   (c) Using the equilibrium price and quantity, calculate the consumer surplus and interpret it.
   (d) Using the equilibrium price and quantity, calculate the producer surplus and interpret it.

4. (a) Sketch possible supply and demand curves where the consumer surplus is greater than the producer surplus.
   (b) Sketch possible supply and demand curves where the consumer surplus is less than the producer surplus.

5. In May 1991 *Car and Driver* described a model Jaguar that sells for $980,000. At that price only 50 have been sold. It is estimated that 350 could have been sold if the price had been $560,000. Assuming that the demand curve is a straight line, and that $560,000 and 350 are the equilibrium price and quantity, find the consumer surplus.

6. Rent controls on apartments are an example of price controls on a commodity. They keep the price of the good artificially low (below the equilibrium price). Sketch a graph of supply and demand curves, and label on it a price p^- below the equilibrium price. What effect does forcing the price down have

   (a) on the producer surplus?
   (b) on the consumer surplus?
   (c) on the total gains from trade (consumer surplus + producer surplus)?

7. In Figure 6.72, page 356, find the regions with the following areas:

   (a) p^*q^* (b) $\displaystyle\int_0^{q^*} D(q)\,dq$ (c) $\displaystyle\int_0^{q^*} S(q)\,dq$

   (d) $\left(\displaystyle\int_0^{q^*} D(q)\,dq\right) - p^*q^*$ (e) $p^*q^* - \displaystyle\int_0^{q^*} S(q)\,dq$ (f) $\displaystyle\int_0^{q^*} (D(q)-S(q))\,dq$

REVIEW PROBLEMS FOR CHAPTER SIX

For each of the functions in Problems 1–5, do the following:
(a) Find f' and f''.
(b) Find the critical points of f.
(c) Find any inflection points of f.
(d) Evaluate f at its critical points and, where appropriate, at the endpoints of the given interval. Identify local and global maxima and minima of f in the interval.
(e) Sketch a graph of f. Indicate clearly where f is increasing or decreasing, and the concavity of its graph.

1. $f(x) = x^3 - 3x^2$ $(-1 \le x \le 3)$ 2. $f(x) = x + \sin x$ $(0 \le x \le 2\pi)$

3. $f(x) = e^{-x} \sin x$ $(0 \le x \le 2\pi)$ 4. $f(x) = 2x^3 - 9x^2 + 12x + 1$

5. $f(x) = x^3 + 3x^2 - 9x - 15$

From the graphs of f' in Problems 6–9 determine:
(a) Over what intervals is f increasing? Decreasing?
(b) Whether f has maxima or minima. If so, which, and where?

6.

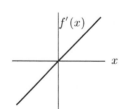

7.

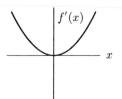

8.

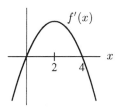

9.

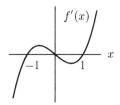

10. Sketch several members of the family $y = x^3 - ax^2$ on the same coordinate plane. Discuss the effect of the parameter a on the graph. Find all critical points for this function.

11. Consider the vase in Figure 6.83. Assume the vase is filled with water at a constant rate (i.e., constant volume per unit time).

   (a) Graph $y = f(t)$, the depth of the water, against time, t. Show on your graph the points at which the concavity changes.

   (b) Where does $y = f(t)$ grow fastest? Slowest? Estimate the ratio between these two growth rates.

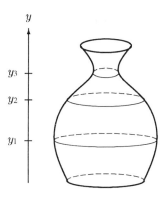

Figure 6.83

12. A developer has recently purchased a laundry and an adjacent factory. For years, the laundry has taken pains to keep the smoke from the factory from soiling the air used by its clothes dryers. Now that the developer owns both the laundry and the factory, she could install filters within the factory's smokestacks to reduce the emission of smoke directly, instead of merely protecting the laundry from it, at a cost which would depend on the number of filters used. The developer faces the schedule of factory's cost of filters versus laundry's cost of protecting against dirt shown in Table 6.5.

TABLE 6.5

# of Filters	Total Expense for Filters	Total Costs of Protecting Laundry from Smoke
0	$0	$127
1	$5	$63
2	$11	$31
3	$18	$15
4	$26	$6
5	$35	$3
6	$45	$0
7	$56	$0

(a) Draw up a table which shows, for each possible number of filters (0 through 7,) the marginal cost of the filter, the average cost of the filters, and the marginal savings protecting the laundry from smoke.

(b) Since the developer wishes to minimize the total costs to both her businesses, what should she do? Use the table from part (a) to explain your answer.

(c) What should the developer do if, in addition to the cost of the filters, the filters must be mounted on a rack which costs $100?

(d) What should the developer do if the rack costs $50?

13. Figure 6.84 shows cost and revenue functions for a product.

(a) Estimate the production level that will maximize profit.

(b) Sketch graphs of marginal revenue and marginal cost for this product on the same coordinate system. Label on this graph the production level that will maximize profits.

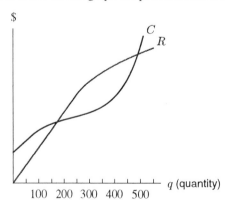

Figure 6.84: Where is profit maximized?

14. A premium ice cream company finds that at a price of $4.00, demand for the product is 4000. For every $0.25 decrease in price, demand increases by 200. Find the price and quantity sold that will maximize revenues.

15. Let $C(q)$ be the total cost of producing a quantity q of a certain good. The average cost is $a(q) = C(q)/q$.

    (a) Interpret $a(q)$ graphically, as the slope of a line in Figure 6.85.
    (b) Find on the graph the quantity q_0 where $a(q)$ is minimal.
    (c) What is the relationship between $a(q_0)$ and the marginal cost, $C'(q_0)$? Explain your result graphically. What does this result mean, in terms of economics?
    (d) Graph $C'(q)$ and $a(q)$ on the same axes. Mark q_0 on the q-axis.

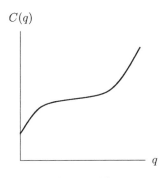

Figure 6.85

16. When birds lay eggs, they do so in clutches of several at a time. When the eggs hatch, each clutch gives rise to a brood of baby birds. We want to determine the clutch size which maximizes the number of birds surviving to adulthood per brood. If the clutch is small, there are few baby birds in the brood; if the clutch is large, there are so many baby birds to feed that most die of starvation. The number of surviving birds per brood as a function of clutch size is shown by the benefit curve in Figure 6.86.[5]

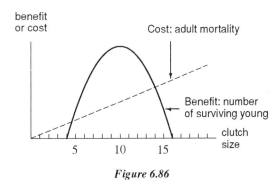

Figure 6.86

    (a) Estimate the clutch size which maximizes the number of survivors per brood.
    (b) Suppose also that there is a biological cost to having a larger clutch: the female survival rate is reduced by large clutches. This cost is represented by the dotted line in Figure 6.86. If we take cost into account by assuming that the optimal clutch size in fact maximizes the vertical distance between the curves, what is the new optimal clutch size?

17. Let $f(v)$ be the amount of energy consumed by a flying bird, measured in joules per second (a joule is a unit of energy), as a function of its speed v (in meters/sec). See Figure 6.87.

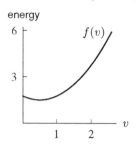

Figure 6.87: Amount of energy consumed by a flying bird

(a) Suggest a reason for the shape of this graph (in terms of the way birds fly).

Now let $a(v)$ be the amount of energy consumed by the same bird, measured in joules *per meter*.

(b) What is the relationship between $f(v)$ and $a(v)$?
(c) Where is $a(v)$ a minimum?
(d) Should the bird try to minimize $f(v)$ or $a(v)$ when it is flying? Why?

18. A bird such as a starling feeds worms to its young. To collect worms, the bird flies to a site where worms are to be found, picks up several in its beak, and flies back to its nest. The *loading curve* in Figure 6.88 shows how the number of worms (the load) a starling collects depends on the time it has been searching for them.[6] The curve is concave down because the bird can pick up worms more efficiently when its beak is empty; when its beak is partly full, the bird becomes much less efficient. The traveling time (from nest to site and back) is represented by the distance PO in Figure 6.88. Suppose the bird wants to maximize the rate at which it brings worms to the nest, where

$$\text{Rate worms arrive at nest} = \frac{\text{Load}}{\text{Traveling time} + \text{Searching time}}$$

(a) Draw a line in Figure 6.88 whose slope is this rate.
(b) Using the graph, estimate the load which maximizes this rate.
(c) If the traveling time is increased, does the optimal load increase or decrease? Why?

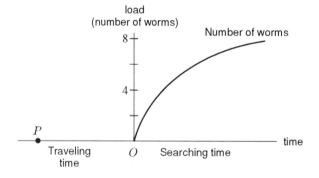

Figure 6.88: Bird's loading curve

---

[6] Alex Kacelnick(1984). Reported by J. R. Krebs and N. B. Davis, *An Introduction to Behavioural Ecology* (Oxford: Blackwell, 1987).

19. Consider a large tank of water, with temperature $W(t)$. The ambient temperature $A(t)$ (i.e., the temperature of the surrounding air), is given by the graph in Figure 6.89.

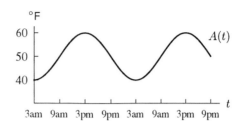

Figure 6.89

The temperature of the water is affected by the temperature of the surrounding air.

(a) How does the temperature of the water change if the water is colder than the surrounding air? What if the water is warmer?

(b) Using your answer in part (a), sketch a possible graph for $W(t)$ on the same axes as $A(t)$.

(c) Explain the relationship between the maxima and minima of $W(t)$ and the points where the two graphs intersect.

(d) What is the relationship between the rate at which the temperature of the water changes and the difference $A(t) - W(t)$?

(e) What is the relationship between the inflection points of $W(t)$ and the points where $A(t) - W(t)$ has a maxima or minima?

(f) Assume the tank is refilled with cold water (35°F) at 3 am. Sketch a possible graph for $W(t)$. Pay attention to the concavity.

20. Find the area under the graph of $f(x) = x^2 + 2$ between $x = 0$ and $x = 6$.

21. Find the average value of the function $f(x) = 5 + 4x - x^2$ between $x = 0$ and $x = 3$.

22. Assume the marginal cost function in producing a certain product is $C'(q) = q^2 - 50q + 700$, for $0 \le q \le 50$. If fixed costs are \$500, find the total cost to produce 50 items.

23. The derivative of a function $f(t)$ is given by $f'(t) = t^3 - 6t^2 + 8t$ for $0 \le t \le 5$. Sketch a graph of this derivative function, and describe how the function $f(t)$ changes over the period $t = 0$ to $t = 5$. When is it increasing and when is it decreasing? When is it at its maximum and when is it at its minimum?

For Problems 24–25 the graph of $f'(x)$ is given. Sketch a possible graph for $f(x)$. Mark the points $x_1 \ldots x_4$ on your graph and label local maxima, local minima and points of inflection.

24.

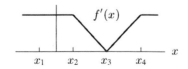

25.

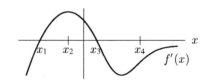

26. Water flows at a constant rate into the left side of the W-shaped container shown in Figure 6.90. Sketch a graph of the height, H, of the water in the left side of the container as a function of time t.

Figure 6.90

27. Throughout much of this century, the yearly consumption of electricity in the US has been increasing exponentially at a continuous rate of 7% per year. Assuming this trend continues, and that the electrical energy consumed in 1900 was 1.4 million megawatt-hours (a megawatt-hour is a measure of electrical energy),

    (a) Write an expression for electricity consumption as a function of time, t, measured in years since 1900.
    (b) Find the average yearly electrical consumption throughout this century.
    (c) During what year was electrical consumption closest to the average for the century?
    (d) Without doing the calculation for part (c), how could you have predicted which half of the century the answer would be in?

28. The width, in feet, at various points along the fairway of a hole on a golf course is given in Figure 6.91. If one pound of fertilizer covers 200 square feet, estimate the amount of fertilizer needed to fertilize the fairway.

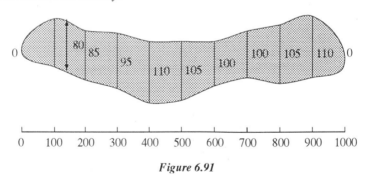

Figure 6.91

29. In 1990 humans generated 1.4×10^{20} joules of energy through the combustion of petroleum. All of the earth's petroleum would generate approximately 10^{22} joules. Assuming the use of energy generated by petroleum combustion will increase by 2% each year, how long will it be before all of our petroleum resources are used up?

30. An economist studying the rate of production, $R(t)$, of oil in a new oil well has proposed the following model:
$$R(t) = 4000 + 1000e^{-t}\sin(2\pi t),$$
where t is the time in years, and $R(t)$ is in barrels of oil per year.

    (a) Find the total amount of oil produced in the first 5 years of operation.

(b) Find the average amount of oil produced per year during the first 5 years.

31. The rate of growth of the height of two species of trees is shown in Figure 6.92, where t is measured in years, and the rate is given in feet per year. If the two trees are the same height at time $t = 0$, which tree is taller after 5 years? After 10 years?

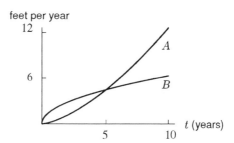

Figure 6.92: Comparing growth rates of two trees

32. The graph of the demand curve and the supply curve for a certain product are given in Figure 6.93. Estimate from this graph the equilibrium price and quantity and the consumer surplus and producer surplus. On sketches similar to Figure 6.93, shade areas corresponding to the consumer surplus and to the producer surplus.

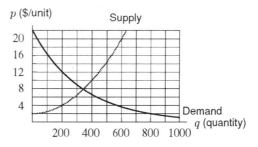

Figure 6.93: Demand and supply curves

33. Draw plasma concentration curves for two drugs A and B if product A has the highest peak concentration, but product B is absorbed more quickly and has greater overall bioavailability.

34. Whether a resource is distributed evenly among members of a population is often an important political or economic question. How can we measure this? How can we decide if the distribution of wealth in this country is becoming more or less equitable over time? How can we measure which country has the most equitable income distribution? This problem describes a way of making such measurements.

    Suppose the resource is distributed evenly. Then any 20% of the population will have 20% of the resource. Similarly, any 30% will have 30% of the resource and so on. If, however, the resource is not distributed evenly, the poorest $P\%$ of the population (in terms of this resource) will not have $P\%$ of the goods. Suppose $F(x)$ represents the fraction of the resources owned by the poorest fraction x of the population. Thus $F(0.4) = 0.1$ means that the poorest 40% of the population owns 10% of the resource.

    (a) What would F be if the resource was distributed evenly?

(b) What must be true of any such F? What must $F(0)$ and $F(1)$ equal? Is F increasing or decreasing? Is the graph of f concave up or concave down?

(c) Gini's index of inequality, G, is one way to measure how evenly the resource is distributed. It is defined by

$$G = 2 \int_0^1 [x - F(x)]\, dx.$$

Show graphically what G represents.

35. Gini's index of inequality is discussed in Problem 34. Graphical representations of this index are given in Figure **??** for two countries. Which country has the more equitable distribution of wealth? Discuss the distribution of wealth in each of the two countries.

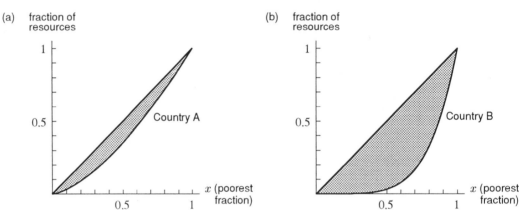

Figure 6.94: Gini's Index of Inequality for two countries

CHAPTER SEVEN

FUNCTIONS OF MANY VARIABLES

Many quantities depend on more than one variable: the amount of food grown depends on the amount of rain and the amount of fertilizer used; the rate of a chemical reaction depends on the temperature and the pressure of the environment in which it proceeds; the quantity of meat purchased depends on the price of meat and the income of the buyer; the rate of fallout from a volcanic eruption depends on the distance from the volcano and the time since the eruption. In this chapter we will see how to extend the concept of the derivative to functions of two or more variables.

7.1 UNDERSTANDING FUNCTIONS OF MANY VARIABLES

Functions of Two Variables

Suppose you are planning to take out a five-year loan to buy a car and you need to calculate what your monthly payment will be; this depends on both the amount of money you borrow and the interest rate. These quantities can vary separately: the loan amount can change while the interest rate remains the same. The interest rate can change while the loan amount remains the same. To calculate your monthly payment you need to know both. If the monthly payment is m, the loan amount is L, and the interest rate is $r\%$, then we express the fact that m is a function of L and r by writing

$$m = f(L, r).$$

This is just like the function notation of one-variable calculus. The variable m is called the dependent variable , and the variables L and r are called the independent variables. The letter f stands for the *function* or rule that gives the value of m corresponding to given values of L and r.

A function of two variables can be represented pictorially by a contour diagram, numerically by a table of values, or algebraically by a formula. In this section we will give examples of each of these three ways of viewing a function.

Graphical Example: A Weather Map

Figure 7.1 shows a weather map from a newspaper.

What information does it convey? It is displaying the predicted high temperature, T, in degrees Fahrenheit (°F), at any point in the US on that day. The curves shown on the map, called *isotherms*, separate the country into zones, according to whether T is in the 60s, 70s, 80s, 90s, or 100s. (*Iso* means same and *therm* means heat.) Notice that the isotherm separating the 80s and 90s zones connects all the points where the temperature is exactly 90°F.

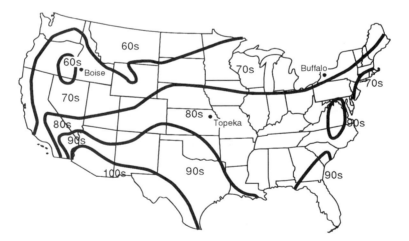

Figure 7.1: Weather map showing predicted high temperatures, T, for June 30, 1992

Example 1   Estimate the value of T in Boise, Idaho; Topeka, Kansas; and Buffalo, New York.

Solution   Boise and Buffalo are in the 70s region, and Topeka is in the 80s region. Thus, the temperature in Boise and Buffalo is between 70 and 80; the temperature in Topeka is between 80 and 90. In fact, we can say more. Although both Boise and Buffalo are in the 70s, Boise is quite close to the $T = 70$ isotherm, whereas Buffalo is quite close to the $T = 80$ isotherm. So we estimate that the temperature is in the low 70s in Boise, and the high 70s in Buffalo. Topeka is more or less halfway between the $T = 80$ isotherm and the $T = 90$ isotherm. Thus, we guess that the temperature in Topeka is in the mid 80s. In fact, the high temperatures for that day were 71°F for Boise, 79°F for Buffalo, and 86°F for Topeka.

The predicted high temperature, T, illustrated by the weather map is a function of (that is, depends on) location. The location of a point in the US is given by two variables, often longitude and latitude, or miles east-west and miles north-south of a fixed point, say, Topeka. Thus, T is a function of two variables and the weather map is one way of visualizing that function. The weather map in Figure 7.1 is called a *contour map* or *contour diagram*. In the next section, we look at contour diagrams in detail.

Numerical Example: Beef Consumption

Suppose you are a beef producer and you want to know how much beef people will buy. This depends on how much money people have and on the price of beef. Thus, the consumption of beef, C (in pounds per week per household) is a function of household income, I (in thousands of dollars per year), and the price of beef, p (in dollars per pound). In function notation, we write

$$C = f(I, p).$$

Table 7.1 contains values of this function. Values of p are shown across the top, values of I are down the left side, and corresponding values of $f(I, p)$ are given in the table.[1] For example, to find the value of $f(40, 3.50)$, we look in the row corresponding to $I = 40$ (for 40 thousand dollars) under $p = 3.50$ (for cost per pound), where we find the number 4.05. Thus,

$$f(40, 3.50) = 4.05.$$

This means that, on average, if a household's income is $40,000 a year and the price of beef is $3.50/lb, the family will buy 4.05 lbs of beef per week.

Notice how this differs from the table of values of a one-variable function, where one row or one column is enough to list the values of the function. Here many rows and columns are needed because the function has a value for every *pair* of values of the independent variables.

TABLE 7.1   *Quantity of beef bought (pounds/household/week)*

$I\backslash p$	3.00	3.50	4.00	4.50
20	2.65	2.59	2.51	2.43
40	4.14	4.05	3.94	3.88
60	5.11	5.00	4.97	4.84
80	5.35	5.29	5.19	5.07
100	5.79	5.77	5.60	5.53

[1] Adapted from Richard G. Lipsey, *An Introduction to Positive Economics 3rd Ed.*, Weidenfeld and Nicolson, London, 1971

Algebraic Examples: Formulas

In both the weather map and beef consumption examples, there is no formula for the underlying function. That is usually the case for functions representing real-life data. On the other hand, for many idealized models in physics, engineering, or economics, there are exact formulas.

Example 2 Give a formula for the function $M = f(B, t)$ where M is the amount of money in a bank account t years after an initial investment of B dollars, if interest accrues at a rate of 5% per year compounded
(a) annually (b) continuously.

Solution (a) Annual compounding means that M increases by a factor of 1.05 every year, so

$$M = f(B, t) = B(1.05)^t.$$

(b) Continuous compounding means that M grows according to the function e^{kt}, with $k = 0.05$, so

$$M = f(B, t) = Be^{0.05t}.$$

Example 3 A car rental company charges $40 a day and 15 cents a mile for its cars.
(a) Write a formula for the cost, C, of renting a car as a function of the number of days, d, and the number of miles driven, m.
(b) If we have $C = f(d, m)$, find $f(5, 300)$ and interpret it.

Solution (a) The total cost in dollars of renting a car is 40 times the numbers of days plus 0.15 times the number of miles, so we have

$$C = 40d + 0.15m.$$

(b) We have

$$f(5, 300) = 40(5) + 0.15(300)$$
$$= 200 + 45$$
$$= 245.$$

We see that $f(5, 300) = 245$. This tells us that if we rent a car for 5 days and drive it 300 miles, it will cost us $245.

Strategy to Investigate Functions of Two Variables: Vary One Variable at a Time

We can learn a great deal about functions of two or more variables by letting one variable vary at a time while holding the others fixed, thus obtaining a function of one variable.

Concentration of a Drug in the Blood

When a drug is injected into muscle tissue, it begins to diffuse into the bloodstream. The concentration of the drug in the blood increases until it reaches its maximum value, and then begins to decrease as the body's metabolic processes remove the drug from the body. The concentration C (in mg per 100 ml) of the drug in the blood is a function of two variables: x, the amount (in 100's of mg) of the drug given in the injection, and t, the time (in hours) since the injection was administered. Suppose we are told that

$$C = f(x, t) = te^{-t(5-x)} \qquad \text{for } 0 \le x \le 5 \text{ and } t \ge 0.$$

Example 4 (a) Explain the significance of $f(4, t)$ in terms of drug concentration in the blood.
(b) Explain the significance of $f(x, 1)$ in terms of drug concentration in the blood.

Solution   (a)   Holding x fixed at 4 means that we are considering an injection of 400 mg of the drug into the muscle; letting t vary means we are watching the effect of this dose as time passes. Thus the function $f(4, t)$ describes the concentration of the drug in the blood resulting from a 400 mg injection as a function of time. Figure 7.2 shows the graph of $f(4, t) = te^{-t}$. Notice that the concentration in the blood from this dose is at a maximum at 1 hour after injection, and that after that time the concentration in the blood goes down and eventually approaches zero.

(b)   Holding t fixed at 1 means that we are focusing on the blood 1 hour after the injection; letting x vary means we are considering the effect of different doses at that instant. Thus the function $f(x, 1)$ gives the concentration of the drug in the blood 1 hour after injection as a function of the amount injected. Figure 7.3 shows the graph of $f(x, 1) = e^{-(5-x)} = e^{x-5}$. Notice that $f(x, 1)$ is an increasing function of x. This makes sense. If we administer more of the drug, the concentration in the bloodstream will be higher.

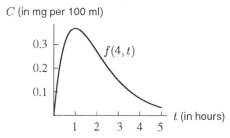

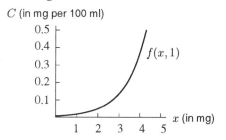

Figure 7.2: The function $f(4, t)$ shows the effect of a 400 mg injection

Figure 7.3: The function $f(x, 1)$ shows the concentration in the blood 1 hour after the injection

Example 5   Sketch the graphs of $f(a, t)$ for $a = 1, 2, 3$ and 4 on the same axes. Describe how the graph changes for larger values of a, and explain what this means in terms of drug concentration in the blood.

Solution   In general, the one-variable function $f(a, t)$ gives the effect of an injection of $100a$ mg into the muscle. Figure 7.4 shows the graphs of the four functions $f(1, t) = te^{-4t}$, $f(2, t) = te^{-3t}$, $f(3, t) = te^{-2t}$, and $f(4, t) = te^{-t}$ corresponding to the injection of 100, 200, 300, and 400 mg of the drug. The general shape of the graph is the same in every case: the concentration in the blood is zero at the time of injection $t = 0$, then increases to a maximum value, and then decreases towards zero again. We see that if a larger dose of the drug is administered, the peak of the graph is later and higher. This makes sense, since a larger dose will take longer to fully diffuse into the bloodstream and will produce a higher concentration when it does.

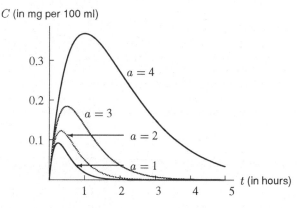

Figure 7.4: Concentration $f(a, t)$ of a drug resulting from $100a$ mg injection

The Beef Data

For a function given by a table of values, such as the beef consumption data, we allow one variable to vary at a time by looking at one row or one column. For example, to hold the income, I, fixed at 40, we look at the row $I = 40$ in table 7.2. This row gives the values of the function $f(40, p)$ and shows how beef consumption varies as the price varies. This data gives us Table 7.2.

TABLE 7.2   *Beef consumption by households making $40,000*

p ($/lb)	3.00	3.50	4.00	4.50
$f(40, p)$ ((lb/household)/week)	4.14	4.05	3.94	3.88

Since I is fixed, we now have a function of one variable that shows how much beef is bought at various prices by people who earn $40,000 a year. Table 7.2 shows that $f(40, p)$ decreases as p increases. The other rows in table 7.1 tell the same story; for each income, I, the consumption of beef goes down as the price, p, increases.

Weather Map

What happens on the weather map in Figure 7.1 when we allow only one variable at a time to vary? For example, suppose we moved along the east-west line through Topeka. Suppose x represents miles east-west of Topeka and y represents miles north-south. We keep y fixed at 0 and let x vary. Along this line, the high temperature T goes from the 60s along the west coast, to the 70s in Nevada and Utah, to the 80s in Topeka, to the 90s just before the east coast, then returns to the 80s. A possible graph is shown in Figure 7.5. Other graphs are possible because we do not know for sure how the temperature varies between contours.

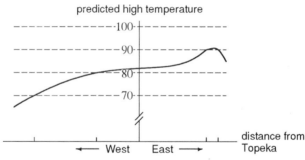

Figure 7.5: Predicted high temperature on an east-west line
through Topeka

Functions of More Than Two Variables

In applications of calculus, functions of any number of variables can arise. The monthly payment on a loan, P, for example, is a function of the amount of the loan, L, the interest rate, k, and the length of time of the loan, t. The monthly payment, P, is a function of three variables. We have

$$P = f(L, k, t).$$

We need to be able to apply calculus to functions of arbitrarily many variables.

The idea of studying a function by keeping one variable fixed works as well for a function of three variables as it does for two.

Example 6 In the beef consumption example, we thought of the consumption of beef as a function of the price, p, of beef and household income, I. In fact, a household's beef consumption depends on many other things as well, such as the prices of competing products. Let us view the consumption, C, of beef as a function of the price q (dollars per pound) of a competing meat, say chicken, as well as the price of beef and household income. That is, $C = f(I, p, q)$. This function can be given by a collection of tables, one for each different value of the price, q, of chicken. Tables 7.3 and 7.4 show two such tables for the consumption function f with q held fixed. Explain how the two tables are related.

TABLE 7.3 *Beef consumption when price of chicken is \$1.50/lb* ($q = 1.50$)

$I \backslash p$	3.00	3.50	4.00
10	2.65	2.59	2.51
60	5.11	5.00	4.97
100	5.79	5.77	5.60

TABLE 7.4 *Beef consumption when price of chicken is \$2.00/lb* ($q = 2.00$)

$I \backslash p$	3.00	3.50	4.00
10	2.75	2.75	2.71
60	5.21	5.12	5.11
100	5.80	5.77	5.60

Solution Comparing tables shows that for households with large incomes (say \$100,000 per year, corresponding to $I = 100$), changes in the price of chicken have little effect on the amount of beef consumed (because price is not a factor in their buying). However for small incomes (say \$10,000 per year, corresponding to $I = 10$), an increase in the price of chicken (from \$1.50/lb to \$2.00/lb) causes families to spend more on beef.

Problems for Section 7.1

Problems 1–3 refer to the weather map in Figure 7.1 on page 372.

1. Give the range of daily high temperatures for
    (a) Pennsylvania   (b) North Dakota   (c) California.

2. Sketch the graph of the predicted high temperature T on a line north-south through Topeka.

3. Sketch the graphs of the predicted high temperature on a north-south line and an east-west line through Boise.

For Problems 4–8 refer to Table 7.1 on page 373, where p is the price of beef and I is annual household income.

4. Make a table showing the amount of money, M, that the average household spends on beef (in dollars per household per week) as a function of the price of beef and household income.

5. Give tables for beef consumption as a function of p, with I fixed at $I = 20$ and $I = 100$. Give tables for beef consumption as a function of I, with p fixed at $p = 3.00$ and $p = 4.00$. Comment on what you see in the tables.

6. How does beef consumption vary as a function of household income if the price of beef is held constant?

7. Make a table of the proportion, P, of household income spent on beef per week as a function of price and income. (Note that P is the fraction of income spent on beef.)

8. Express P, the proportion of household income spent on beef per week, in terms of the original function $f(I, p)$ which gave consumption as a function of p and I.

9. Sketch the graph of the bank account function f in Example 2(a) on page 374, holding B fixed at three different values and letting only t vary. Then sketch the graph of f, holding t fixed at three different values and letting only B vary. Explain what you see.

10. Consider a function giving the number, n, of new cars sold in a year as a function of the price of new cars, c, and of the average price of gas, g.

    (a) If c is held constant, is n an increasing or decreasing function of g? Why?
    (b) If g is held constant, is n an increasing or decreasing function of c? Why?

11. The total sales of a product, S, can be expressed as a function of the price p charged for the product and the amount, a, spent on advertising. Therefore, $S = f(p, a)$. Is f an increasing or decreasing function of p? Is S an increasing or decreasing function of a? Explain your answers.

12. Values of a function $f(x, y)$ are given in Table 7.5. Is f an increasing or decreasing function of x? Is f an increasing or decreasing function of y?

TABLE 7.5 *Values of a function $f(x, y)$.*

$x\backslash y$	0	1	2	3	4	5
0	102	107	114	123	135	150
20	96	101	108	117	129	144
40	90	95	102	111	123	138
60	85	90	97	106	118	133
80	81	86	93	102	114	129

13. The cost of renting a car from a certain company is $40 a day and 15 cents a mile. We saw in Example 3 that we can express the cost, C, of renting the car as a function of the number of days, d, and the number of miles, m, as follows:

$$C = f(d, m) = 40d + 0.15m.$$

    (a) Find $f(3, 200)$ and interpret it.
    (b) Explain the significance of $f(3, m)$ in terms of rental car costs. Sketch a graph of this function, with C as a function of m.
    (c) Explain the significance of $f(d, 100)$ in terms of rental car costs. Sketch a graph of this function, with C as a function of d.

14. In Example 3, we saw that the function

$$C = f(d, m) = 40d + 0.15m.$$

gives the cost, C, of renting a car as a function of the number of days, d, and the number of miles, m. Make a table of values for C, using $d = 1, 2, 3, 4$ and $m = 100, 200, 300, 400$. You should have 16 values in your table.

15. You are planning a long driving trip and your principal cost will be gasoline.

    (a) Make a table showing how the daily fuel cost varies as a function of the price of gasoline (in dollars per gallon) and the number of gallons you buy each day.

    (b) If your car goes 30 miles on each gallon of gasoline, make a table showing how your daily fuel cost varies as a function of your daily travel distance and the price of gas.

16. The *temperature adjusted for wind-chill* is a temperature which tells you how cold it feels, as a result of the combination of wind and temperature. Table 7.6 shows the temperature adjusted for wind-chill as a function of wind speed and temperature.

TABLE 7.6   *Temperature adjusted for wind-chill ($°F$)*

Wind Speed (mph)	Temperature ($°F$)							
	35	30	25	20	15	10	5	0
5	33	27	21	16	12	7	0	−5
10	22	16	10	3	−3	−9	−15	−22
15	16	9	2	−5	−11	−18	−25	−31
20	12	4	−3	−10	−17	−24	−31	−39
25	8	1	−7	−15	−22	−29	−36	−44

    (a) If the temperature is 0°F and the wind speed is 15 mph, how cold does it feel?
    (b) If the temperature is 35°F, what wind speed makes it feel like 22°F?
    (c) If the temperature is 25°F, what wind speed makes it feel like 20°F?
    (d) If the wind is blowing at 15 mph, what temperature feels like 0°F?

17. Using Table 7.6, make tables of the temperature adjusted for wind-chill as a function of wind speed for temperatures of 20°F and 0°F.

18. Using Table 7.6, make tables of the temperature adjusted for wind-chill as a function of temperature for wind speeds of 5 mph and 20 mph.

19. Table 7.7 shows the heat index as a function of temperature and humidity. The heat index is a temperature which tells you how hot it feels as a result of the combination of the two. Heat exhaustion is likely to occur when the heat index reaches 105.

TABLE 7.7   *Heat index ($°F$)*

Humidity (%)	Temperature ($°F$)									
	70	75	80	85	90	95	100	105	110	115
0	64	69	73	78	83	87	91	95	99	103
10	65	70	75	80	85	90	95	100	105	111
20	66	72	77	82	87	93	99	105	112	120
30	67	73	78	84	90	96	104	113	123	135
40	68	74	79	86	93	101	110	123	137	151
50	69	75	81	88	96	107	120	135	150	
60	70	76	82	90	100	114	132	149		

    (a) If the temperature is 80°F and the humidity is 50%, how hot does it feel?

(b)   At what humidity does 90°F feel like 90°F?

(c)   Make a table showing the approximate temperature at which heat exhaustion becomes a danger, as a function of humidity.

(d)   Explain why the heat index is sometimes above the actual temperature and sometimes below it.

20.   Using Table 7.7, graph the heat index as a function of humidity with temperature fixed at 70°F and at 100°F. Explain the features of each graph and the difference between them in common sense terms.

21.   An airport can be cleared of fog by heating the air. The amount of heat required to do the job depends on the air temperature and the wetness of the fog. Figure 7.6 shows the heat H required (in calories per cubic meter of fog) as a function of the temperature T (in degrees Celsius) and the water content w (in grams per cubic meter of fog). We have $H = f(T, w)$. Note that figure 7.6 is several graphs of H against T as w is held fixed at different values.

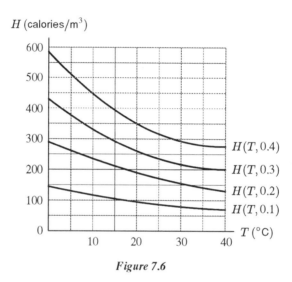

Figure 7.6

(a)   Estimate $f(20, 0.3)$ and explain what information it gives us.

(b)   Make a table of values for $H = f(T, w)$. Use $T = 0, 10, 20, 30,$ and 40, and $w = 0.05, 0.1, 0.2, 0.3, 0.4, 0.5$.

In Problems 22–26 we investigate how the demand for coffee depends on the price of coffee and the price of tea. Suppose that the demand, Q, in thousands of pounds per week, for a certain brand of coffee depends on the price of coffee, c, and the price of tea, t, both in dollars per pound, according to the formula

$$Q = f(c, t) = 100\frac{t}{c}.$$

22.   If the price of tea is \$1 per pound, the demand curve for coffee is given by $Q = f(c, 1) = 100/c$. Sketch a graph of this demand curve, with Q as a function of c, for $0 \le c \le 6$.

23.   In Problem 22, we sketched the demand curve for coffee when $t = 1$. On the same axes, sketch demand curves for coffee with $t = 1, 2, 3, 4, 5$, and label each curve with the corresponding value of t. What do you observe? How does the demand curve for coffee change as the price of tea increases? Explain in terms of the demand for coffee why this is reasonable.

24. Is Q an increasing or decreasing function of c? Is Q an increasing or decreasing function of t? Explain in terms of the demand for coffee why this is so.

25. Make a table showing the value of Q when $t = 1, 2, 3, 4$, and $c = 1, 2, 3, 4$. (You should have 16 values of Q in your table.) Use your table to check your answers to Problem 24.

26. We can also think of the demand for tea as a function of the price of coffee and the price of tea. Give a possible formula for the demand for tea as a function of c and t.

Problems 27–30 concern a vibrating guitar string. Suppose you pluck a guitar string and watch it vibrate. If you take snapshots of the guitar string at millisecond intervals, you might get something like Figure 7.7.

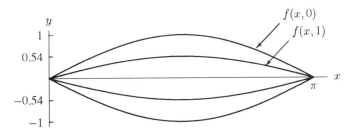

Figure 7.7: A vibrating guitar string: $f(x, t) = \cos t \sin x$

We can analyze the motion of the guitar string using a function of two variables. Think of the guitar string stretched tight along the x-axis from $x = 0$ to $x = \pi$. Each point on the string has an x-value, $0 \le x \le \pi$. As the string vibrates, each point on the string moves back and forth on either side of the x-axis. The ends of the string at $x = 0$ and $x = \pi$ remain stationary, while the point at the middle of the string moves the most. Let $f(x, t)$ be the displacement at time t of the point on the string located x units from the left end. Then a possible formula for $f(x, t)$ is

$$f(x, t) = \cos t \sin x, \quad 0 \le x \le \pi, \quad t \text{ in milliseconds.}$$

27. (a) Sketch graphs of y versus x for fixed t values, $t = 0, \pi/4, \pi/2, 3\pi/4, \pi$.
    (b) Use your graphs to explain why this function could represent a vibrating guitar string.

28. Explain what the functions $f(x, 0)$ and $f(x, 1)$ represent in terms of the vibrating string.

29. Explain what the functions $f(0, t)$ and $f(1, t)$ represent in terms of the vibrating string.

30. Describe the motion of the guitar strings whose displacements are given by the following:
    (a) $y = g(x, t) = \cos 2t \sin x$ (b) $y = h(x, t) = \cos t \sin 2x$

7.2 CONTOUR DIAGRAMS

How can we visualize a function of two variables? Functions of two variables are often represented by *contour diagrams* such as the weather map on page 372.

Topographical Maps

One of the most common examples of a contour diagram is a topographical map like that shown in Figure 7.8. It gives the elevation in the region and is a good way of getting an overall picture of the terrain such as where the mountains and flat areas are. Such topographical maps are frequently colored green at the lower elevations and brown, red, or even white at the higher elevations.

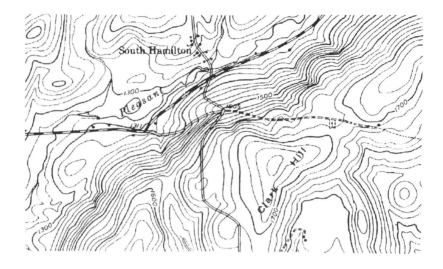

Figure 7.8: A topographical map showing the region around South Hamilton, NY

Example 1 Explain why the topographical map shown in Figure 7.9 corresponds to terrain such as that shown in Figure 7.10.

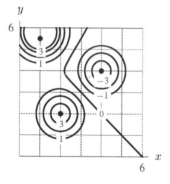

Figure 7.9: A topographical map

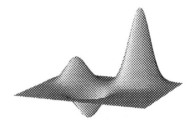

Figure 7.10: Terrain corresponding to the topographical map in Figure 7.9

Solution We see from the topographical map in Figure 7.9 that there are two hills, one with height about 5, and the other with height about 4. Most of the terrain is around height 0, and there is one valley with height about −4. This matches the terrain in Figure 7.10 since there are two hills (one taller than the other) and one valley.

The curves on a topographical map that separate lower elevations from higher elevations are called *contour lines* because they outline the contour or shape of the land.[2] Because every point

---

[2]In fact they are usually not straight lines, but curves.

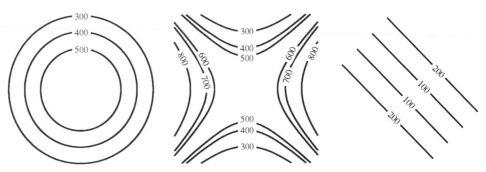

Figure 7.11: Mountain peak

Figure 7.12: Pass between two mountains

Figure 7.13: Long valley

along the same contour has the same elevation, contour lines are also called *level curves* or *level sets*. The more closely spaced the contours, the steeper the terrain; the more widely spaced the contours, the flatter the terrain. (Provided, of course, that the elevation between contours varies by a constant amount.) Certain features have distinctive characteristics. A mountain peak is typically surrounded by contour lines like those in Figure 7.11. A pass in a range of mountains may have contours that look like Figure 7.12. A long valley has approximately parallel contour lines indicating the rising elevations on both sides of the valley (see Figure 7.13); a long ridge of mountains has the same type of contour lines, only the elevations decrease on both sides of the ridge. Notice that the elevation numbers on the contour lines are just as important as the curves themselves.

There are some things contour lines cannot do. Two contours corresponding to different elevations cannot cross each other as shown in Figure 7.14. If they did, the point of intersection of the two curves would have two different elevations, which is impossible (assuming the terrain has no overhangs). We will usually follow the convention of drawing contours for equally spaced values of z. Thus, if a diagram has contours for $z = 1, 2,$ and 4, then it will usually (but not always) have a contour for $z = 3$ as well.

Figure 7.14: Impossible contour lines

Using Contour Diagrams

Consider how to represent the effect of different weather conditions on US corn production. What would happen if the average temperature were to increase (due to global warming, for example)? What would happen if the rainfall were to decrease (due to a drought)? One way of estimating the effect of these climatic changes is to use Figure 7.15. This map is a contour diagram giving the corn production $f(R, T)$ in the United States as a function of the total rainfall, R, in inches and average temperature, T, in degrees Fahrenheit, during the growing season[3]. Suppose at the present time, $R = 15$ inches and $T = 76°$. Production is measured as a percentage of the present production; thus, the contour through $R = 15, T = 76$ has value 100, that is, $f(15, 76) = 100$.

[3] Adapted from S. Beaty and R. Healy, *The Future of American Agriculture*, Scientific American, Vol. 248, No.2, February, 1983

Example 2   Use Figure 7.15 to evaluate $f(18, 78)$ and $f(12, 76)$ and explain the answers in terms of corn production.

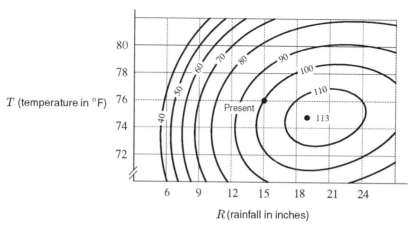

T (temperature in °F)

R (rainfall in inches)

Figure 7.15: Corn production as a function of rainfall and temperature

Solution   The point with R-coordinate 18 and T-coordinate 78 is on the contour with value 100, so $f(18, 78) = 100$. This means that if the annual rainfall were 18 inches and the temperature were 78°F, the country would produce about the same amount of corn as at present, although it would be wetter and warmer than it is now.

The point with R-coordinate 12 and T-coordinate 76 is between the value 80 contour and the value 90 contour, and slightly closer to the value 90 contour, so $f(12, 76) \approx 86$. This means that if the rainfall dropped to 12 inches and the temperature stayed at 76°, then corn production would drop to about 86% of what it is now.

Example 3   Describe in words how corn production changes as a function of rainfall if temperature is held constant at the point representing present conditions in Figure 7.15. Describe in words how corn production changes as a function of temperature if rainfall is held constant at present conditions. Give common sense explanations for your answers.

Solution   To see what happens to corn production if the temperature stays fixed at the present value of 76°F but the rainfall changes, look along the horizontal line $T = 76$. As we stay on the $T = 76$ line and move to the left, the values on the contour lines go down. In other words, if there is a drought, corn production decreases. Conversely, if rainfall increases, that is, as we move to the right from $R = 15, T = 76$, corn production increases, reaching a maximum of more than 110% when $R = 19$ inches, and then decreases (too much rainfall floods the fields).

If, instead, rainfall remains at the present value and temperature increases, we move up the vertical line $R = 15$. Under these circumstances corn production decreases; a 2°F increase causes a 10% drop in production. This makes sense since hotter temperatures lead to greater evaporation and hence drier conditions, even with rainfall constant at 15 inches. Similarly, a decrease in temperature leads to a very slight increase in production, reaching a maximum of around 102% of the current level when the average temperature is 74°F, followed by a decrease in production for lower temperatures (the corn won't grow if it is too cold).

Example 4   Antibiotics can be toxic in large doses, and this can cause problems if the antibiotics are not excreted normally, as can happen in a patient with some degree of kidney failure. If repeated doses of the antibiotic are to be given, it is important to closely monitor the rate at which the medicine is excreted. One measure of kidney function is the glomerular filtration rate, or GFR, which measures the amount of material crossing the outer (or glomerular) membrane of the kidney, in milliliters per minute. A normal GFR is about 125 ml/min. Figure 7.16 gives a contour diagram of the percent, P, of a dose of mezlocillin (an antibiotic) excreted, as a function of the patient's GFR and the time t in hours since the dose was administered. [4]

(a) In a patient with a GFR of 40, approximately how long will it take for 30% of the dose to be excreted?

(b) In a patient with a GFR of 60, approximately what percent of the dose has been excreted after 6 hours?

(c) Explain how we can tell from the graph that, for a patient with a fixed GFR, the amount excreted changes very little after 12 hours.

(d) Is the percent excreted an increasing or decreasing function of time? Explain why this makes sense.

(e) Is the percent excreted an increasing or decreasing function of GFR? Explain why this makes sense.

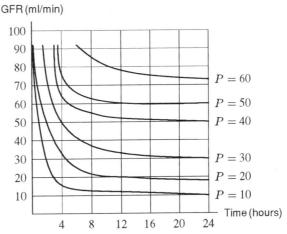

Figure 7.16: Excretion of mezlocillin in patients with kidney disease; percent of dose excreted

Solution   (a) If we look across the horizontal line at GFR = 40, we see that the $P = 30$ contour crosses it at about $t = 6$. In such a patient, it takes about 6 hours to excrete 30% of the dose.

(b) When GFR = 60 and $t = 6$, we are approximately on the 40 contour. Forty percent of the dose has been excreted after 6 hours.

(c) If we look across a horizontal line at any value of the GFR, we see that the contour changes very little between $t = 12$ and $t = 24$. This means that the percent excreted changes very little during this time.

(d) If we hold GFR fixed and increase t, looking from left to right across a horizontal line we see that the values of the contours increase. The percent excreted is an increasing function of time. This makes sense because as time goes by, more of the drug will pass through the patient's system.

[4]Peter G. Welling and Francis L. S. Tse, *Pharmacokinetics of Cardiovascular, Central Nervous System, and Antimicrobial Drugs*, The Royal Society of Chemistry, 1985, p. 316

(e)   If we hold time fixed and increase GFR, we look from bottom to top along a vertical line. The values of the contours increase as we do this and so the percent excreted is an increasing function of GFR. Lower GFR values correspond to sicker patients. We see that in sicker patients, less of the dose is excreted. This is why, in patients with kidney disease, physicians must be careful not to administer doses of antibiotics too frequently.

Example 5   Suppose the contour diagram in Figure 7.17 shows your happiness as a function of love and money. Describe in words your happiness:
(a)   As a function of money, with love fixed.
(b)   As a function of love, with money fixed.
(c)   Draw the graphs of two different cross-sections with love fixed and two different cross-sections with money fixed.

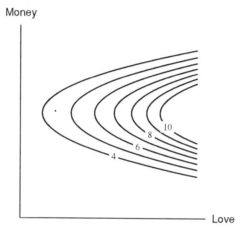

Figure 7.17: Happiness as a function of love and money

Solution   (a)   As money increases, with love fixed, your happiness goes up, reaches a maximum and then goes back down. Evidently, there is such a thing as too much money.
(b)   On the other hand, as love increases, with money fixed, your happiness keeps going up.
(c)   A cross-section with love fixed will show your happiness as money increases; the curve goes up to a maximum then back down, as in Figure 7.18. The higher cross-section, showing more overall happiness, corresponds to a larger amount of love, because as love increases so does happiness. Figure 7.19 shows two cross-sections with money fixed. Happiness increases as love increases. We cannot say, however, which cross-section corresponds to a larger fixed amount of money, because as money increases happiness can either increase or decrease.

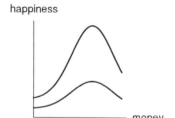

Figure 7.18: Cross-sections with love fixed

Figure 7.19: Cross-sections with money fixed

Contour Diagrams and Tables

Table 7.8 shows the heat index as a function of temperature and humidity. (This table was also used in the exercises for the previous section.) We can also display this function using a contour diagram. Scales for the two independent variables (temperature and humidity) go on the axes. The heat indices shown range from 64 to 151, so we will draw contours at values of 70, 80, 90, 100, 110, 120, 130, 140, and 150. How do we know where the contour for 70 goes? We see in Table 7.8 that, when humidity is 0%, a heat index of 70 occurs between 75°F and 80°F, so the contour will go approximately through the point $(0, 76)$. We see that it also goes through the point $(10, 75)$. Continuing in this way, we can sketch the approximate path of the 70 contour. See Figure 7.20. We can construct all contours in a similar way. The finished contour diagram is shown in Figure 7.21.

TABLE 7.8   *Heat index (°F)*

Humidity (%)	Temperature (°F)									
	70	75	80	85	90	95	100	105	110	115
0	64	69	73	78	83	87	91	95	99	103
10	65	70	75	80	85	90	95	100	105	111
20	66	72	77	82	87	93	99	105	112	120
30	67	73	78	84	90	96	104	113	123	135
40	68	74	79	86	93	101	110	123	137	151
50	69	75	81	88	96	107	120	135	150	
60	70	76	82	90	100	114	132	149		

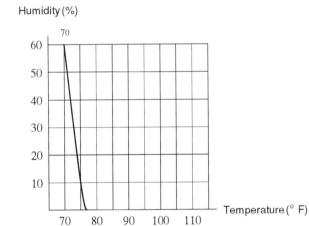

Figure 7.20: The contour for a heat index of 70

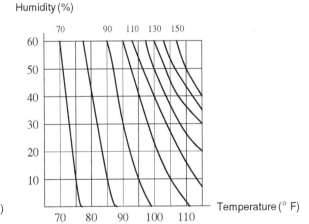

Figure 7.21: Contour diagram for the heat index

Example 6   The contour diagram for the heat index is given in Figure 7.21. Heat exhaustion is likely to occur where the heat index is 105 or higher. On the contour diagram, shade in the region where heat exhaustion is likely to occur.

Solution   The region where the heat index is above 105 is shaded in Figure 7.22, and at these values of temperature and humidity, heat exhaustion is likely.

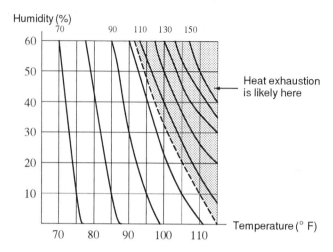

Figure 7.22: Where is heat exhaustion likely?

Finding Contours Algebraically

Algebraic equations for the contours of a function f are easy to find if we have a formula for $f(x, y)$. Suppose the function has equation

$$z = f(x, y).$$

A contour is obtained by finding all points where $z = c$. Thus, the equation for the contour of value c is given by

$$f(x, y) = c.$$

Example 7   In Example 3 of Section 7.1, we saw that if a car rental company charges $40 a day and 15 cents a mile, the cost, C, to rent a car is given by

$$C = 40d + 0.15m,$$

where d is the number of days and m is the number of miles driven. Draw a contour diagram for $C = f(d, m)$. Include contours for $C = 50, 100, 150, 200$.

Solution   The contour for $C = 50$ is given by

$$40d + 0.15m = 50.$$

This is the equation of a line with intercepts $d = 50/40 = 1.25$ and $m = 50/0.15 \approx 333$. See Figure 7.23. The contour for $C = 100$ is given by

$$40d + 0.15m = 100.$$

This is the equation of a line with intercepts $d = 100/40 = 2.5$ and $m = 100/0.15 \approx 667$. We can draw the contours for $C = 150$ and $C = 200$ similarly. The contour diagram is shown in Figure 7.23.

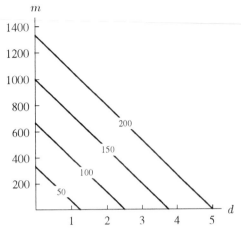

Figure 7.23: A contour diagram for
$$C = 40d + 0.15m$$

Problems for Section 7.2

1. Figure 7.24 shows cardiac output in patients suffering from shock as a function of right atrial pressure (in mm Hg) and the time in hours since hemorrhagic shock began. [5] Cardiac output is measured in liters per minute.

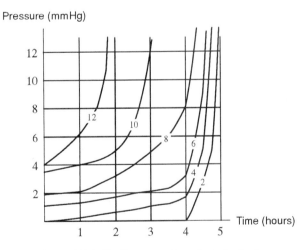

Figure 7.24: Cardiac output in patients in shock

(a) In a patient with right atrial pressure of 4 mm Hg, what is cardiac output when the patient first goes into shock? Estimate cardiac output three hours later. How much time has passed when cardiac output is reduced to 50% of the value at $t = 0$?

[5] Arthur C. Guyton and John E. Hall, *Textbook of Medical Physiology, Ninth Edition* (Philadelphia: W. B. Saunders,1996)
288

(b)   Is cardiac output an increasing or decreasing function of right atrial pressure in patients suffering from shock?

(c)   Is cardiac output an increasing or decreasing function of time since the patient went into shock? Why is this reasonable?

(d)   If right atrial pressure is 3 mm Hg, explain how cardiac output changes as a function of time. In particular, does it change rapidly or slowly during the first two hours of shock? During hours 2 to 4? During the last hour of the study? Explain why this information might be useful to a physician treating a patient for shock.

2.   A contour diagram is given in Figure 7.25 for a function $z = f(x, y)$. Is z an increasing or a decreasing function of x? Is z an increasing or a decreasing function of y?

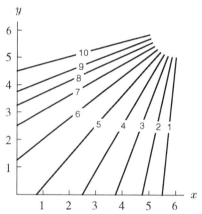

Figure 7.25: Is z an increasing or a decreasing function of x? Of y?

3.   The demand for orange juice is a function of the price of orange juice and the price of apple juice. A possible contour diagram for this demand function is given in Figure 7.26. Which axis corresponds to the price of orange juice? Which axis corresponds to the price of apple juice? Explain.

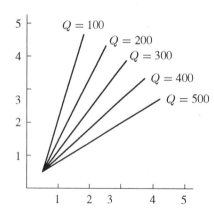

Figure 7.26: Demand for orange juice

4. The total sales of a product is a function of the price of the product and the amount spent on advertising. A contour diagram for the number of sales is given in Figure 7.27. Which axis corresponds to the amount spent on advertising? Explain.

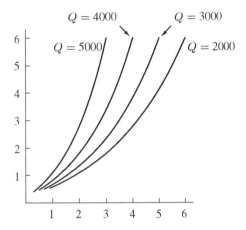

Figure 7.27: Total sales of a product

5. A topographic map is given in Figure 7.28. How many hills are there? Give the x- and y-coordinates of the tops of the hills. Which hill is the highest? There is a river running through the valley; in which direction is it flowing? Assume north is in the positive y-direction.

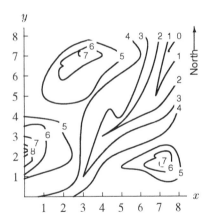

Figure 7.28: A topographic map

6. Table 7.6 on page 379 shows the wind-chill factor as a function of wind speed and temperature. Draw a possible contour diagram for this function. Include contours at wind chills of $20°$, $0°$, and $-20°$.

7. We saw in Section 7.1 that the concentration, C, of a drug in the blood could be given by

$$C = f(x, t) = te^{-t(5-x)},$$

where x is the amount of drug injected and t is the number of hours since the injection. The contour diagram of $f(x, t)$ is given in Figure 7.29. Explain the diagram by varying one variable at a time: describe f as a function of x if t is held fixed, and then describe f as a function of t if x is held fixed.

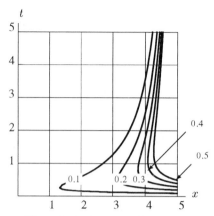

Figure 7.29: A contour diagram for concentration of a drug in the blood

8. Figure 7.30 is a contour diagram of the monthly payment on a 5-year car loan as a function of the interest rate and the amount you borrow. Suppose that the interest rate is 13% and that you decide to borrow $6000.

   (a) What is your monthly payment?

   (b) If interest rates drop to 11%, how much more can you borrow without increasing your monthly payment?

   (c) Make a table of how much you can borrow, without increasing your monthly payment, as a function of the interest rate.

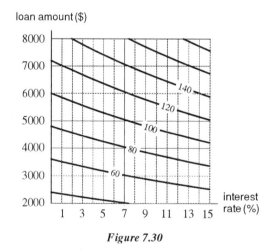

Figure 7.30

For the functions in Problems 9–14, sketch a contour diagram with at least four labeled contours.

9. $f(x, y) = x + y$ 10. $f(x, y) = x + y + 1$ 11. $f(x, y) = 3x + 3y$

12. $f(x, y) = -x - y$ 13. $f(x, y) = 2x - y$ 14. $f(x, y) = y - x^2$

15. The map in Figure 7.31 is from the undergraduate senior thesis of Professor Robert Cook, Director of Harvard's Arnold Arboretum. It shows level curves of the function giving the species density of breeding birds at each point in the US, Canada, and Mexico.

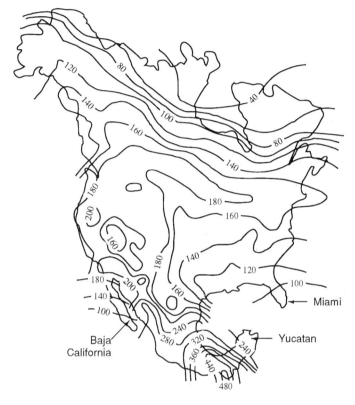

Figure 7.31

Using the map in Figure 7.31, decide if the the following statements are true or false. Explain your answers.

(a) Moving from south to north across Canada, the species density increases.

(b) The species density in the area around Miami is over 100.

(c) In general, peninsulas (for example, Florida, Baja California, the Yucatan) have lower species densities than the areas around them.

(d) The greatest rate of change in species density with distance is in Mexico. If you think this is true, mark the point and direction which give the maximum rate of change and explain why you picked the point and direction you did.

16.

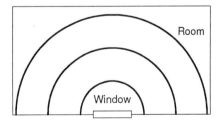

Figure 7.32

Figure 7.32 shows the level curves of the temperature H in a room near a recently opened window. Label the three level curves with reasonable values of H if the house is in the following locations.

(a)   Minnesota in winter (where winters are harsh).

(b)   San Francisco in winter (where winters are mild).

(c)   Houston in summer (where summers are hot).

(d)   Oregon in summer (where summers are mild).

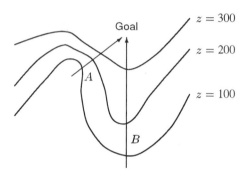

Figure 7.33

17.   Figure 7.33 shows a contour map of a hill with two paths, A and B.

(a)   On which path, A or B, will you have to climb more steeply?

(b)   On which path, A or B, will you probably have a better view of the surrounding countryside? (Assuming trees do not block your view.)

(c)   Alongside which path is there more likely to be a stream?

18.   Each of the contour diagrams in Figure 7.34 shows population density in a certain region of a city. Choose the contour diagram that best corresponds to each of the following situations. Many different matchings are possible. Pick a reasonable one and justify your choice.

(a)   The middle contour line is a highway.

(b)   The middle contour line is an open sewage canal.

(c)   The middle contour line is a railroad line.

(I) 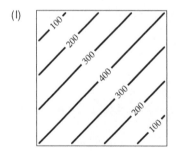 (II) 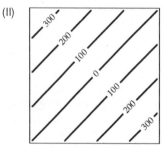 (III)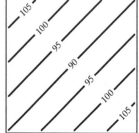

Figure 7.34

19. Each of the contour diagrams in Figure 7.35 shows population density in a certain region. Choose the contour diagram that best corresponds to each of the following situations. Many different matchings are possible. Pick any reasonable one and justify your choice.

    (a) The center of the diagram is a city.
    (b) The center of the diagram is a lake.
    (c) The center of the diagram is a power plant.

    (I) 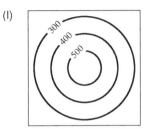   (II) 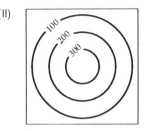   (III)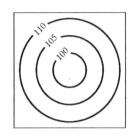

    Figure 7.35

20. Match the following descriptions of a company's success with the contour diagrams of success as a function of money and work in Figure 7.36.

    (a) Our success is measured in dollars, plain and simple. More hard work won't hurt, but it also won't help.
    (b) No matter how much money or hard work we put into the company, we just can't make a go of it.
    (c) Although we aren't always totally successful, it seems that the amount of money invested doesn't matter. As long as we put hard work into the company our success will increase.
    (d) The company's success is based on both hard work and investment.

    (I) 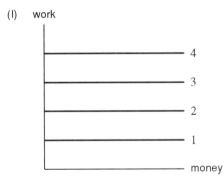   (II)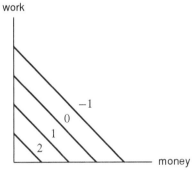

    (III) 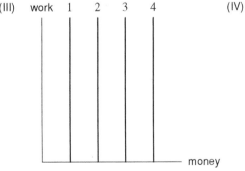   (IV)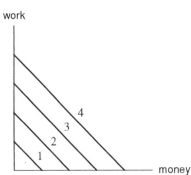

    Figure 7.36

21. You like pizza and you like cola. Which of the contour diagrams in Figure 7.37, of happiness as a function of the number of pizzas and the number of colas, represents your happiness as a function of how many pizzas and how much cola you have if

    (a) There is no such thing as too many pizzas and too much cola?
    (b) There is such a thing as too many pizzas or too much cola?
    (c) There is such a thing as too much cola but no such thing as too many pizzas?

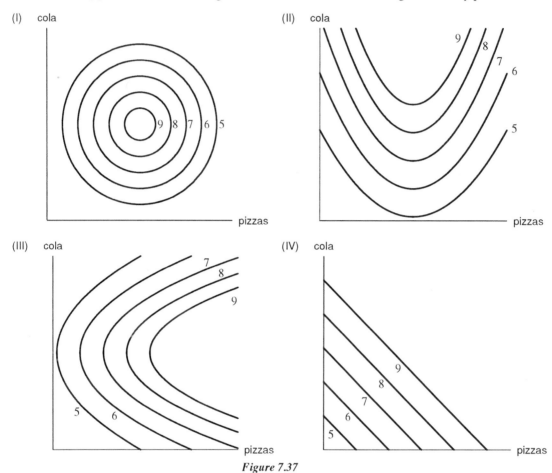

Figure 7.37

22. For each of the contour diagrams I-IV in Problem 21 draw:

    (a) a graph of happiness as a function of cola, if pizza is fixed
    (b) a graph of happiness as a function of pizza, if cola is fixed.

23. Match Tables 7.9–7.12 with the contour diagrams (I) - (IV) in Figure 7.38.

TABLE 7.9

$y \backslash x$	−1	0	1
−1	2	1	2
0	1	0	1
1	2	1	2

TABLE 7.10

$y \backslash x$	−1	0	1
−1	0	1	0
0	1	2	1
1	0	1	0

TABLE 7.11

$y \backslash x$	−1	0	1
−1	2	0	2
0	2	0	2
1	2	0	2

TABLE 7.12

$y \backslash x$	−1	0	1
−1	2	2	2
0	0	0	0
1	2	2	2

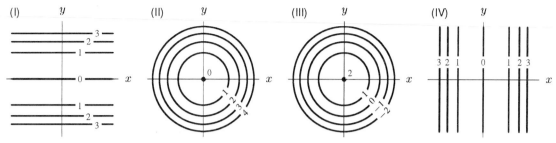

Figure 7.38

24. In Problems 22– 26 in Section 7.1, we looked at the demand for coffee, Q, in thousands of pounds per week, as a function of the price of coffee, c, and the price of tea, t. We have

$$Q = f(c, t) = 100\frac{t}{c}.$$

   (a) Draw a contour diagram for this function. Include contours for $Q = 25$, $Q = 50$, $Q = 100$, and $Q = 200$.
   (b) Determine from the contour diagram whether Q is an increasing or a decreasing function of c. Is Q an increasing or a decreasing function of t?
   (c) Demand is called *elastic* if small increases in the price of the product cause relatively large changes in the demand for the product. If the price of tea is fixed at $t = 1$, is the demand for coffee more elastic at low prices or high prices? Explain your answers using the contour diagram.

25. A city on an island has a large central park. Draw a possible contour diagram showing light intensity at night as a function of position. Label your contours with values between 0 and 1, where 0 represents total darkness and 1 represents maximum artificial illumination.

26. Figure 7.39 shows the density of the fox population P (in foxes per square kilometer) for southern England. Draw two different graphs of the fox population as a function of miles north, with miles east fixed at two different values, and draw two different graphs of the fox population as a function of miles east, with miles north fixed at two different values.

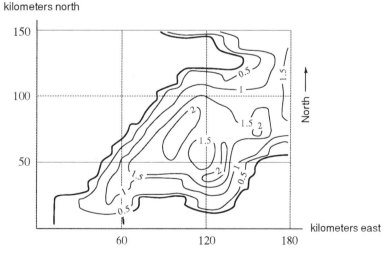

Figure 7.39: Population density of foxes in southwestern England

27. A manufacturer sells two goods, one at a price of $3000 a unit and the other at a price of $12,000 a unit. Suppose a quantity q_1 of the first good and q_2 of the second good are sold at a total cost of $4000 to the manufacturer.

    (a) Express the manufacturer's profit, P, as a function of q_1 and q_2.
    (b) Sketch the contour diagram for profit P as a function of q_1 and q_2. Include contours at $P = 10,000$, $P = 20,000$, and $P = 30,000$ and the break-even curve $P = 0$.

28. Figure 7.40 shows the contours of the temperature along one wall of a heated room through one winter day, with time indicated as on a 24-hour clock. The room has a heater located at the left-most corner of the wall and one window in the wall. The heater is controlled by a thermostat about 2 feet from the window.

    (a) Where is the window?
    (b) When is the window open?
    (c) When is the heat on?
    (d) Draw graphs of the temperature along the wall of the room at 6 am, at 11 am, at 3 pm (15 hours) and at 5 pm (17 hours).
    (e) Draw a graph of the temperature as a function of time at the heater, at the window and midway between them.
    (f) The temperature at the window at 5 pm (17 hours) is less than at 11 am. Why do you think this might be?
    (g) To what temperature do you think the thermostat is set? How do you know?
    (h) Where is the thermostat?

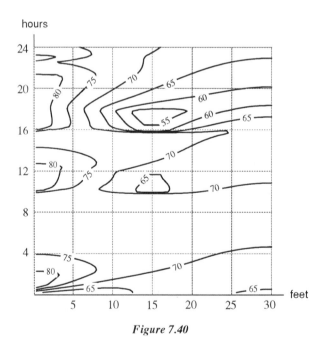

Figure 7.40

29. Figure 7.41 shows contour diagrams of temperature in degrees Celsius in a room at three different times. Describe the heat flow in the room. What could be causing this?

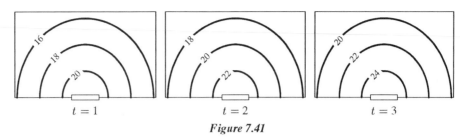

Figure 7.41

7.3 THE PARTIAL DERIVATIVE

In one-variable calculus we saw how the derivative measures the rate of change of a function. We begin by reviewing this idea.

Formaldehyde Toxicity in Rats: a One-Variable Problem

About 120 female rats were exposed to formaldehyde vapor in a concentration of 6 ppm (parts per million) for a period of 24 months in a study to examine the toxicity of formaldehyde. The percent of rats surviving, P, was recorded each month, and the data is given in Table 7.13. Figure 7.42 shows a graph of this data, and we see that the percent surviving goes down very slowly at first, and then more rapidly near the end of the study.

TABLE 7.13 *Percent of rat population surviving after exposure to formaldehyde vapor in 6 ppm*

t (months)	0	2	4	6	8	10	12	14	16	18	20	22	24
P (%)	100	100	100	99	99	98	96	96	95	93	90	86	80

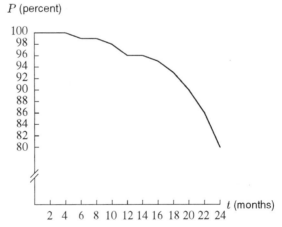

Figure 7.42: Percent surviving after exposure to formaldehyde vapor in 6 ppm

Example 1   Let $P = g(t)$ be the function given in Table 7.13, and estimate the derivative $g'(18)$. Explain what the answer means in terms of formaldehyde toxicity.

Solution   The derivative $g'(18)$ is the rate of change of P as a function of t, which can be approximated by $\Delta P/\Delta t$, the change in P divided by the change in t, for a small change in t. Using the data in the table, we take $\Delta t = 2$. We estimate the derivative to be

$$g'(18) \approx \frac{\Delta P}{\Delta t} = \frac{g(20) - g(18)}{20 - 18} = \frac{90 - 93}{20 - 18} = -1.5 \text{ \%/month.}$$

The fact that $g'(18) \approx -1.5$ means that the percent surviving is decreasing at a rate of 1.5 per month. In other words, in the next month (the nineteenth) we estimate that about another 1.5% of the original rat population will die.

Formaldehyde Toxicity in Rats: a Two-Variable Problem

The data given in Table 7.13 is not very helpful, because we have no way of knowing how many of the deaths were caused by formaldehyde and how many were natural or from other causes. The actual study involved four groups of female rats (each containing about 120 rats). Each group was exposed to a different concentration of formaldehyde vapor: one at 15 ppm, one (seen in Table 7.13) at 6 ppm, one at 2 ppm, and the fourth was a control group with no formaldehyde exposure. [6] The results of the study are given in Table 7.14. Notice that the percent surviving, P, is a function of time, t, in months since the start of the study and concentration, c, of the formaldehyde vapor. We have

$$P = f(t, c)$$

TABLE 7.14   *Percent of rat population surviving after exposure to formaldehyde vapor*

Concentration c (ppm)	time t (months)												
	0	2	4	6	8	10	12	14	16	18	20	22	24
0	100	100	100	100	100	100	100	100	100	100	99	97	95
2	100	100	100	100	100	100	100	100	99	98	97	95	92
6	100	100	100	99	99	98	96	96	95	93	90	86	80
15	100	100	100	99	99	99	99	96	93	82	70	58	36

How does the percent surviving, P, vary when $t = 18$ months and $c = 6$ ppm? Since we know how to measure the rate of change of a function of a single variable, we vary just one variable at a time. If we fix c at 6 ppm, we can find the rate of change of percent surviving, P, with respect to t. Notice that the $c = 6$ row of Table 7.14 is exactly the same as Table 7.13. Thus, $f(t, 6) = g(t)$.

What is the meaning of the derivative $g'(18)$ in this context? It is the rate of change of percent surviving, P, *in the time t-direction* at the point $(18, 6)$. We denote this rate of change by $f_t(18, 6)$, so that

$$f_t(18, 6) = g'(18) \approx -1.5 \text{ \%/month.}$$

We call $f_t(18, 6)$ the *partial derivative of f with respect to t at the point* $(18, 6)$. The fact that it is negative means that P is decreasing as we read across the $c = 6$ row of the table in the direction of increasing t (that is, horizontally from left to right in Table 7.14).

[6] James E. Gibson, *Formaldehyde Toxicity*, Hemisphere Publishing Company, McGraw-Hill, 1983, p. 125.

Example 2   Using Table 7.14, estimate the rate of change of P in the c-direction at the point $(18, 6)$, and interpret your answer in terms of formaldehyde toxicity.

TABLE 7.15   *Percent surviving 18 months*

c (ppm)	P (%)
0	100
2	98
6	93
15	82

Solution   Since we want the rate of change of P *in the c-direction*, we fix t at 18, and look at how P changes as we move in the direction of increasing c (that is, from top to bottom in Table 7.14.) This column is reproduced in Table 7.15. If we let $h(c)$ represent the function $f(18, c)$, and denote the rate of change of P in the c-direction at $(18, 6)$ by $f_c(18, 6)$, we have

$$f_c(18, 6) = h'(6) \approx \frac{\Delta P}{\Delta c} = \frac{82 - 93}{15 - 6} = -1.22\%/\text{ppm}.$$

We call $f_c(18, 6)$ *the partial derivative of f with respect to c at the point* $(18, 6)$. The rate of change of P as c increases is -1.22% per ppm. This means that as the concentration increases by 1 ppm from 6 ppm, we expect the percent surviving 18 months to decrease by about 1.22%. The partial derivative is negative because fewer rats survive this long when the concentration of formaldehyde is greater (P goes down as c goes up).

Definition of the Partial Derivative

For any function $f(x, y)$ we study the influence of x and y separately on the value $f(x, y)$, by holding one fixed and letting the other vary. The method of the previous example allows us to calculate the rates of change of $f(x, y)$ with respect to x and y, giving us the following definitions.

Partial Derivatives of f With Respect to x and y

For all points at which the limits exist, define the **partial derivative at the point** $(\mathbf{a}, \mathbf{b})$

$$f_x(a, b) = \begin{array}{c} \text{Rate of change of } f \text{ with respect to } x \\ \text{at the point } (a, b) \end{array} = \lim_{h \to 0} \frac{f(a + h, b) - f(a, b)}{h}$$

$$f_y(a, b) = \begin{array}{c} \text{Rate of change of } f \text{ with respect to } y \\ \text{at the point } (a, b) \end{array} = \lim_{h \to 0} \frac{f(a, b + h) - f(a, b)}{h}.$$

If we think of a and b as variables, $a = x$ and $b = y$, we have the **partial derivative functions** $f_x(x, y)$ and $f_y(x, y)$.

Just as with ordinary derivatives, there is an alternative notation:

Alternative Notation for Partial Derivatives

If $z = f(x, y)$ we can write

$$f_x(x, y) = \frac{\partial z}{\partial x} \quad \text{and} \quad f_y(x, y) = \frac{\partial z}{\partial y}$$

$$f_x(a, b) = \frac{\partial z}{\partial x}\bigg|_{(a,b)} \quad \text{and} \quad f_y(a, b) = \frac{\partial z}{\partial y}\bigg|_{(a,b)}$$

We use the symbol ∂ to distinguish partial derivatives from ordinary derivatives. In cases where the independent variables have names different from x and y, we adjust the notation accordingly. For example, the partial derivatives of $f(u, v)$ are denoted by f_u and f_v.

Estimating Partial Derivatives from a Table

Example 3   To avoid flying planes with too many empty seats, airlines sell some tickets at full price and some at a discount. Table 7.16 shows an airline's revenue R in dollars, from tickets sold on a particular route, as a function of the number of full-price tickets sold (x) and the number of discount tickets sold (y). We write $R = f(x, y)$.

(a) Evaluate $f(200, 400)$, and interpret your answer.

(b) Is $f_x(200, 400)$ positive or negative? Is $f_y(200, 400)$ positive or negative? Explain.

(c) Estimate each of the partial derivatives in part (b). Give units with your answers and interpret them in terms of airline revenue.

TABLE 7.16   *Revenue from ticket sales (dollars)*

	Number of full-price tickets, x			
Number of discount tickets, y	100	**200**	300	400
200	39,700	**63,600**	87,500	111,400
400	**55,500**	**79,400**	**103,300**	**127,200**
600	71,300	**95,200**	119,100	143,000
800	87,100	**111,000**	134,900	158,800
1000	102,900	**126,800**	150,700	174,600

Solution   (a) $f(200, 400) = 79,400$. This means that sales of 200 full-price tickets and 400 discount tickets generate $79,400 in revenue.

(b)   The notation $f_x(200, 400)$ indicates the rate of change of f as we fix y at 400 and increase x from 200. What happens to the revenue as we look right along the row $y = 400$ in the table? Revenue increases, and so $f_x(200, 400)$ is positive.

The notation $f_y(200, 400)$ indicates the rate of change of f as we fix x at 200 and increase y from 400. What happens to the revenue as we look down the column $x = 200$ in the table? Revenue increases, and so $f_y(200, 400)$ is positive.

Both partial derivatives are positive. This makes sense, because revenue will go up if more of either type of ticket is sold.

(c) To estimate $f_x(200, 400)$, we calculate $\Delta R / \Delta x$ as x increases from 200 to 300, and y is held constant. We have

$$f_x(200, 400) \approx \frac{\Delta R}{\Delta x} = \frac{103{,}300 - 79{,}400}{300 - 200} = 239 \text{ dollars/ticket.}$$

The partial derivative of f with respect to x is 239 dollars per full-price ticket. The price of one full-price ticket is \$239.

To estimate $f_y(200, 400)$, we calculate $\Delta R / \Delta y$ as y increases from 400 and 600, and x is held constant. We have

$$f_y(200, 400) \approx \frac{\Delta R}{\Delta y} = \frac{95{,}200 - 79{,}400}{600 - 400} = 79 \text{ dollars/ticket.}$$

The partial derivative of f with respect to y is 79 dollars per discount ticket. The price of one discount ticket is \$79.

Using Partial Derivatives to Estimate Values of the Function

Example 4   In Example 3 the revenue is \$79,400 when 200 full-price tickets and 400 discount tickets are sold; that is, $f(200, 400) = 79{,}400$. Use this fact and the partial derivatives $f_x(200, 400) = 239$ and $f_y(200, 400) = 79$ to find the revenue when

(a) $x = 201$ and $y = 400$.
(b) $x = 200$ and $y = 405$.
(c) $x = 203$ and $y = 406$.

Solution   (a) The partial derivative $f_x = 239$ tells us that R goes up by 239 as x goes up by 1. Thus $f(201, 400) = f(200, 400) + 239 = 79{,}400 + 239 = \$79{,}639$. Sales of 201 full-price and 400 discount tickets will bring in a revenue of \$79,639.

(b) The partial derivative $f_y = 79$ tells us that R goes up by 79 as y goes up by 1. Since y is going up by 5, we have $f(200, 405) = f(200, 400) + 5(79) = \$79{,}795$.

(c) Here x is going up by 3 and y is going up by 6, and we have $f(203, 406) = f(200, 400) + 3(239) + 6(79) = \$80{,}591$.

Example 5   Use Table 7.14 and partial derivatives to estimate the percent of rats surviving after 24 months if they are exposed to a formaldehyde concentration of 18 ppm.

Solution   We have percent surviving, $P = f(t, c)$, where t is time in months and c is formaldehyde concentration, and we wish to evaluate $f(24, 18)$. The closest entry to this in Table 7.14 is $f(24, 15) = 36$. We want to keep t fixed at 24 and increase c from 15 to 18. We need to estimate the rate of change in P as c changes – this is f_c. We see from Table 7.14 that

$$f_c(24, 15) \approx \frac{\Delta P}{\Delta c} = \frac{36 - 80}{15 - 6} = -4.89 \text{ \%/ppm.}$$

The percent surviving 24 months goes down from 36% by about 4.89% for every unit increase above 15 ppm in the formaldehyde concentration. We have:

$$f(24, 18) \approx 36 - (3)4.89 = 21.33\%.$$

We estimate that only about 21% of the rats would survive for 24 months if they were exposed to formaldehyde as strong as 18 ppm. Since this figure is an extrapolation from the available data, we should use it with caution.

Estimating Partial Derivatives from a Contour Diagram

If we move parallel to one of the axes on a contour diagram, the partial derivative is the rate of change of the value of the function on the contours. For example, if the values on the contours are increasing, then the partial derivative must be positive.

Example 6   Figure 7.43 shows the contour diagram for the temperature $H(x, t)$ (in °F) in a room as a function of distance x (in feet) from a heater and time t (in minutes) after the heater has been turned on. What are the signs of $H_x(10, 20)$ and $H_t(10, 20)$? Estimate these partial derivatives and explain the answers in practical terms.

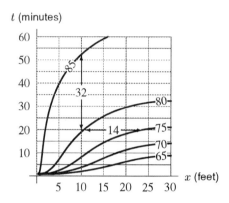

Figure 7.43: Temperature in a heated room

Solution   The point $(10, 20)$ is on the $H = 80$ contour. As x increases, we move towards the $H = 75$ contour, so H is decreasing and $H_x(10, 20)$ is negative. This makes sense because as we move further from the heater, the temperature drops. On the other hand, as t increases, we move towards the $H = 85$ contour, so H is increasing and $H_t(10, 20)$ is positive. This also makes sense, because it says that as time passes, the room warms up.

To estimate the partial derivatives, use a difference quotient. Looking at the contour diagram, we see there is a point on the $H = 75$ contour about 14 units to the right of $(10, 20)$. Hence, H decreases by 5 when x increases by 14, so the rate of change of H with respect to x is about $\Delta H / \Delta x = -5/14 \approx -0.36$. Thus, we find

$$H_x(10, 20) \approx -0.36° \text{ F/ft.}$$

This means that near the point 10 ft from the heater, after 20 minutes the temperature drops about 1/3 of a degree for each foot we move away from the heater.

To estimate $H_t(10, 20)$, we look again at the contour diagram and notice that the $H = 85$ contour is about 32 units directly above the point $(10, 20)$. So H increases by 5 when t increases by 32. Hence,

$$H_t(10, 20) \approx \frac{\Delta H}{\Delta t} = \frac{5}{32} \approx 0.16°\text{F/min.}$$

This means that after 20 minutes the temperature is going up about 1/6 of a degree each minute at the point 10 ft from the heater.

Using Units to Interpret Partial Derivatives

The meaning of a partial derivative can often be explained using units.

Example 7 Suppose that your weight w in pounds is a function $f(c, n)$ of the number c of calories you consume daily and the number n of minutes you exercise daily. Using the units for w, c and n, interpret in everyday terms the statements

$$\frac{\partial w}{\partial c}(2000, 15) = 0.02 \quad \text{and} \quad \frac{\partial w}{\partial n}(2000, 15) = -0.025.$$

Solution The units of $\partial w/\partial c$ are pounds per calorie. The statement

$$\frac{\partial w}{\partial c}(2000, 15) = 0.02$$

means that if you are presently consuming 2000 calories daily and exercising 15 minutes daily, you will weigh 0.02 pounds more for each extra calorie you consume daily, or about 2 pounds for each extra 100 calories per day.

The units of $\frac{\partial w}{\partial n}$ are pounds per minute. The statement

$$\frac{\partial w}{\partial n}(2000, 15) = -0.025$$

means for the same calorie consumption and number of minutes of exercise, you will weigh 0.025 pounds less for each extra minute you exercise daily, or about 1 pound less for each extra 40 minutes per day.

Problems for Section 7.3

1. The demand for coffee, Q, in pounds sold per week, can be thought of as a function of the price of coffee, c, in dollars per pound, and the price of tea t, in dollars per pound. Thus, we have $Q = f(c, t)$.

   (a) Do you expect f_c to be positive or negative? What about f_t? Explain.

   (b) Interpret each of the following statements:

      (i) $f(3, 2) = 780$

      (ii) $f_c(3, 2) = -60$

      (iii) $f_t(3, 2) = 20$

2. A drug is injected into a patient's blood vessel. The function $c = f(x, t)$ represents the concentration of the drug at a distance x mm in the direction of the blood flow measured from the point of injection and at time t seconds since the injection. What are the units of the following partial derivatives? What are their practical interpretations? What do you expect their signs to be?

   (a) $\partial c/\partial x$ (b) $\partial c/\partial t$

3. The monthly mortgage payment in dollars, P, for a house is a function of three variables

$$P = f(A, r, N),$$

where A is the amount borrowed in dollars, r is the interest rate, and N is the number of years before the mortgage is paid off.

(a) $f(92000, 14, 30) = 1090.08$. What does this tell you, in financial terms?

(b) $\dfrac{\partial P}{\partial r}(92000, 14, 30) = 72.82$. What is the financial significance of the number 72.82?

(c) Would you expect $\partial P/\partial A$ to be positive or negative? Why?

(d) Would you expect $\partial P/\partial N$ to be positive or negative? Why?

4. Suppose you borrow $\$A$ at an interest rate of $r\%$ (per month) and pay it off over t months by making monthly payments of $\$P$, as determined by the function $P = g(A, r, t)$. In financial terms, what do the following statements tell you?

(a) $g(8000, 1, 24) = 376.59$

(b) $\dfrac{\partial g}{\partial A}(8000, 1, 24) = 0.047$

(c) $\dfrac{\partial g}{\partial r}(8000, 1, 24) = 44.83$

5. Suppose P is your monthly car payment in dollars and $P = f(P_0, t, r)$, where $\$P_0$ is the amount you borrowed, t is the number of months it takes to pay off the loan, and $r\%$ is the interest rate. What are the units, the financial meanings, and the signs of $\partial P/\partial t$ and $\partial P/\partial r$?

6. Table 7.17 gives a table of values for a positive function $z = f(x, y)$.

TABLE 7.17

	$x = 0$	$x = 10$	$x = 20$	$x = 30$
$y = 0$	89	80	74	71
$y = 2$	93	85	80	76
$y = 4$	98	91	85	81
$y = 6$	104	98	92	88
$y = 8$	112	105	99	94

(a) Is f_x positive or negative? Is f_y positive or negative? Explain.

(b) Find $f_x(10, 6)$ and $f_y(10, 6)$.

(c) Use partial derivatives to estimate $f(30, 9)$. Explain your reasoning.

(d) Use partial derivatives to estimate $f(34, 8)$. Explain your reasoning.

7. Using the contour diagram for $z = f(x, y)$ given in Figure 7.44, determine whether each of these partial derivatives is positive, negative, or zero.

(a) $f_x(4, 1)$

(b) $f_y(4, 1)$

(c) $f_x(5, 2)$

(d) $f_y(5, 2)$

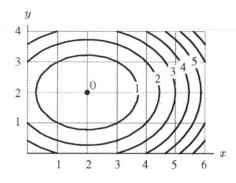

Figure 7.44

8. According to the contour diagram for $z = f(x, y)$ given in Figure 7.44, which is larger: $f_x(3, 1)$ or $f_x(5, 2)$? Explain.

9. Figure 7.45 shows a contour diagram for the monthly payment P as a function of the interest rate, $r\%$, and the amount, L, of a 5-year loan. Estimate $\partial P/\partial r$ and $\partial P/\partial L$ at the point where $r = 8$ and $L = 5000$. Give the units and the everyday meaning of your answer.

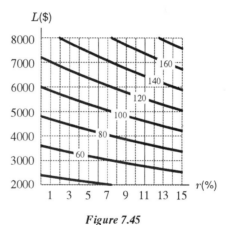

Figure 7.45

10. Table 7.1 on page 373 gives the quantity of beef bought, C, as a function of household income, I, and the price of beef, p. Thus we have $C = f(I, p)$.
    (a) Find $f_p(80, 4.0)$ and interpret it in terms of beef consumption.
    (b) Find $f_I(80, 4.0)$ and interpret it in terms of beef consumption.
    (c) Use partial derivatives to estimate the quantity of beef bought by a family with a household income of \$110,000 if the price of beef is \$4.00 per pound. Explain your reasoning.

11. Figure 7.15 on page 384 gives a contour diagram of corn production as a function of rainfall, R, in inches and temperature, T, in degrees Fahrenheit. Corn production is measured as a percentage of the present production, and is equal to $f(R, T)$.
    (a) Find $f_R(15, 76)$. Give units with your answer and interpret it in terms of corn production.
    (b) Find $f_T(15, 76)$. Give units with your answer and interpret it in terms of corn production.

12. Suppose for a function $f(x, y)$ we know that $f(100, 20) = 2750$, $f_x(100, 20) = 4$, and $f_y(100, 20) = 7$. Estimate $f(105, 21)$. Justify your answer.

13. Suppose for a function $f(r, s)$ that $f(50, 100) = 5.67$, $f_r(50, 100) = 0.60$, and $f_s(50, 100) = -0.15$. Estimate $f(52, 108)$. Justify your answer.

For Problems 14–16, refer to Table 7.6 on page 379 giving the temperature adjusted for wind-chill, C, in °F, as a function $f(w, T)$ of the wind speed, w, in mph, and the temperature, T, in °F. The temperature adjusted for wind-chill tells you how cold it feels, as a result of the combination of wind and temperature.

14. Estimate $f_w(10, 25)$. What does your answer mean in practical terms?

15. Estimate $f_T(5, 20)$. What does your answer mean in practical terms?

16. From Table 7.6 you can see that when the temperature is 20°F, the temperature adjusted for wind-chill drops by an average of about 2.6°F with every 1 mph increase in wind speed from 5 mph to 10 mph. Which partial derivative is this telling you about?

For Problems 17–18 refer to Table 7.8 on page 387 giving the heat index, I, in °F, as a function $f(H, T)$ of the relative humidity, H, and the temperature, T, in °F. The heat index is a temperature which tells you how hot it feels as a result of the combination of humidity and temperature.

17. Estimate $\partial I / \partial H$ and $\partial I / \partial T$ for typical weather conditions in Tucson in summer ($H = 10$, $T = 100$). What do your answers mean in practical terms for the residents of Tucson?

18. Answer the question in Problem 17 for Boston in summer ($H = 50$, $T = 80$).

19. Table 7.14 on page 400 gives the percent of rats surviving, P, as a function of time, t, in months and concentration of formaldehyde, c, in ppm. Thus we have $P = f(t, c)$. Use partial derivatives to estimate the percent surviving after 26 months, if the concentration is 15. Explain your reasoning.

20. Suppose that x is the average price of a new car and that y is the average price of a gallon of gasoline. Then q_1, the number of new cars bought in a year, depends on both x and y, so $q_1 = f(x, y)$. Similarly, if q_2 is the quantity of gas bought in a year, then $q_2 = g(x, y)$.

    (a) What do you expect the signs of $\partial q_1 / \partial x$ and $\partial q_2 / \partial y$ to be? Explain.

    (b) What do you expect the signs of $\partial q_1 / \partial y$ and $\partial q_2 / \partial x$ to be? Explain.

21. Estimate $z_x(1, 0)$ and $z_x(0, 1)$ and $z_y(0, 1)$ from the contour diagram for $z(x, y)$ in Figure 7.46.

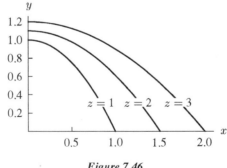

Figure 7.46

22. Figure 7.47 gives a contour diagram for the number n of foxes per square kilometer in southwestern England. Estimate $\partial n / \partial x$ and $\partial n / \partial y$ at the points marked A, B, and C, where x is kilometers east-west and y is kilometers north-south.

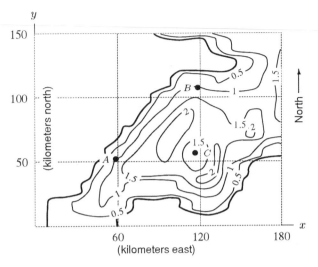

Figure 7.47

23. Use the contour diagram from Problem 21, page 380 to estimate $H_T(T, w)$ for $T = 10, 20, 30$ and $w = 0.1, 0.2, 0.3$. What is the practical meaning of these partial derivatives?

24. Repeat Problem 23 for $H_w(T, w)$ at $T = 10, 20,$ and 30 and $w = 0.1, 0.2,$ and 0.3. What is the practical meaning of these partial derivatives?

25. The cardiac output, represented by c, is the volume of blood flowing through a person's heart, per unit time. The systemic vascular resistance (SVR), represented by s, is the resistance to blood flowing through veins and arteries. Let p be a person's blood pressure. Then p is a function of c and s, so $p = f(c, s)$.

   (a) What does $\partial p / \partial c$ represent?

   Suppose now that $p = kcs$, where k is a constant.

   (b) Sketch the level curves of p. What do they represent? Label your axes.
   (c) For a person with a weak heart, it is desirable to have the heart pumping against less resistance, while maintaining the same blood pressure. Such a person may be given the drug Nitroglycerine to decrease the SVR and the drug Dopamine to increase the cardiac output. Represent this on a graph showing level curves. Put a point A on the graph representing the person's state before drugs are given and a point B for after.
   (d) Right after a heart attack, a patient's cardiac output drops, thereby causing the blood pressure to drop. A common mistake made by medical residents is to get the patient's blood pressure back to normal by using drugs to increase the SVR, rather than by increasing the cardiac output. On a graph of the level curves of p, put a point D representing the patient before the heart attack, a point E representing the patient right after the heart attack, and a third point F representing the patient after the resident has given the drugs to increase the SVR.

7.4   COMPUTING PARTIAL DERIVATIVES ALGEBRAICALLY

The partial derivative $f_x(x, y)$ is the ordinary derivative of the function $f(x, y)$ with respect to x with y held constant, and the partial derivative $f_y(x, y)$ is the ordinary derivative of $f(x, y)$ with respect to y with x held constant. Thus, we can use all the techniques for differentiation from single-variable calculus to find partial derivatives.

Example 1   Let $f(x, y) = x^2 + 5y^2$. Find $f_x(3, 2)$ and $f_y(3, 2)$ algebraically.

Solution   We use the fact that $f_x(3, 2)$ equals the derivative of $f(x, 2)$ at $x = 3$. To find f_x, we hold y fixed, so we can substitute 2 for y:

$$f(x, 2) = x^2 + 5(2^2) = x^2 + 20.$$

Differentiating with respect to x, we have

$$f_x(x, 2) = 2x \qquad \text{and so} \qquad f_x(3, 2) = 2(3) = 6.$$

Similarly, $f_y(3, 2)$ equals the derivative of $f(3, y)$ at $y = 2$. To find f_y, we hold x fixed, so we can substitute 3 for x:

$$f(3, y) = 3^2 + 5y^2 = 9 + 5y^2.$$

Differentiating with respect to y, we have

$$f_y(3, y) = 10y \qquad \text{and so} \quad f_y(3, 2) = 10(2) = 20.$$

Example 2   The concentration C of bacteria in the blood following injection of an antibiotic can be viewed as a function of the dose x (in units of antibiotic) injected and the time t (in hours) since the injection. Suppose we are told that $C = f(x, t) = te^{-xt}$.
(a)   Evaluate $f_x(1, 2)$ and explain what it means in practical terms.
(b)   Evaluate $f_t(1, 2)$ and explain what it means in practical terms.

Solution   (a)   Since

$$f(x, 2) = 2e^{-2x}$$

we have

$$f_x(x, 2) = \frac{d(2e^{-2x})}{dx} = -4e^{-2x}$$

In particular, $f_x(1, 2) = -4e^{-2} \approx -0.54$. To see what $f_x(1, 2)$ means, think about the function $f(x, 2)$ of which it is the derivative. The graph of $f(x, 2)$ in Figure 7.48 gives the concentration of bacteria two hours after the injection as a function of the dose. The derivative $f_x(1, 2)$ is the slope of this graph at the point $x = 1$, and it is negative because a larger dose results in a reduced bacterial population. More precisely, the partial derivative $f_x(1, 2)$ gives the rate of change of bacterial concentration at that time with respect to the dose injected, namely a decrease of 0.54 units of bacteria concentration per unit of antibiotic injected.

(b)   Since

$$f(1, t) = te^{-t}$$

we have, using the product rule,

$$f_t(1, t) = -te^{-t} + e^{-t}$$

In particular, $f_t(1, 2) = -2e^{-2} + e^{-2} \approx -0.135$. To see what $f_t(1, 2)$ means, think about the function $f(1, t)$ of which it is the derivative. The graph of $f(1, t)$ in Figure 7.49 gives the concentration of bacteria at time t if the dose of antibiotic is 1 unit. The derivative $f_t(1, 2)$ is the slope of the graph at the point $t = 2$, and it is negative because after 2 hours the concentration of bacteria is already going down. More precisely, the partial derivative $f_t(1, 2)$ gives the rate at which the bacterial concentration is changing at that time, namely a decrease of 0.135 units of bacterial concentration per hour.

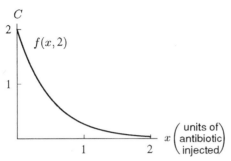

Figure 7.48: Bacterial concentration after 2 hours

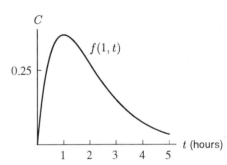

Figure 7.49: Bacterial concentration if 1 unit of antibiotic is injected

Example 3 Let $f(x, y) = x^2 + 5y^2$ as in Example 1. Find f_x and f_y as functions of x and y.

Solution To find f_x, we think of y as a constant. Thus $5y^2$ is a constant, and the derivative with respect to x of this term is 0. We have
$$f_x(x, y) = 2x + 0 = 2x.$$
To find f_y, we think of x as a constant and so the derivative of x^2 with respect to y is zero. We have
$$f_y(x, y) = 0 + 10y = 10y.$$

Example 4 Find both partial derivatives of each of the following functions:
(a) $f(x, y) = 3x + e^{-5y}$ (b) $f(x, y) = x^2 y$ (c) $f(u, v) = u^2 e^{2v}$

Solution (a) To find f_x, we treat y as a constant and so the entire term e^{-5y} is a constant, and the derivative of this term is zero. Likewise, to find f_y, we treat x as a constant. We have
$$f_x = 3 + 0 = 3 \quad \text{and} \quad f_y = 0 + (-5)e^{-5y} = -5e^{-5y}.$$

(b) To find f_x, we treat y as a constant and so the function is treated as a constant times x^2. The derivative of a constant times x^2 is the constant times $2x$, and so we have
$$f_x = (2x)y = 2xy \quad \text{Similarly,} \quad f_y = (x^2)(1) = x^2.$$

(c) To find f_u, we treat v as a constant, and to find f_v, we treat u as a constant. We have
$$f_u = (2u)(e^{2v}) = 2ue^{2v} \quad \text{and} \quad f_v = u^2(2e^{2v}) = 2u^2 e^{2v}.$$

Cobb-Douglas Production Functions

Suppose you are running a small printing business, and decide to expand because you have more orders than you can handle. How should you expand? Should you start a night shift and hire more workers? Should you buy more expensive but faster computers which will enable the current staff to keep up with the work? Or should you do some combination of the two?

Obviously, the way such a decision is made in practice involves many other considerations — such as whether you could get a suitably trained night shift, or whether there are any faster computers available that your current staff could use. Nevertheless, you might model the quantity, P, of work produced by your business as a function of two variables: your total number, N, of workers, and the total value, V, of your equipment.

How would you expect such a production function to behave? In general, having more equipment and more workers enables you to produce more. However, increasing equipment without increasing the number of workers will increase production a bit, but not beyond a point. (If equipment is already lying idle, having more of it won't help.) Similarly, increasing the number of workers without increasing equipment will increase production, but not past the point where the equipment is fully utilized, as any new workers would have no equipment available to them.

Example 5   Explain why the contour diagram in Figure 7.50 does not model the behavior expected of the production function, whereas the contour diagram in Figure 7.51 does.

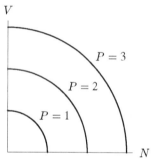

Figure 7.50: Incorrect contours for printing production

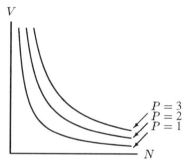

Figure 7.51: Correct contours for printing production

Solution   Production P should be an increasing function of N and an increasing function of V. We see that both contour diagrams (in Figure 7.50 and Figure 7.51) satisfy this condition. Which of the contour diagrams has production increasing in the correct way? First look at the contour diagram in Figure 7.50. Fixing V at a particular value and letting N increase means moving to the right on the contour diagram. As you do so, you cross contours with larger and larger P values, meaning that production increases indefinitely. On the other hand, in Figure 7.51, as you move in the same direction you eventually find yourself moving nearly parallel to the contours, crossing them less and less frequently. Therefore, production increases more and more slowly as N increases while V is held fixed. Similarly, if you hold N fixed and let V increase, the contour diagram in Figure 7.50 shows production increasing at a steady rate, whereas Figure 7.51 shows production increasing, but at a decreasing rate. Thus, Figure 7.51 fits the expected behavior of the production function best.

Formula for a Production Function

Production functions with the qualitative behavior we want are often approximated by formulas of the form

$$P = f(N, V) = cN^\alpha V^\beta$$

where P is the total quantity produced and c, α, and β are positive constants with $0 < \alpha < 1$ and $0 < \beta < 1$. Notice that this is the product of two power functions.

Example 6   Let's return to the small printing business we discussed earlier. Recall that N is the number of workers, V is the value of the equipment (in units of \$25,000), and P is the production output, or quantity of work produced by the business, measured in thousands of pages processed per day. Suppose the production function for this company is given by

$$P = 2N^{0.6}V^{0.4}.$$

(a)   If this company has a labor force of 100 workers and 200 units worth of equipment, what is the production output of the company?

(b)   If we write $P = f(N, V)$, find $f_N(100, 200)$. Interpret your answer in terms of production.

(c)   Find $f_V(100, 200)$. Interpret your answer in terms of production.

Solution   (a)   We have $N = 100$ and $V = 200$ so production

$$P = 2(100)^{0.6}(200)^{0.4} = 263.9 \text{ thousand pages per day.}$$

(b)   To find f_N, we treat V as a constant. Since we are finding $f_N(100, 200)$, we can fix V at 200 to obtain

$$f(N, 200) = 2N^{0.6}(200^{0.4}) = 16.65N^{0.6}.$$

We differentiate to obtain

$$f_N(N, 200) = 16.65(0.6N^{-0.4}) = 9.99N^{-0.4},$$

and so

$$f_N(100, 200) = 9.99(100^{-0.4}) = 1.583 \text{ thousand pages/worker.}$$

This tells us that if we have 200 units of equipment and increase the number of workers by 1 from 100 to 101, the production output will go up by about 1.58 units, or 1580 pages per day.

(c)   Similarly, to find $f_V(100, 200)$, we substitute $N = 100$ to obtain

$$f(100, V) = 2(100^{0.6})V^{0.4} = 31.70V^{0.4}.$$

We differentiate to obtain

$$f_V(100, V) = 31.70(0.4V^{-0.6}) = 12.68V^{-0.6},$$

and so

$$f_V(100, 200) = 12.68(200^{-0.6}) \approx 0.53 \text{ thousand pages/unit of equipment.}$$

This tells us that if we have 100 workers and increase the value of the equipment by 1 unit (\$25,000) from 200 units to 201 units, the production will go up by about 0.53 units, or 530 pages per day.

The Cobb-Douglas Production Model

In 1928, Cobb and Douglas used a similar function to model the production of the entire US economy in the first quarter of this century. Using government estimates of P, the total yearly production between 1899 and 1922, and of K, the total capital investment over the same period, and of L, the total labor force, they found that P was well approximated by the following function:

The Cobb-Douglas Production Function

$$P = 1.01 L^{0.75} K^{0.25}$$

where P is production, K is capital and L is labor.

This function turned out to model the US economy surprisingly accurately, both for the period on which it was based, and for some time afterwards.

Problems for Section 7.4

1. If $f(x, y) = x^3 + 3y^2$, then find the three quantities $f(1, 2)$, $f_x(1, 2)$, and $f_y(1, 2)$.
2. If $f(u, v) = 5uv^2$, then find the three quantities $f(3, 1)$, $f_u(3, 1)$, and $f_v(3, 1)$.

Find the indicated partial derivatives for Problems 3–14.

3. f_x and f_y if $f(x, y) = x^2 + 2xy + y^3$

4. $\dfrac{\partial z}{\partial x}$ if $z = x^2 e^y$

5. f_x and f_y if $f(x, y) = 2x^2 + 3y^2$

6. $\dfrac{\partial Q}{\partial p}$ if $Q = 5a^2 p - 3ap^3$

7. $\dfrac{\partial P}{\partial r}$ if $P = 100e^{rt}$

8. f_t if $f(t, a) = 5a^2 t^3$

9. f_x and f_y if $f(x, y) = 100x^2 y$

10. f_x and f_y if $f(x, y) = 10x^2 e^{3y}$

11. z_x if $z = x^2 y + 2x^5 y$

12. f_u and f_v if $f(u, v) = u^2 + 5uv + v^2$

13. $\dfrac{\partial A}{\partial h}$ if $A = \frac{1}{2}(a + b)h$

14. $\dfrac{\partial}{\partial m}\left(\dfrac{1}{2}mv^2\right)$

15. The amount of money, $\$B$, in a bank account earning interest at a continuous rate, r depends on the amount deposited, $\$P$, and the time, t, it has been in the bank, where

$$B = Pe^{rt}.$$

Find $\partial B/\partial t$ and $\partial B/\partial r$ and $\partial B/\partial P$ and interpret each in financial terms.

16. Recall that if the cost of renting a car from a certain company is $40 per day plus 15 cents per mile, then we can express the cost, C, of renting the car as a function of the number of days, d, and the number of miles, m, as follows:

$$C = f(d, m) = 40d + 0.15m.$$

(a) Find $\dfrac{\partial C}{\partial d}$. Give units with your answer and explain why this answer makes sense.

(b) Find $\dfrac{\partial C}{\partial m}$. Give units with your answer and explain why this answer makes sense.

17. A manufacturing company produces two items, and the quantities of the two items produced are denoted by q_1 and q_2, respectively. The company's total production costs are given by the expression

$$C = f(q_1, q_2) = 16 + 1.2q_1 + 1.5q_2 + 0.2q_1q_2.$$

Find $f(500, 1000)$, $f_{q_1}(500, 1000)$, and $f_{q_2}(500, 1000)$. Give units with your answers and interpret each of your answers in terms of production cost.

18. Consider the function $f(x, y) = x^2 + y^2$.

(a) Estimate $f_x(2, 1)$ and $f_y(2, 1)$ using the contour diagram for f in Figure 7.52.

(b) Estimate $f_x(2, 1)$ and $f_y(2, 1)$ from a table of values for f with $x = 1.9, 2, 2.1$ and $y = 0.9, 1, 1.1$.

(c) Compare your estimates in parts (a) and (b) with the exact values of $f_x(2, 1)$ and $f_y(2, 1)$ found algebraically.

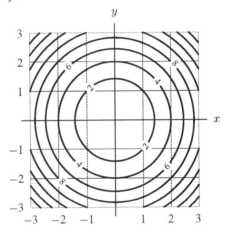

Figure 7.52

19. Suppose you are in a stadium where the audience is doing the wave. This is a ritual in which members of the audience stand up and down in such a way as to create a wave that moves around the stadium. Normally a single wave travels all the way around the stadium, but we will assume there is a continuous sequence of waves. Let $h(x, t) = 5 + \cos(0.5x - t)$ be the function describing this stadium wave. The value of $h(x, t)$ gives the height (in feet) of the head of the spectator in seat x at time t seconds. Evaluate $h_x(2, 5)$ and $h_t(2, 5)$ and interpret each in terms of the wave.

20. Is there a function f which has the following partial derivatives? If so what is it? Are there any others?

$$f_x(x, y) = 4x^3y^2 - 3y^4,$$
$$f_y(x, y) = 2x^4y - 12xy^3.$$

21. Let's return to the small printing business we discussed at the beginning of this section. Recall that N is the number of workers, V is the value of the equipment (in units of $25,000), and P is the production output, or quantity of work produced by the business, measured in thousands of pages processed per day. Suppose the production function for this company is given by

$$P = 2N^{0.6}V^{0.4}.$$

(a) If this company has a labor force of 100 workers and 200 units worth of equipment, what is the production output of the company?

(b) If the labor force is doubled (to 200 workers), how does production change?

(c) If the company purchases enough equipment to double the value of its equipment (to 400 units), how does production change?

(d) If both N and V are doubled from the values given in part (a), how does production change?

22. Suppose that the quantity, Q, produced of a certain good depends on the number of units of labor, L, and of capital, K, according to the function $Q = 900L^{\frac{1}{2}}K^{\frac{2}{3}}$.

(a) If $L = 70$ and $K = 50$, what quantity is produced?

(b) Find Q when $L = 140$ and $K = 100$. If L and K are doubled, discuss the effects on Q.

23. Figure 7.53 gives contour diagrams for different Cobb-Douglas production functions $F(L, K)$. Match each contour diagram with the correct statement.

(A) Tripling each input triples output.

(B) Quadrupling each input doubles output.

(C) Doubling each input almost triples output.

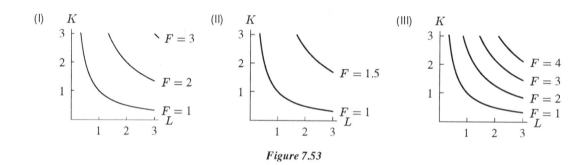

Figure 7.53

24. A company's production output, P, is given in tons, and is a function of the number of workers, N, and the value of the equipment, V, in units of \$25,000. The production function for the company is

$$P = f(N, V) = 5N^{0.75}V^{0.25}.$$

The company currently employs 80 workers, and has equipment worth \$750,000. What are N and V? Find the values of f, f_N, and f_V at these values of N and V. Give units with your answers, and explain what each means in terms of production.

25. Suppose the Cobb-Douglas production function for a product is given by

$$Q = 25K^{0.75}L^{0.25},$$

where Q is the quantity produced given a capital investment of K dollars and a labor investment of L.

(a) Find Q_K and Q_L.

(b) Find the values of Q, Q_K and Q_L given that $K = 60$ and $L = 100$.

(c) Interpret each of the values you found in part (b) in terms of production.

7.5 CRITICAL POINTS AND OPTIMIZATION

To optimize a function means to find the largest (or smallest) possible value of the function. If the function represents profit, we want to find the conditions that will maximize profit. On the other hand, if the function represents cost, we may want to find the conditions that minimize cost. In Chapter 6, we learned how to optimize a function of one variable by investigating critical points. In this section, we will see how to extend the notions of critical points and local extrema to a function of more than one variable.

Review of the One-Variable Case

We begin by reviewing how to optimize a function of one variable. The critical points of a function of one variable are the points where the derivative is zero or undefined. See Figure 7.54. A local maximum or minimum will occur at a critical point, so the first thing we do is to find all critical points. Once we have found the critical points, we must determine whether each is a local maximum, a local minimum, or neither. One way to do this is to investigate the behavior of the function near each critical point. We follow the same steps when dealing with a function of more than one variable.

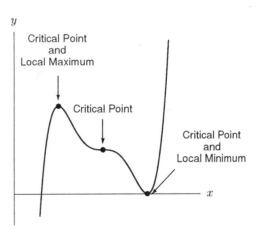

Figure 7.54: Critical points for a function of one variable

Local and Global Extrema for Functions of Two Variables

Functions of several variables, like functions of one variable, can have *local and global extrema.* (That is, local and global maxima and minima.) A function has a local extremum at a point where it takes on the largest or smallest value in a small region around the point. Global extrema are the largest or smallest value anywhere on the set on which the function is defined.

Estimating Extrema From a Table or Contour Diagram

Example 1   Table 7.18 gives a table of values for a function $f(x, y)$. Estimate the location and value of any maxima or minima.

TABLE 7.18   *Where are the extreme points of this function $f(x, y)$?*

$y \backslash x$	0	0.2	0.4	0.6	0.8	1.0
0	80	84	82	76	71	65
5	86	90	88	73	77	71
10	91	95	93	88	82	76
15	87	91	89	84	78	72
20	82	86	84	79	73	67

Solution   The maximum value of the function appears to be 95 at the point $(0.2, 10)$. Since the table only gives certain values, we cannot be sure that this is exactly the maximum. (The function might have a larger value at, for example $(0.3, 11)$.) The minimum value of this function on the points given is 65 at the point $(1, 0)$.

Example 2   Figure 7.55 gives a contour diagram for a function $f(x, y)$. Estimate the location and value of any local maxima or minima. Are any of these global maxima or minima?

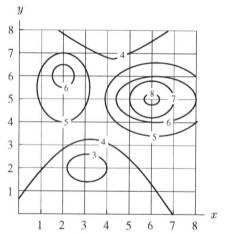

Figure 7.55: Where are the local and global extreme points of this function?

Solution   We see in Figure 7.55 that there is a local maximum of about 8.5 near the point $(6, 5)$, a local maximum of about 6.5 near the point $(2, 6)$, and a local minimum of about 2.5 near the point $(3, 2)$. The value 8.5 appears to be a global maximum and the value 2.5 appears to be a global minimum. Note that, as in Example 1, we can't be certain of the exact location or value of the extreme points.

In Example 1 and Example 2, we can estimate the location and value of extreme points of a function, but we do not have enough information to find them exactly. This is usually true when we are given a table of values or a contour diagram. To find local or global extrema exactly, we usually need to have a formula for the function, and we need to analyze the behavior of the function analytically.

Finding a Local Maximum or Minimum Analytically

In one-variable calculus, the local extrema of a function occur at critical points, namely points where the derivative is zero or undefined. How does this generalize to the case of functions of two or more variables? Suppose that a function $f(x, y)$ has a local maximum at a point (x_0, y_0) which is not on the boundary of the domain of f. If the partial derivative $f_x(x_0, y_0)$ were defined and positive, then we could increase f by increasing x, and if $f_x(x_0, y_0) < 0$, then we could increase f by decreasing x. Similarly, if $f_y(x_0, y_0)$ were positive, we could increase f by increasing y, and if $f_y(x_0, y_0) < 0$, then we could increase f by decreasing y. Since f has a local maximum at (x_0, y_0), there can be no direction in which f is increasing, so we must have $f_x(x_0, y_0) = 0$, and $f_y(x_0, y_0) = 0$. The case in which $f(x, y)$ has a local minimum is similar. Therefore, we arrive at the following conclusion:

If a function $f(x, y)$ has a local maximum or minimum at a point (x_0, y_0), not on the boundary of the domain of f, then either

$$f_x(x_0, y_0) = 0 \quad \text{and} \quad f_y(x_0, y_0) = 0$$

or (at least) one partial derivative is undefined at the point (x_0, y_0).

Points where the partial derivatives are either zero or undefined are called *critical points* of the function. Note that just because (x_0, y_0) is a critical point, such as where $f_x(x_0, y_0) = f_y(x_0, y_0) = 0$, it does not necessarily mean that f has a maximum or a minimum there. It is possible for both partial derivatives to be zero at a point where f does not have an extremum.

Notice that the definition of critical points of a function is easy to extend to functions of three or more variables.

How Do We Find Critical Points?

To find critical points of a function f, we find the points where all the partial derivatives of f equal zero. We must also look for the points where one or more of the partial derivatives is undefined.

Example 3   Find and analyze the critical points of $f(x, y) = x^2 - 2x + y^2 - 4y + 5$.

Solution   Since

$$f_x = 2x - 2 \quad \text{and} \quad f_y = 2y - 4$$

we solve $2x - 2 = 0$, and $2y - 4 = 0$. We see that $x = 1$ and $y = 2$ is the only solution. Hence, f has only one critical point, namely $(1, 2)$. What is the behavior of f near $(1, 2)$? Look at the values of the function in Table 7.19.

Table 7.19 suggests that the function has a local minimum value of 0 at the point $(1, 2)$.

TABLE 7.19   *Values of $f(x, y)$ near the point $(1, 2)$*

$y \backslash x$	0.8	0.9	1.0	1.1	1.2
1.8	0.08	0.05	0.04	0.05	0.08
1.9	0.05	0.02	0.01	0.02	0.05
2.0	0.04	0.01	0.00	0.01	0.04
2.1	0.05	0.02	0.01	0.02	0.05
2.2	0.08	0.05	0.04	0.05	0.08

Example 4   A manufacturing company produces two products which are sold in two separate markets. The company's economists analyze the two markets and determine that the quantities, q_1 and q_2, demanded by consumers and the prices, p_1 and p_2 (in dollars), of each item are related by the equations

$$p_1 = 600 - 0.3q_1 \quad \text{and} \quad p_2 = 500 - 0.2q_2.$$

Thus, if the price for either item increases, the demand for it decreases. The company's total production cost is given by

$$C = 16 + 1.2q_1 + 1.5q_2 + 0.2q_1q_2.$$

If the company wants to maximize its total profits, how much of each product should it produce? What will be the maximum profit? [7]

Solution   The total revenue R is the sum of the revenues, p_1q_1 and p_2q_2, from each market. Substituting for p_1 and p_2, we get

$$\begin{aligned} R &= p_1q_1 + p_2q_2 \\ &= (600 - 0.3q_1)q_1 + (500 - 0.2q_2)q_2 \\ &= 600q_1 - 0.3q_1^2 + 500q_2 - 0.2q_2^2. \end{aligned}$$

Thus the total profit π is given by

$$\begin{aligned} \pi &= R - C \\ &= 600q_1 - 0.3q_1^2 + 500q_2 - 0.2q_2^2 - (16 + 1.2q_1 + 1.5q_2 + 0.2q_1q_2) \\ &= -16 + 598.8q_1 - 0.3q_1^2 + 498.5q_2 - 0.2q_2^2 - 0.2q_1q_2. \end{aligned}$$

To maximize π, we first compute partial derivatives:

$$\frac{\partial \pi}{\partial q_1} = 598.8 - 0.6q_1 - 0.2q_2,$$

$$\frac{\partial \pi}{\partial q_2} = 498.5 - 0.4q_2 - 0.2q_1.$$

Since the partial derivatives are defined everywhere, the only critical points of π are those where the partial derivatives of π are both equal to zero. Thus, we solve the equations

$$598.8 - 0.6q_1 - 0.2q_2 = 0,$$
$$498.5 - 0.4q_2 - 0.2q_1 = 0,$$

for q_1 and q_2, and find that

$$q_1 = 699.1 \approx 699 \quad \text{and} \quad q_2 = 896.7 \approx 897.$$

To see that this is a maximum point, we look at a table of values of profit π around this point. We see in Table 7.20 that profit is greatest at $(699, 897)$. The company should produce 699 units of the first product priced at \$390.30 per unit, and 897 units of the second product priced at \$320.60 per unit. The maximum profit will be $\pi(699, 897) = \$432{,}797$.

TABLE 7.20   *Does this profit function have a maximum at $(699, 897)$?*

$q_2\backslash q_1$	698	699	700
896	432,796.4	432,796.9	432,796.8
897	432,796.7	432,797.0	432,796.7
898	432,796.6	432,796.7	432,796.2

[7] Adapted from M. Rosser, *Basic Mathematics for Economists*, p. 316, (New York: Routledge, 1993).

Problems for Section 7.5

1. By looking at the weather map in Figure 7.1 on page 372, find the maximum and minimum daily temperatures in the states of Mississippi, Alabama, Pennsylvania, New York, California, Arizona, and Massachusetts.

2. By looking at the table of UV exposure as a function of latitude and year in Table 7.21, find the latitude and year that will have the worst UV exposure between 1970 and the year 2010.

TABLE 7.21 *Ultraviolet exposure.*

Latitude \ Year	1970	1975	1980	1985	1990	1995	2000	2005	2010
90	0.00	0.00	0.00	0.00	0.00	0.00	0.00	0.00	0.00
80	0.00	0.00	0.00	0.00	0.00	3.03	5.79	6.37	6.52
70	0.00	0.00	0.00	0.00	0.00	0.02	3.14	6.80	8.21
60	0.01	0.01	0.01	0.01	0.01	0.01	0.12	1.78	3.41
50	0.08	0.08	0.08	0.08	0.08	0.08	0.08	0.08	0.33
40	0.25	0.25	0.25	0.25	0.25	0.25	0.25	0.25	0.25
30	0.49	0.49	0.49	0.49	0.49	0.49	0.49	0.49	0.49
20	0.74	0.74	0.74	0.74	0.74	0.74	0.74	0.74	0.74
10	0.93	0.93	0.93	0.93	0.93	0.93	0.93	0.93	0.93
0	1.00	1.00	1.00	1.00	1.00	1.00	1.00	1.00	1.00
−10	0.93	0.93	0.93	0.93	0.93	0.93	0.93	0.93	0.93
−20	0.74	0.74	0.74	0.74	0.74	0.74	0.74	0.74	0.74
−30	0.49	0.49	0.49	0.49	0.49	0.49	0.49	0.49	0.49
−40	0.25	0.25	0.25	0.25	0.25	0.25	0.25	0.25	0.25
−50	0.08	0.08	0.08	0.08	0.08	0.08	0.08	0.08	0.33
−60	0.01	0.01	0.01	0.01	0.01	0.01	0.12	1.78	3.41
−70	0.00	0.00	0.00	0.00	0.00	0.02	3.14	6.80	8.21
−80	0.00	0.00	0.00	0.00	0.00	3.03	5.79	6.37	6.52
−90	0.00	0.00	0.00	0.00	0.00	0.00	0.00	0.00	0.00

For Problems 3–6,
(a) find all critical points of the function, and
(b) for each critical point, make a small table of values to determine whether the critical point is a local maximum, a local minimum, or neither.

3. $f(x, y) = x^2 + y^2 + 6x - 10y + 8$
4. $f(x, y) = x^2 + 4x + y^2$
5. $f(x, y) = x^2 + xy + 3y$
6. $f(x, y) = y^3 - 3xy + 6x$

7. Figure 7.56 shows a contour diagram of a function $f(x, y)$. List x, y, and the value of the function at any local maximum and local minimum points, and identify which are which. Are any of these local extrema also global extrema? If so, which ones?

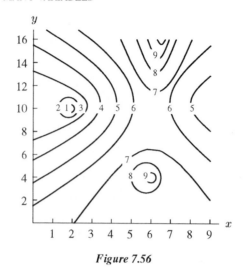

Figure 7.56

8. Figure 7.57 shows a contour diagram of a function $f(x, y)$. List x, y, and the value of the function at any local maximum and local minimum points, and identify which are which. Are any of these local extrema also global extrema? If so, which ones?

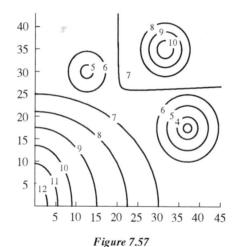

Figure 7.57

9. Suppose $f(x, y) = A - (x^2 + Bx + y^2 + Cy)$. What values of A, B, and C give $f(x, y)$ a local maximum value of 15 at the point $(-2, 1)$?

10. A company sells two products which are partial substitutes for each other, such as coffee and tea. If the price of one product rises, then the demand for the other product rises. The quantities demanded, q_1 and q_2, are given as a function of the prices, p_1 and p_2, by

$$q_1 = 517 - 3.5p_1 + 0.8p_2 \quad \text{and} \quad q_2 = 770 - 4.4p_2 + 1.4p_1.$$

(a) Write total sales revenue as a function of p_1 and p_2.
(b) What prices should the company charge in order to maximize the total sales revenue? [8]

[8] Adapted from M. Rosser, *Basic Mathematics for Economists*, p. 318 (New York: Routledge, 1993).

7.6 LAGRANGE MULTIPLIERS

Many, perhaps most, real optimization problems are constrained by external circumstances. For example, a city wanting to build a public transportation system has only a limited number of tax dollars it can spend on the project. Any nation trying to maintain its balance of trade must spend less on imports than it earns on exports. In this section, we will see how to find an optimum value under such constraints.

A Constrained Optimization Problem

Let's consider the example of trying to maximize the production of a firm under a budget constraint. Suppose production, f, is a function of two variables, x and y, which could be quantities of two raw materials, or labor and capital, or the number of two different types of workers (doctors and nurses, for example). We suppose that

$$f(x,y) = x^{2/3}y^{1/3}.$$

Suppose that x and y are purchased at prices of p_1 and p_2 dollars per unit. What is the maximum production f that can be obtained with a budget of c dollars?

If we want to maximize f without regard to the budget, we simply increase x and y as far as we can. However, the budget will prevent us from increasing x and y beyond a certain point. Exactly how does the budget constrain us? Suppose that x and y each cost \$100 per unit, and suppose that the total budget is \$378,000. The amount spent on x and y together is given by $g(x,y) = 100x + 100y$, and since we can't spend more than the budget allows, we have the following constraint equation:

$$100x + 100y = 378,000.$$

The goal is to make the function

$$f(x,y) = x^{2/3}y^{1/3}$$

as large as possible, subject to the constraint

$$100x + 100y = 378,000.$$

Example 1   Suppose a company has budget constraint $100x + 100y = 378,000$ as described above, and wants to maximize the production function $f(x,y) = x^{2/3}y^{1/3}$.
(a) If the company spends \$100,000 on x, how much can it spend on y? What is the value of the production function in this case?
(b) If the company spends \$200,000 on x, how much can it spend on y? What is the value of the production function in this case?
(c) Which of the two options above is the better choice for the company? Do you think this is the best of all possible options?

Solution   (a) If the company spends \$100,000 on x, then it has \$278,000 left to spend on y. In this case, we have $100x = 100,000$, so $x = 1000$, and $100y = 278,000$, so $y = 2780$. We have

$$f(1000, 2780) = (1000)^{2/3}(2780)^{1/3} = 1406.$$

The company will produce 1406 units.

(b)   If the company spends $200,000 on x, then it has $178,000 left to spend on y. We have $x = 2000$ and $y = 1780$, and so

$$f(2000, 1780) = (2000)^{2/3}(1780)^{1/3} = 1924.$$

Production will be 1924 units.

(c)   Of these two options, the company is better off spending $200,000 on x and $178,000 on y, since its production is larger in this case. This is almost certainly not optimal, however, since there are many other options we could check.

A Graphical Approach to Maximizing Production Subject to a Budget Constraint

In Example 1 we used trial and error to find values of x and y that would give us a large value for the production function. How can we find the maximum value? We begin by examining the problem graphically. On a graph, the budget constraint is represented by the line in Figure 7.58. Any point on or below the line represents a pair of values of x and y that we can afford. A point on the line completely exhausts the budget, a point below the line represents values of x and y which can be bought without using up the budget. Any point above the line represents a pair of values that breaks the budget — we cannot afford such a combination.

Figure 7.58 also shows some contours of the production function f. Since we want to maximize f, we want to find the point which lies on the level curve with the largest possible f value *and* which lies within the budget. The point we are looking for must lie on the budget constraint line because we should clearly spend all the available money if we want to maximize f. Unless we are at the point where the budget constraint is tangent to the contour $f = 2000$, we can always increase f by moving along the line representing the budget constraint in Figure 7.58. For example, if we are on the line to the left of the point of tangency, moving right will increase f; if we are on the line to the right of the point of tangency, moving left will increase f. Thus, the maximum value of f on the budget constraint line occurs at the point where the budget constraint line is tangent to the contour $f = 2000$.

In theory, we could find the values of x and y giving this maximum by reading them off the graph. In practice, however, we often use the method of Lagrange multipliers.

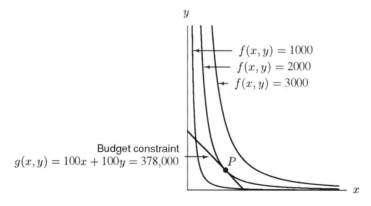

Figure 7.58: Budget constraint and level curves of production

The Method of Lagrange Multipliers

Suppose we wish to optimize an *objective function* $f(x, y)$ subject to a *constraint* $g(x, y) = c$. We saw above that the optimum value is achieved at the point, P, where the constraint is tangent to a contour line of the objective function. Since we are dealing with tangent lines and slopes, it should not surprise you that this graphical condition translates into a condition on the partial derivatives of f and g. It turns out that the optimum value (if it exists) occurs at the values of x and y satisfying the following system of equations:

$$f_x = \lambda g_x$$
$$f_y = \lambda g_y$$
$$g(x, y) = c,$$

where the constant λ is called the *Lagrange multiplier*.

The method of Lagrange multipliers

To optimize $f(x, y)$ subject to the constraint $g(x, y) = c$, we solve the following system of three equations

$$f_x(x, y) = \lambda g_x(x, y),$$
$$f_y(x, y) = \lambda g_y(x, y),$$
$$g(x, y) = c,$$

for the three unknowns x, y, λ.
If f has a constrained global maximum or minimum, then it must occur at one of the solutions (x_0, y_0) to this system.

Example 2   Maximize $f(x, y) = x^{2/3} y^{1/3}$ subject to $100x + 100y = 378{,}000$.

Solution   We have

$$f_x = \frac{2}{3} x^{-1/3} y^{1/3} \qquad \text{and} \qquad f_y = \frac{1}{3} x^{2/3} y^{-2/3}$$

and

$$g_x = 100 \qquad \text{and} \qquad g_y = 100,$$

leading to the equations

$$\frac{2}{3} x^{-1/3} y^{1/3} = \lambda(100)$$
$$\frac{1}{3} x^{2/3} y^{-2/3} = \lambda(100)$$
$$100x + 100y = 378{,}000.$$

Combining the first two equations gives

$$\frac{2}{3} x^{-1/3} y^{1/3} = \frac{1}{3} x^{2/3} y^{-2/3}.$$

We can rewrite this as

$$\frac{2y^{1/3}}{3x^{1/3}} = \frac{x^{2/3}}{3y^{2/3}},$$

and so

$$2y^{1/3}(3y^{2/3}) = x^{2/3}(3x^{1/3})$$
$$6y = 3x$$
$$2y = x.$$

Since we must also satisfy the constraint $100x + 100y = 378{,}000$, we have

$$100(2y) + 100y = 378{,}000$$
$$300y = 378{,}000$$
$$y = 1260.$$

Since $x = 2y$, we have $x = 2520$. The optimum value occurs at $x = 2520$ and $y = 1260$. For these values,

$$f(2520, 1260) = (2520)^{2/3}(1260)^{1/3} \approx 2000.1 \approx 2000.$$

Thus, we see that the maximum value of f is 2000; we also learn that this maximum occurs at $x = 2520$ and $y = 1260$.

The Meaning of λ

In the previous example, we never found (or needed) the value of λ. However, λ does have a practical interpretation.

Let's look back at the production problem (Example 2) where we wanted to maximize

$$f(x, y) = x^{2/3}y^{1/3}$$

subject to the constraint

$$g(x, y) = 100x + 100y = 378{,}000.$$

We solved the equations

$$\frac{2}{3}x^{-1/3}y^{1/3} = 100\lambda,$$
$$\frac{1}{3}x^{2/3}y^{-2/3} = 100\lambda,$$
$$100x + 100y = 378{,}000,$$

to get $x = 2520, y = 1260$. Continuing to find λ gives us

$$\lambda \approx 0.0053.$$

Suppose now we do another, apparently unrelated calculation. Suppose our budget is increased slightly, from 378,000 to 379,000, giving a new budget constraint of

$$100x + 100y = 379{,}000.$$

Then the corresponding solution is at $x = 2527$ and $y = 1263$ and the new maximum value (instead of $f = 2000.1$) is

$$f = (2527)^{2/3}(1263)^{1/3} \approx 2005.4.$$

Notice that the additional \$1000 in the budget increased the production level f by 5.3 units, or by $5.3/1000 = 0.0053$ units per dollar, which is the value of λ. The value of λ represents the extra production achieved by increasing the budget by one dollar — in other words, the extra 'bang' you get for an extra 'buck' of budget.

If we solve for λ in either of the equations $f_x = \lambda g_x$ or $f_y = \lambda g_y$, we see that we can think of the Lagrange multiplier in general terms as the change in f over the change in g, or the expected change in f if g increases by 1 unit:

$$\lambda \approx \frac{\Delta f}{\Delta g}$$

In summary, we have:

- The value of λ is approximately the increase in the optimum value of f when the value of the constraint is increased by 1 unit.

More precisely:

- The value of λ represents the rate of change of the optimum value of f as the constraint increases.

Example 3   Suppose the quantity of goods produced according to the function $P(x,y) = x^{2/3}y^{1/3}$ is maximized subject to the constraint $100x + 100y = 378{,}000$. What price must the product sell for if it is to be worth an increased budget for its production?

Solution   Previously we found that $\lambda = 0.0053$, and therefore increasing the budget by \$1 increases production by about 0.0053 unit. In order to make the increase in budget profitable, the extra goods produced must sell for more than \$1. If the price is P, we must have $P(0.0053) > 1$. Thus, the price per unit must be at least $1/0.0053 \approx \$189$.

Example 4   The quantity, Q, of a product manufactured by a company is given by

$$Q = xy,$$

where x and y are the quantities of raw materials used. Assume that x costs \$20 per unit, y costs \$10 per unit, and the budget is \$10,000.

(a)   How many units of x and y should be purchased in order to maximize production?
(b)   How many units are produced at the maximum value?
(c)   Find the value of λ and interpret it.

Solution   (a)   We maximize $f(x,y) = xy$ subject to the constraint $20x + 10y = 10{,}000$ where $g(x,y) = 20x + 10y$. We have the following partial derivatives:

$$f_x = y, \qquad f_y = x, \qquad \text{and} \qquad g_x = 20, \qquad g_y = 10.$$

The method of Lagrange multipliers gives the following equations:

$$y = 20\lambda$$
$$x = 10\lambda$$
$$20x + 10y = 10{,}000.$$

If we substitute the values of x and y from the first two equations into the third equation, we have

$$20(10\lambda) + 10(20\lambda) = 10{,}000$$
$$400\lambda = 10{,}000$$
$$\lambda = 25.$$

Substituting 25 for λ in the first two equations above, we see that $x = 250$ and $y = 500$. The company should purchase 250 units of x and 500 units of y.

(b)   The maximum value of the production function is

$$f(250, 500) = (250)(500) = 125,000 \text{ units.}$$

(c)   We saw above that $\lambda = 25$. This tells us that if the budget is increased by \$1, we expect production to go up by about 25 units. (If the budget goes up to \$1000, maximum production will increase about 25,000 to a total of roughly 150,000 units.)

Problems for Section 7.6

In Problems 1–5, use Lagrange multipliers to find the maximum or minimum values of $f(x, y)$ subject to the given constraints.

1. $f(x, y) = xy$, $5x + 2y = 100$

2. $f(x, y) = x^2 + 3y^2 + 100$, $8x + 6y = 88$

3. $f(x, y) = x^2 + 4xy$, $x + y = 100$

4. $f(x, y) = 5xy$, $x + 3y = 24$

5. $f(x, y) = x + y$, $x^2 + y^2 = 1$

6.   Assume that the production function for a company is given by

$$P = 24K^{0.6}L^{0.4},$$

where P is the amount produced by the company, L is the amount spent on labor, and K is the value of the equipment, or capital. The total amount spent on L and K together cannot exceed \$1000. If the goal is to maximize production, how much should be spent on labor and how much on capital?

(a)   Set up the equations to solve this problem using Lagrange multipliers.
(b)   What are the optimal values of L and K?
(c)   How many units are produced at this level of production?
(d)   Find the value of λ and interpret it.

7.   The director of a neighborhood health clinic has an annual budget of \$600,000. He wants to allocate his budget so as to maximize the number of patient visits, V, which is given as a function of the number of doctors, D, and the number of nurses, N, by

$$V = 1000D^{0.6}N^{0.3}.$$

Doctors receive a salary of \$40,000, while nurses get \$10,000.

(a)   Set up the director's constrained optimization problem.
(b)   Solve the problem formulated in part (a).
(c)   Find the value of the Lagrange multiplier and interpret its meaning in this problem.

8.   Figure 7.59 shows a contour diagram of the function $f(x, y)$ and the constraint equation $g(x, y) = c$. Approximately what values of x and y will maximize $f(x, y)$ subject to the constraint equation? What is the value of f at this maximum?

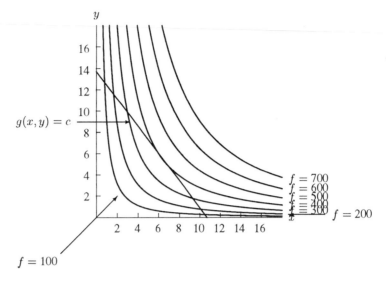

Figure 7.59

9. Suppose that for a certain company with two inputs, x and y, the *production function $P(x, y)$* gives the number of units that can be produced for given values of x and y, while the *cost function $C(x, y)$* gives the cost of production for given values of x and y.

    (a) If the company wishes to maximize production at a cost of $50,000, what is the objective function f? What is the constraint equation? What is the meaning of λ in this situation?

    (b) If instead the company wishes to minimize the costs at a fixed production level of 2000 units, what is the objective function f? What is the constraint equation? What is the meaning of λ in this situation?

10. Suppose that the quantity, Q, manufactured of a certain product depends on the quantity of labor, L, and of capital, K, used according to the function

$$Q = 900L^{1/2}K^{2/3}.$$

Suppose that labor costs $100 per unit and that capital costs $200 per unit. What combination of labor and capital should be used to produce 36,000 units of the goods at minimum cost? What is that minimum cost?

11. The quantity, Q, of a product manufactured by a company is given by

$$Q = aK^{0.6}L^{0.4},$$

where a is a positive constant, K is the quantity of capital and L is the quantity of labor used. Capital costs are $20 per unit, labor costs are $10 per unit, and the company wants costs for capital and labor combined to be no higher than $150. Suppose you are asked to consult for the company, and learn that 5 units each of capital and labor are being used.

    (a) What do you advise? Should the plant use more or less labor? More or less capital? If so, by how much?

    (b) Write a one sentence summary that could be used to sell your advice to the board of directors.

12. Suppose the quantity, q, of a product manufactured depends on the number of workers, W, and the amount of capital invested, K, and is represented by the Cobb-Douglas function

$$q = 6W^{3/4}K^{1/4}.$$

In addition, labor costs are \$10 per worker and capital costs are \$20 per unit, and the budget is \$3000.

   (a) What are the optimum number of workers and the optimum number of units of capital?
   (b) Recompute the optimum values of W and K when the budget is increased by one dollar. Check that increasing the budget by \$1 allows the production of λ extra units of the good, where λ is the Lagrange multiplier.

13. Suppose that the minimum value of the cost function $f(x, y)$, given the constraint that production must equal 50, is $f(33, 87) = 1200$, with $\lambda = 15$. Estimate the additional production cost or savings if the production quota is (a) raised to 51; (b) lowered to 49.

14. A company manufactures x units of one item and y units of another. The total cost in dollars, C, of producing these two items is approximated by the function

$$C = 5x^2 + 2xy + 3y^2 + 800.$$

   (a) If the production quota for the total number of items (both types combined) is 39, find the minimum production cost.
   (b) Estimate the additional production cost or savings if the production quota is raised to 40 or lowered to 38.

15. Consider a firm which manufactures a commodity at two different factories. The total cost of manufacturing depends on the quantities, q_1 and q_2, supplied by each factory, and is expressed by the *joint cost function*, $C = f(q_1, q_2)$. Suppose the joint cost function is approximated by

$$f(q_1, q_2) = 2q_1^2 + q_1q_2 + q_2^2 + 500$$

and that the company's objective is to produce 200 units, at the same time minimizing production costs. How many units should be supplied by each factory?

16. Each person tries to balance his or her time between leisure and work. The tradeoff is that as you work less your income falls. Therefore each person has *indifference curves* which connect the number of hours of leisure, l, and income, s. If, for example, you are indifferent between 0 hours of leisure and an income of \$1125 a week on the one hand, and 10 hours of leisure and an income of \$750 a week on the other hand, then the points $l = 0$, $s = 1125$, and $l = 10$, $s = 750$ both lie on the same indifference curve. Table 7.22 gives information on three indifference curves, I, II, and III.

TABLE 7.22

Weekly Income			Weekly Leisure Hours		
I	II	III	I	II	III
1125	1250	1375	0	20	40
750	875	1000	10	30	50
500	625	750	20	40	60
375	500	625	30	50	70
250	375	500	50	70	90

(a) Sketch the three indifference curves on graph paper.

(b) Suppose you have 100 hours a week available for work and leisure combined, and that you earn $10/hour. Write an equation in terms of l and s which represents this constraint.

(c) On the same graph paper, sketch a graph of this constraint.

(d) Estimate from the graph what combination of leisure hours and income you would choose under these circumstances. Give the corresponding number of hours per week you would work. Explain how you made this estimate.

17. A mountain climber at the summit of a mountain wants to descend to a lower altitude as fast as possible. Suppose the altitude of the mountain is given approximately by

$$h(x, y) = 3000 - \frac{1}{10000}(5x^2 + 4xy + 2y^2) \quad \text{meters,}$$

where x, y are horizontal coordinates on the earth (in meters), with the mountain summit located above the origin. In thirty minutes, the climber can reach any point (x, y) on a circle of radius 1000 m. In which direction should she travel in order to descend as far as possible?

18. A large automobile manufacturing plant currently employs 1500 workers and has capital investment equal to 4 million dollars per month. The production function for this factory is

$$Q = x^{0.4}y^{0.6},$$

where Q is the number of cars produced per month, x represents the units of labor measured by the number of workers, and y is capital investment measured in thousands of dollars. Each worker's salary, i.e., their "cost", is $2,100 per month and each unit of capital costs $1,000 per month (since y is measured in thousands of dollars).

(a) How many cars is the factory currently assembling each month?

(b) Due to the sluggish economy, the factory decides to decrease production to 2000 cars per month. The factory wants to minimize the cost of producing these 2000 cars. How many workers will need to be laid off? By what amount should monthly investment be decreased?

(c) Give the value of the Lagrange multiplier, λ, and interpret it in terms of the car factory.

REVIEW PROBLEMS FOR CHAPTER SEVEN

1. Table 7.21 on page 421 shows the predictions of one simple model on how average yearly ultraviolet (UV) exposure might vary with the year and the latitude.

(a) Graph UV exposure against latitude for the years 1970, 1996 and 2000.

(b) Produce a table showing what latitude has the most severe exposure to UV, as a function of the year.

(c) What do you notice in your answer to part (b)? What is a possible explanation for this phenomenon?

2. Suppose you are in a room 30 feet long with a heater at one end. In the morning the room is 65°F. You turn on the heater, which quickly warms up to 85°F. Let $H(x, t)$ be the temperature x feet from the heater, t minutes after the heater is turned on. Figure 7.60 shows the contour

diagram for H. How warm is it 10 feet from the heater 5 minutes after it was turned on? 10 minutes after it was turned on?

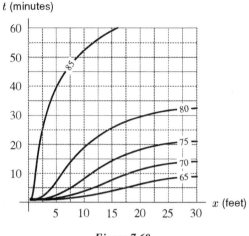

Figure 7.60

3. Using the contour diagram in Figure 7.60, sketch the graphs of the one-variable functions $H(x, 5)$ and $H(x, 20)$. Interpret the two graphs in practical terms, and explain the difference between them.

Problems 4–5 refer to the fallout, V, (in pounds per cubic meter) from a volcanic explosion. The fallout depends on the distance from the volcano, d, and the time since the explosion, t, and is given by

$$V = f(d, t) = \sqrt{t}e^{-d}.$$

4. On the same axes, graph cross-sections of f with $t = 1$, and $t = 2$. Describe the shape of the graphs: As distance from the volcano changes, how does the fallout change? Look at the relationship between the graphs: How is the fallout changing over time? Explain why your answers make sense in terms of volcanoes.

5. On the same axes, graph cross-sections of f with $d = 0$, $d = 1$, and $d = 2$. Describe the shape of the graphs: As time since explosion changes, how does the fallout change? Look at the relationship between the graphs: How is the fallout changing over time? Explain why your answers make sense in terms of volcanoes.

For Problems 6–7, make a contour plot in the region $-2 < x < 2$ and $-2 < y < 2$. Describe the shape of the contour lines.

6. $z = 3x - 5y + 1$ 7. $z = 2x^2 + y^2$

8. You are an anthropologist observing a native ritual. Sixteen people arrange themselves with their backs to you along a bench; all but the three on the far left side are seated. The first person on the far left is standing with her hands at her side, the second is standing with his hands raised and the third is standing with her hands at her side. At some unseen signal, the first one sits down, and everyone else copies what his neighbor to the left was doing one second earlier. Every second that passes, this behavior is repeated until all are once again seated.

   (a) Draw graphs at several different times showing how the height depends upon the distance along the bench.

(b)   Graph the location of the raised hands as a function of time.

(c)   What US ritual is most closely related to what you have observed?

9.   The cornea is the front surface of the eye. Corneal specialists use a TMS, or Topographical Modeling System, to produce a "map" of the curvature of the eye's surface. A computer analyzes light reflected off the eye and draws level curves joining points of constant curvature. The regions between these curves are colored different colors.

   The first two pictures in Figure 7.61 are cross-sections of eyes with constant curvature, the smaller being about 38 units and the larger about 50 units. For contrast, the third eye has varying curvature.

(a)   Describe in words how the TMS map of an eye of constant curvature will look.

(b)   Draw the TMS map of an eye with the cross-section in Figure 7.62. Assume the eye is circular when viewed from the front, and the cross-section is the same in every direction. Put reasonable numeric labels on your level curves.

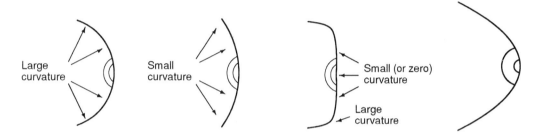

Figure 7.61: Pictures of eyes with different curvature   *Figure 7.62*

10.   Figure 7.63 shows the contour diagram for the vibrating string function from page 381:

$$f(x, t) = \cos t \sin x, \quad 0 \le x \le \pi.$$

Using the diagram, describe in words the cross-sections of f with t fixed and the cross-sections of f with x fixed. Explain what you see in terms of the behavior of the string.

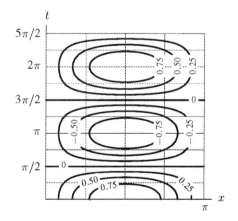

Figure 7.63

11.   The Cobb-Douglas production function for a certain product is given by

$$P = 5L^{0.8}K^{0.2}$$

where P is the quantity produced, L is the size of the labor force, and K is the amount of total equipment. Assume that each unit of labor costs $300, each unit of equipment costs $100, and the total budget is $15,000.

(a)   Make a table of different possible values for L and K and find the production level, P, for each.

(b)   Use the method of Lagrange multipliers to find the optimal way to spend the budget.

For problems 12–17, find the indicated partial derivatives. Assume the variables are restricted to a domain in which the function is defined.

12. f_x and f_y if $f(x, y) = x^2 + xy + y^2$ 13. $\dfrac{\partial Q}{\partial p_1}$ and $\dfrac{\partial Q}{\partial p_2}$ if $Q = 50p_1p_2 - p_2^2$

14. P_a and P_b if $P = a^2 - 2ab^2$ 15. $\dfrac{\partial f}{\partial x}$ and $\dfrac{\partial f}{\partial t}$ if $f = 5xe^{-2t}$

16. $\dfrac{\partial P}{\partial K}$ and $\dfrac{\partial P}{\partial L}$ if $P = 10K^{0.7}L^{0.3}$ 17. f_x and f_y if $f(x, y) = \sqrt{x^2 + y^2}$

18.   Figure 7.64 shows a contour diagram for the temperature T (in °C) along a wall in a heated room as a function of distance x along the wall and time t in minutes. Estimate $\partial T/\partial x$ and $\partial T/\partial t$ at the given points. Give units for your answers and say what the answers mean.

(a) $x = 15, t = 20$ (b) $x = 5, t = 12$

t (minutes)

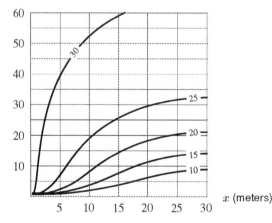

Figure 7.64

19.   The quantity Q (in pounds) of beef that a certain community buys during a week is a function $Q = f(b, c)$ of the prices of beef, b, and chicken, c, during the week. Do you expect $\partial Q/\partial b$ to be positive or negative? What about $\partial Q/\partial c$?

20.   Suppose the cost of producing one unit of a certain product is given by

$$c = a + bx + ky,$$

where x is the amount of labor used (in man hours) and y is the amount of raw material used (by weight) and a and b and k are constants. What does $\partial c/\partial x = b$ mean? What is the practical interpretation of b?

21. People commuting to a city can choose to go either by bus or by train. The number of people who choose either method depends in part upon the price of each. Let $f(P_1, P_2)$ be the number of people who take the bus when P_1 is the price of a bus ride and P_2 is the price of a train ride. What can you say about the signs of $\partial f/\partial P_1$ and $\partial f/\partial P_2$? Explain your answers.

22. A soft drink company is interested in seeing how the demand for its products is affected by prices. The company believes that the quantity, q, of its soft drinks sold depends on p_1, the average price of the company's soft drinks, p_2, the average price of competing soft drinks, and p_3, the average amount of money spent by the company on advertising. Suppose also that

$$q = C - 8 \cdot 10^6 p_1 + 4 \cdot 10^6 p_2 + 2p_3.$$

   (a) What does the constant C represent in terms of soft drink sales?
   (b) Find the marginal demand for soft drinks with respect to changes in p_1, p_2, and p_3. Explain why the signs and relative magnitudes of your answers are reasonable. [Note: The marginal demand is the rate of change of quantity demanded with price.]

23. Suppose that x is the price of one brand of gasoline and y is the price of a competing brand. Then q_1, the quantity of the first brand sold in a fixed time period, depends on both x and y, so $q_1 = f(x, y)$. Similarly, if q_2 is the quantity of the second brand sold during the same period, $q_2 = g(x, y)$. What do you expect the signs of the following quantities to be? Explain.
   (a) $\partial q_1/\partial x$ and $\partial q_2/\partial y$ (b) $\partial q_1/\partial y$ and $\partial q_2/\partial x$

24. In the 1940s the quantity, q, of beer sold each year in Britain was found to depend on I (the aggregate personal income, adjusted for taxes and inflation), p_1 (the average price of beer), and p_2 (the average price of all other goods and services). Would you expect $\partial q/\partial I, \partial q/\partial p_1, \partial q/\partial p_2$ to be positive or negative? Give reasons for your answers.

25. The quantity of a product demanded by consumers is a function of its price. The quantity of one product demanded may also depend on the price of other products. For example, the demand for tea is affected by the price of coffee; the demand for cars is affected by the price of gas. Suppose the quantities demanded, q_1 and q_2, of two products depend on the prices, p_1 and p_2, as follows

$$q_1 = 150 - 2p_1 - p_2$$
$$q_2 = 200 - p_1 - 3p_2.$$

   (a) What does the fact that the coefficients of p_1 and p_2 are negative tell you? Give an example of two products that might be related this way.
   (b) Suppose one manufacturer sells both of these products. How should the manufacturer set prices to earn the maximum possible revenue? What is that maximum possible revenue?

26. Find all critical points of $f(x, y) = x^2 + 3y^2 - 4x + 6y + 10$.

27. Find all critical points of $f(x, y) = x^3 - 3x + y^2$. Make a table of values to determine if each critical point is a local minimum, a local maximum, or neither.

28. Show that the Cobb-Douglas function

$$Q = bK^\alpha L^{1-\alpha} \quad \text{where} \quad 0 < \alpha < 1$$

satisfies the equation

$$K\frac{\partial Q}{\partial K} + L\frac{\partial Q}{\partial L} = Q.$$

APPENDICES

A ROOTS AND ACCURACY

It is often necessary to find the zeros of a polynomial or the points of intersection of two curves. So far, you have probably used algebraic methods, such as the quadratic formula, to solve such problems. Unfortunately, however, mathematicians' search for formulas for the solutions to equations, such as the quadratic formula, has not been all that successful. The formulas for the solutions to third- and fourth-degree equations are so complicated that you'd never want to use them. Early in the nineteenth century, it was proved that there is no algebraic formula for the solutions to equations of degree 5 and higher. Most nonpolynomial equations cannot be solved using a formula either.

However, we can still find roots of equations, provided we use approximation methods, not formulas. In this section we will discuss three ways to find roots: algebraic, graphical, and numerical. Of these, only the algebraic method gives exact solutions.

First, let's get some terminology straight. Given the equation $x^2 = 4$, we call $x = -2$ and $x = 2$ the *roots*, or *solutions of the equation*. If we are given the function $f(x) = x^2 - 4$, then -2 and 2 are called the *zeros of the function*; that is, the zeros of the function f are the roots of the equation $f(x) = 0$.

The Algebraic Viewpoint: Roots by Factoring

If the product of two numbers is zero, then one or the other or both must be zero, that is, if $AB = 0$, then $A = 0$ or $B = 0$. This observation lies behind finding roots by factoring. You may have spent a lot of time factoring polynomials. Here you will also factor expressions involving trigonometric and exponential functions.

Example 1 Find the roots of $x^2 - 7x = 8$.

Solution Rewrite the equation as $x^2 - 7x - 8 = 0$. Then factor the left side: $(x + 1)(x - 8) = 0$. By our observation about products, either $x + 1 = 0$ or $x - 8 = 0$, so the roots are $x = -1$ and $x = 8$.

Example 2 Find the roots of $\dfrac{1}{x} - \dfrac{x}{(x + 2)} = 0$.

Solution Rewrite the left side with a common denominator:

$$\frac{x + 2 - x^2}{x(x + 2)} = 0.$$

Whenever a fraction is zero, the numerator must be zero. Therefore we must have

$$x + 2 - x^2 = (-1)(x^2 - x - 2) = (-1)(x - 2)(x + 1) = 0.$$

We conclude that $x - 2 = 0$ or $x + 1 = 0$, so 2 and -1 are the roots. They can be checked by substitution.

Example 3   Find the roots of $e^{-x}\sin x - e^{-x}\cos x = 0$.

Solution   Factor the left side: $e^{-x}(\sin x - \cos x) = 0$. The factor e^{-x} is never zero; it is impossible to raise e to a power and get zero. Therefore, the only possibility is that $\sin x - \cos x = 0$. This equation is equivalent to $\sin x = \cos x$. If we divide both sides by $\cos x$, we get

$$\frac{\sin x}{\cos x} = \frac{\cos x}{\cos x} \quad \text{so} \quad \tan x = 1.$$

The roots of this equation are

$$\ldots, \frac{-7\pi}{4}, \frac{-3\pi}{4}, \frac{\pi}{4}, \frac{5\pi}{4}, \frac{9\pi}{4}, \frac{13\pi}{4}, \ldots.$$

Warning: Using factoring to solve an equation only works when one side of the equation is 0. It is not true that if, say, $AB = 7$ then $A = 7$ or $B = 7$. For example, you *cannot* solve $x^2 - 4x = 2$ by factoring $x(x - 4) = 2$ and then assuming that either x or $x - 4$ equals 2.

The problem with factoring is that factors are not easy to find. For example, the left side of the quadratic equation $x^2 - 4x - 2 = 0$ does not factor, at least not into "nice" factors with integer coefficients. For the general quadratic equation:

$$ax^2 + bx + c = 0$$

there is the quadratic formula for the roots:

$$x = \frac{-b \pm \sqrt{b^2 - 4ac}}{2a}.$$

Thus the roots of $x^2 - 4x - 2 = 0$ are $(4 \pm \sqrt{24})/2$, or $2 + \sqrt{6}$ and $2 - \sqrt{6}$.

Notice that in each of these examples, we have found the roots exactly.

The Graphical Viewpoint: Roots by Zooming

To find the roots of an equation $f(x) = 0$, it helps to draw the graph of f. The roots of the equation, that is the zeros of f, are *the values of x where the graph of f crosses the x-axis*. Even a very rough sketch of the graph can be useful in determining how many zeros there are and their approximate values. If you have a computer or graphing calculator, then finding solutions by graphing is the easiest method, especially if you use the zoom feature. However, a graph can never tell you the exact value of a root, only an approximate one.

Example 4   Find the roots of $x^3 - 4x - 2 = 0$.

Solution   Attempting to factor the left side with integer coefficients will convince you it cannot be done, so we cannot easily find the roots by algebra. We know the graph of $f(x) = x^3 - 4x - 2$ will have the usual cubic shape; see Figure 0.1. There are clearly three roots: one between $x = -2$ and $x = -1$, another between $x = -1$ and $x = 0$, and a third between $x = 2$ and $x = 3$. Zooming in on the largest root with a graphing calculator or computer shows that it lies in the following interval:

$$2.213 < x < 2.215.$$

Thus, the root is $x = 2.21$, accurate to two decimal places. Zooming in on the other two roots shows them to be $x = -1.68$ and $x = -0.54$, accurate to two decimal places.

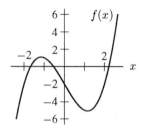

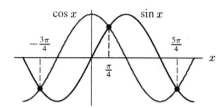

Figure 0.1: The cubic
$f(x) = x^3 - 4x - 2$

Figure 0.2: Finding roots of
$\sin x - \cos x = 0$

Useful trick: Suppose you want to solve the equation $\sin x - \cos x = 0$ graphically. Instead of graphing $f(x) = \sin x - \cos x$ and looking for zeros, you may find it easier to rewrite the equation as $\sin x = \cos x$ and graph $g(x) = \sin x$ and $h(x) = \cos x$. (After all, you already know what these two graphs look like. See Figure 0.2.) The roots of the original equation are then precisely the x coordinates of the points of intersection of the graphs of $g(x)$ and $h(x)$.

Example 5   Find the roots of $2 \sin x - x = 0$.

Solution   Rewrite the equation as $2 \sin x = x$, and graph both sides. Since $g(x) = 2 \sin x$ is always between -2 and 2, there are no roots of $2 \sin x = x$ for $x > 2$ or for $x < -2$. Thus, we need only consider the graphs between -2 and 2 (or between $-\pi$ and π, which makes graphing the sine function easier). Figure 0.3 shows the graphs. There are three points of intersection: one appears to be at $x = 0$, one between $x = \pi/2$ and $x = \pi$, and one between $x = -\pi/2$ and $x = -\pi$. You can tell that $x = 0$ is the exact value of one root because it satisfies the original equation exactly. Zooming in shows that there is a second root $x \approx 1.9$, and the third root is $x \approx -1.9$ by symmetry.

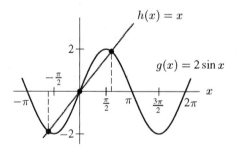

Figure 0.3: Finding roots of $2 \sin x - x = 0$

The Numerical Viewpoint: Roots by Bisection

We now look at a numerical method of approximating the solutions to an equation. This method depends on the idea that if the value of a function $f(x)$ changes sign in an interval, and if we believe there is no break in the graph of the function there, then there is a root of the equation $f(x) = 0$ in that interval.

Let's go back to the problem of finding the root of $f(x) = x^3 - 4x - 2 = 0$ between 2 and 3. To locate the root, we close in on it by evaluating the function at the midpoint of the interval, $x = 2.5$. Since $f(2) = -2$, $f(2.5) = 3.625$, and $f(3) = 13$, the function changes sign between $x = 2$ and $x = 2.5$, so the root is between these points. Now we look at $x = 2.25$.

Since $f(2.25) = 0.39$, the function is negative at $x = 2$ and positive at $x = 2.25$, so there is a root between 2 and 2.25. Now we look at 2.125. We find $f(2.125) = -0.90$, so there is a root between 2.125 and 2.25, ... and so on. (You may want to round the decimals as you work.) See Figure 0.4. The intervals containing the root are listed in Table 0.1 and show that the root is $x = 2.21$ to two decimal places.

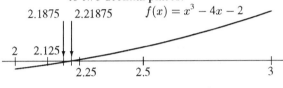

Figure 0.4: Locating a root of $x^3 - 4x - 2 = 0$

TABLE 0.1 *Intervals containing root to* $x^3 - 4x - 2 = 0$ *(Note that* [2, 3] *means* $2 \le x \le 3$*)*

[2, 3]
[2, 2.5]
[2, 2.25]
[2.125, 2.25]
[2.1875, 2.25] .... So $x = 2.2$ rounded to one decimal place
[2.1875, 2.21875]
[2.203125, 2.21875]
[2.2109375, 2.21875]
[2.2109375, 2.2148438] So $x = 2.21$ rounded to two decimal places

This method of finding roots is called the **Bisection Method**:

- To solve an equation $f(x) = 0$ using the bisection method, we need two starting values for x, say, $x = a$ and $x = b$, such that $f(a)$ and $f(b)$ have opposite signs and f is continuous on $[a, b]$.

- Evaluate f at the midpoint of the interval $[a, b]$, and decide in which half-interval the root lies.

- Repeat, using the new half-interval instead of $[a, b]$.

There are some problems with the bisection method:

- The function may not change signs near the root. For example, $f(x) = x^2 - 2x + 1 = 0$ has a root at $x = 1$, but $f(x)$ is never negative because $f(x) = (x - 1)^2$, and a square cannot be negative. (See Figure 0.5.)

- The function f must be continuous between the starting values $x = a$ and $x = b$.

- If there is more than one root between the starting values $x = a$ and $x = b$, the method will find only one of the roots. For example, if we had tried to solve $x^3 - 4x - 2 = 0$ starting at $x = -12$ and $x = 10$, the bisection method would zero in on the root between $x = -2$ and $x = -1$, not the root between $x = 2$ and $x = 3$ that we found earlier. (Try it! Then see what happens if you use $x = -10$ instead of $x = -12$.)

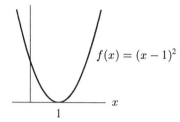

Figure 0.5: f doesn't change sign at
the root

- The bisection method is slow and not very efficient. Applying bisection three times in a row only traps the root in an interval $(\frac{1}{2})^3 = \frac{1}{8}$ as large as the starting interval. Thus, if we initially know that a root is between, say, 2 and 3, then we would need to apply the bisection method at least four times to know the first digit after the decimal point.

There are much more powerful methods available for finding roots, such as Newton's method, which is more complicated, but which avoids some of these difficulties.

Example 6   Find all the roots of $xe^x = 5$ to at least one decimal place.

Solution   If we rewrite the equation as $e^x = 5/x$ and graph both sides, as in Figure 0.6, it is clear that there is exactly one root, and it is somewhere between 1 and 2. Table 0.2 shows the intervals obtained by the bisection method. After five iterations, we have the root trapped between 1.3125 and 1.34375, so we can say the root is $x = 1.3$ to one decimal place.

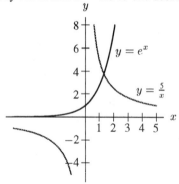

Figure 0.6: Intersection of $y = e^x$ and
$y = 5/x$

TABLE 0.2 *Bisection method for*
$f(x) = xe^x - 5 = 0$ *(Note that*
$[1, 2]$ *means the interval* $1 \leq x \leq 2$)

Interval Containing Root
$[1, 2]$
$[1, 1.5]$
$[1.25, 1.5]$
$[1.25, 1.375]$
$[1.3125, 1.375]$
$[1.3125, 1.34375]$

Iteration

Both zooming in and bisection as discussed here are examples of *iterative* methods, in which a sequence of steps is repeated over and over again, using the results of one step as the input for the next. We can use the method to locate a root to any degree of accuracy. In bisection, each iteration traps the root in an interval that is half the length of the previous one. Each time you zoom in on a calculator, you trap the root in a smaller interval; how much smaller depends on the settings on the calculator.

Accuracy and Error

In the previous discussion, we used the phrase "accurate to 2 decimal places." For an iterative process where we get closer and closer estimates for some quantity, we take a common-sense approach to accuracy: we watch the numbers carefully, and when a digit stays the same for a few iterations, we assume it has stabilized and is correct, especially if the digits to the right of that digit also stay the same. For example, suppose 2.21429 and 2.21431 are two successive estimates for a zero of $f(x) = x^3 - 4x - 2$. Since these two estimates agree to the third digit after the decimal point, we probably have at least 3 decimal places correct.

There is a problem with this, however. Suppose we are finding a root whose true value is 1, and the estimates are converging to the value from below — say, 0.985, 0.991, 0.997 and so on. In this case, not even the first decimal place is "correct," even though the difference between the estimates and the actual answer is very small — much less than 0.1. To avoid this difficulty, we say that an estimate a for some quantity r is *accurate to p decimal places* if the error, which is the absolute value of the difference between a and r, or $|r - a|$, is as follows:

Accuracy to p decimal places	means	Error less than
$p = 1$		0.05
2		0.005
3		0.0005
$\vdots$		$\vdots$
n		$0.\underbrace{000\ldots0}_{n}5$

This is the same as saying that r must lie in an interval of length twice the maximum error, centered on a. For example, if a is accurate to 1 decimal place, r must lie in the following interval:

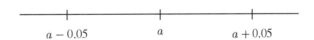

$$a - 0.05 \qquad a \qquad a + 0.05$$

Since both the graphing calculator and the bisection method give us an interval in which the root is trapped, this definition of decimal accuracy is a natural one for these processes.

Example 7   Suppose the numbers $\sqrt{10}$, 22/7, and 3.14 are given as approximations to $\pi = 3.1415\ldots$. To how many decimal places is each approximation accurate?

Solution   Using $\sqrt{10} = 3.1622\ldots$,

$$|\sqrt{10} - \pi| = |3.1622\ldots - 3.1415\ldots| = 0.0206\ldots < 0.05,$$

so $\sqrt{10}$ is accurate to one decimal place. Similarly, using $22/7 = 3.1428\ldots$,

$$\left|\frac{22}{7} - \pi\right| = |3.1428\ldots - 3.1415\ldots| = 0.0013\ldots < 0.005,$$

so $22/7$ is accurate to two decimal places. Finally,

$$|3.14 - 3.1415\ldots| = 0.0015\ldots < 0.005,$$

so 3.14 is accurate to two decimal places.

Warning:

- Saying that an approximation is accurate to, say, 2 decimal places does *not* guarantee that its first two decimal places are "correct," that is, that the two digits of the approximation are the same as the corresponding two digits in the true value. For example, an approximate value of 5.997 is accurate to 2 decimal places if the true value is 6.001, but neither of the 9s in the approximation agrees with the 0s in the true value (nor does the digit 5 agree with the digit 6).

- The number of decimal places of accuracy refers to the number of digits that have stabilized in the root, r. It does *not* refer to the number of digits of $f(r)$ that are zero. For example, Table 0.1 on page 441 shows that $x = 2.2$ is a root of $f(x) = x^3 - 4x - 2 = 0$, accurate to one decimal place. Yet, $f(2.2) = -0.152$, so $f(2.2)$ does not have one zero after the decimal point. Similarly, $x = 2.21$ is the root accurate to two decimal places, but $f(2.21) = -0.046$ does not have two zeros after the decimal point.

Example 8   Is $x = 2.2143$ a zero of $f(x) = x^3 - 4x - 2$ accurate to four decimal places?

Solution   We want to know whether r, the exact value of the zero, lies in the interval

$$2.2143 - 0.00005 < r < 2.2143 + 0.00005$$

which is the same as

$$2.21425 < r < 2.21435.$$

Since $f(2.21425) < 0$ and $f(2.21435) > 0$, the zero does lie in this interval, and so $r = 2.2143$ is accurate to four decimal places.

How to Write a Decimal Answer

The graphing calculator and bisection method naturally give an interval for a root or zero. Other numerical techniques, however, do not give a pair of numbers bounding the true value, but rather a single number near the true value. What should you do if you want to give a single number, rather than an interval, for an answer?

When giving a single number as an answer and interpreting it, be careful about giving rounded answers. For example, suppose you have computed a value to be 0.84, and you know the true value is between 0.81 and 0.87. It would be wrong to round 0.84 to 0.8 and say that the answer is 0.8 accurate to one decimal place; the true value could be 0.86, which is not within 0.05 of 0.8. The right thing to say is that the answer is 0.84 accurate to one decimal place. Similarly, to give an answer accurate to, say, 2 decimal places, you may have to show 3 or more decimal places in your answer.

Problems for Section A

1. Use a calculator or computer graph of $f(x) = 13 - 20x - x^2 - 3x^4$ to determine:
    (a) The range of this function;
    (b) The number of zeros of this function.

For Problems 2–12, determine the roots or points of intersection to an accuracy of one decimal place.

2. (a) The root of $x^3 - 3x + 1 = 0$ between 0 and 1
    (b) The root of $x^3 - 3x + 1 = 0$ between 1 and 2
    (c) The third root of $x^3 - 3x + 1 = 0$

3. The root of $x^4 - 5x^3 + 2x - 5 = 0$ between -2 and -1

4. The root of $x^5 + x^2 - 9x - 3 = 0$ between -2 and -1

5. The largest real root of $2x^3 - 4x^2 - 3x + 1 = 0$

6. All real roots of $x^4 - x - 2 = 0$

7. All real roots of $x^5 - 2x^2 + 4 = 0$

8. The first positive root of $x \sin x - \cos x = 0$

9. The first positive point of intersection between $y = 2x$ and $y = \cos x$

10. The first positive point of intersection between $y = 1/2^x$ and $y = \sin x$

11. The point of intersection between $y = e^{-x}$ and $y = \ln x$

12. All roots of $\cos t = t^2$

13. Estimate all real zeros of the following polynomials, accurate to 2 decimal places:
    (a) $f(x) = x^3 - 2x^2 - x + 3$
    (b) $f(x) = x^3 - x^2 - 2x + 2$

14. Find the largest zero of
$$f(x) = 10xe^{-x} - 1$$
to two decimal places, using the bisection method. Make sure to demonstrate that your approximation is as good as you claim.

15. (a) Find the smallest positive value of x where the graphs of $f(x) = \sin x$ and $g(x) = 2^{-x}$ intersect.
    (b) Repeat with $f(x) = \sin 2x$ and $g(x) = 2^{-x}$.

16. Use a graphing calculator to sketch $y = 2\cos x$ and $y = x^3 + x^2 + 1$ on the same set of axes. Find the positive zero of $f(x) = 2\cos x - x^3 - x^2 - 1$. A friend claims there is one more real zero. Is your friend correct? Explain.

17. Use the table below to investigate the zeros of the function
$$f(\theta) = (\sin 3\theta)(\cos 4\theta) + 0.8$$
in the interval $0 \le \theta \le 1.8$.

θ	0	0.2	0.4	0.6	0.8	1.0	1.2	1.4	1.6	1.8
$f(\theta)$	0.80	1.19	0.77	0.08	0.13	0.71	0.76	0.12	-0.19	0.33

(a) Decide how many zeros the function has in the interval $0 \le \theta \le 1.8$.

(b)   Locate each zero, or a small interval containing each zero.

(c)   Are you sure you have found all the zeros in the interval $0 \leq \theta \leq 1.8$? Graph the function on a calculator or computer to decide.

18. (a)   Use the accompanying table to locate approximate solution(s) to

$$(\sin 3x)(\cos 4x) = \frac{x^3}{\pi^3}$$

in the interval $1.07 \leq x \leq 1.15$. Give an interval of length 0.01 in which each solution lies.

(b)   Make an estimate for each solution accurate to two decimal places.

x	x^3/π^3	$(\sin 3x)(\cos 4x)$
1.07	0.0395	0.0286
1.08	0.0406	0.0376
1.09	0.0418	0.0442
1.10	0.0429	0.0485
1.11	0.0441	0.0504
1.12	0.0453	0.0499
1.13	0.0465	0.0470
1.14	0.0478	0.0417
1.15	0.0491	0.0340

B   COMPOUND INTEREST

If you have some money, you may decide to invest it to earn interest. The interest can be paid in many different ways — for example, once a year or many times a year. If the interest is paid more frequently than once per year and the interest is not withdrawn, there is a benefit to the investor since the interest earns interest. This effect is called *compounding*. You may have noticed banks offering accounts that differ both in interest rates and in compounding methods. Some offer interest compounded annually, some quarterly, and others daily. Some even offer continuous compounding.

What is the difference between a bank account advertising 8% compounded annually (once per year) and one offering 8% compounded quarterly (four times per year)? In both cases 8% is an annual rate of interest. The expression 8% *compounded annually* means that at the end of each year, 8% of the current balance is added. This is equivalent to multiplying the current balance by 1.08. Thus, if $100 is deposited, the balance, P, in dollars, will be

$$P = 100(1.08) \qquad \text{after one year,}$$
$$P = 100(1.08)^2 \qquad \text{after two years,}$$
$$P = 100(1.08)^t \qquad \text{after } t \text{ years.}$$

The expression 8% *compounded quarterly* means that interest is added four times per year (every three months) and that $\frac{8}{4} = 2\%$ of the current balance is added each time. Thus, if $100 is deposited, at the end of one year, four compoundings will have taken place and the account will contain $100(1.02)^4$. Thus, the balance, P, in dollars, will be

$$P = 100(1.02)^4 \qquad \text{after one year,}$$
$$P = 100(1.02)^8 \qquad \text{after two years,}$$
$$P = 100(1.02)^{4t} \qquad \text{after } t \text{ years.}$$

Note that 8% is *not* the rate used for each three month period; the annual rate is divided into four 2% payments. Calculating the total balance after one year under each method shows that

$$\text{Annual compounding:} \quad P = 100(1.08) = 108.00$$
$$\text{Quarterly compounding:} \quad P = 100(1.02)^4 = 108.24$$

Thus, more money is earned from quarterly compounding, because the interest earns interest as the year goes by. In general, the more often interest is compounded, the more money will be earned (although the increase may not be very large).

We can measure the effect of compounding by introducing the notion of *effective annual yield*. Since $100 invested at 8% compounded quarterly grows to $108.24 by the end of one year, we say that the *effective annual yield* in this case is 8.24%. We now have two interest rates which describe the same investment: the 8% compounded quarterly and the 8.24% effective annual yield. Banks call the 8% the *annual percentage rate*, or *APR*. We may also call the 8% the *nominal rate* (nominal means "in name only"). However, it is the effective yield which tells you exactly how much interest the investment really pays. Thus, to compare two bank accounts, simply compare the effective annual yields. The next time that you walk by a bank, look at the advertisements, which should (by law) include both the APR, or nominal rate, and the effective annual yield. We will often abbreviate *annual percentage rate* to *annual rate*.

Using the Effective Annual Yield

Example 1   Which is better: Bank X paying a 7% annual rate compounded monthly or Bank Y offering a 6.9% annual rate compounded daily?

Solution   We will find the effective annual yield for each bank.
Bank X: There are 12 interest payments in a year, each payment being $0.07/12 = 0.005833$ times the current balance. If the initial deposit were $100, then the balance P will be

$$P = 100(1.005833) \quad \text{after one month,}$$
$$P = 100(1.005833)^2 \quad \text{after two months,}$$
$$P = 100(1.005833)^t \quad \text{after } t \text{ months.}$$

To find the effective annual yield, we need to look at one year, or 12 months, which gives $P = 100(1.005833)^{12} = 100(1.072286)$, so the effective annual yield $\approx 7.23\%$.

Bank Y: There are 365 interest payments in a year (assuming it is not a leap year), each being $0.069/365 = 0.000189$ times the current balance. Then the balance P in the account is

$$P = 100(1.000189) \quad \text{after one day,}$$
$$P = 100(1.000189)^2 \quad \text{after two days,}$$
$$P = 100(1.000189)^t \quad \text{after } t \text{ days.}$$

so at the end of one year we have multiplied the initial deposit by

$$(1.000189)^{365} = 1.0714$$

so the effective annual yield for Bank Y $\approx 7.14\%$.

Comparing effective annual yields for the banks, we see that Bank X is offering a better investment, by a small margin.

Example 2   If $1000 is invested in each bank in Example 1, write an expression for the balance in each bank after t years.

Solution   For Bank X, the effective annual yield $\approx 7.23\%$, so after t years the balance, in dollars, will be

$$P = 1000(1.0723)^t.$$

For Bank Y, the effective annual yield $\approx 7.14\%$, so after t years the balance, in dollars, will be

$$P = 1000(1.0714)^t.$$

(Again, we are ignoring leap years.)

> If interest at an annual rate of r is compounded n times a year, then r/n times the current balance is added n times a year. Thus, with an initial deposit of $\$P_0$, the balance t years later is
>
> $$P = P_0\left(1 + \frac{r}{n}\right)^{nt}$$
>
> Note that r is the nominal rate and that, for example, $r = 0.05$ when the annual rate is 5%.

Increasing the Frequency of Compounding: Continuous Compounding

---

Example 3   Find the effective annual yield for a 7% annual rate compounded
(a)   1000 times a year.   (b)   10,000 times a year.

Solution   (a)   The balance is given by

$$\left(1 + \frac{0.07}{1000}\right)^{1000} \approx 1.0725056,$$

giving an effective annual yield of about 7.25056%.
(b)   The balance is

$$\left(1 + \frac{0.07}{10,000}\right)^{10,000} \approx 1.0725079,$$

giving an effective annual yield of about 7.25079%.

You can see that there's not a great deal of difference between compounding 1000 times each year (about three times per day) and 10,000 times each year (about 30 times per day). What happens if we compound more often still? Every minute? Every second? You may be surprised that the effective annual yield does not increase indefinitely, but tends to a finite value. The benefit of increasing the frequency of compounding becomes negligible beyond a certain point.

For example, if you were to compute the effective annual yield on a 7% investment compounded n times per year for values of n larger than 100,000, you would find that

$$\left(1 + \frac{0.07}{n}\right)^n \approx 1.0725082.$$

So the effective annual yield is about 7.25082%. Even if you take $n = 1{,}000{,}000$ or $n = 10^{10}$, the effective annual yield will not change appreciably. The value 7.25082% is an upper bound which is approached as the frequency of compounding increases.

When the effective annual yield is at this upper bound, we say that the interest is being *compounded continuously*. (The word *continuously* is used because the upper bound is approached by compounding more and more frequently.) Thus, when a 7% nominal annual rate is compounded so frequently that the effective annual yield is 7.25082%, we say that the 7% is compounded *continuously*. This represents the most one can get from a 7% nominal rate.

The Number e

In Chapter 1 we discuss general exponential functions $y = a^x$ and in Chapter 4 we discuss the particular exponential function $y = e^x$, with base $e = 2.71828....$ One of the reasons that the number e is important is that many of the formulas of calculus are simplified if e is used as the base.

In addition, it turns out that e is intimately connected to continuous compounding. To see this, use your calculator to check that $e^{0.07} \approx 1.0725082$, which is the same number we obtained when we compounded 7% a large number of times. So you have discovered that for very large n

$$\left(1 + \frac{0.07}{n}\right)^n \approx e^{0.07}.$$

As n gets larger, the approximation gets better and better, which we write as

$$\left(1 + \frac{0.07}{n}\right)^n \longrightarrow e^{0.07}$$

, meaning that as n increases, the value of $\left(1 + 0.07/n\right)^n$ gets closer and closer to $e^{0.07}$.

Thus, if $\$P_0$ is deposited at an annual rate of 7% compounded continuously, the balance, $\$P$, will be given by

$$P = P_0(1.0725082) = P_0 e^{0.07} \qquad \text{after one year,}$$
$$P = P_0(1.0725082)^2 = P_0 \left(e^{0.07}\right)^2 = P_0 e^{(0.07)2} \qquad \text{after two years,}$$
$$P = P_0(1.0725082)^t = P_0 \left(e^{0.07}\right)^t = P_0 e^{0.07t} \qquad \text{after } t \text{ years.}$$

If interest on an initial deposit of $\$P_0$ is *compounded continuously* at an annual rate r, the balance t years later can be calculated using the formula

$$P = P_0 e^{rt}.$$

Again, r is the nominal rate, and, for example, $r = 0.05$ when the annual rate is 5%.

In solving a problem involving compound interest, it is important to be clear whether interest rates are nominal rates or effective yields, as well as whether compounding is continuous or not.

Example 4   Suppose that a bank advertises a nominal annual interest rate of 8%. If you deposit $5000, how much will be in the account three years later if the interest is compounded
(a) Annually?   (b) Monthly?   (c) Continuously?

Solution   (a)   If interest is compounded annually, the amount in dollars in the account after three years is

$$P = P_0(1 + r)^t = 5000(1.08)^3 = 6298.56.$$

(b)   If interest is compounded monthly, the amount in dollars in the account after three years is

$$P = P_0 \left(1 + \frac{r}{n}\right)^{nt} = 5000 \left(1 + \frac{0.08}{12}\right)^{(12)(3)} = 5000(1.006667)^{36} = 6351.18.$$

As expected, we make more money if interest is compounded monthly rather than annually.

(c)   If interest is compounded continuously, the amount in dollars in the account after three years is

$$P = P_0 e^{rt} = 5000e^{(0.08)(3)} = 6356.25.$$

Notice that we make still more money with continuous compounding.

Problems for Section B

1.   A bank pays a nominal annual rate of 6%. If you deposit $1000, how much is in the bank after 10 years if interest is compounded (a)   Annually?   (b)   Quarterly?   (c)   Monthly?

2.   A bank pays a nominal annual rate of 7%. If you deposit $8000, how much is in the bank after 5 years if interest is compounded (a)   Annually?   (b)   Quarterly?   (c)   Monthly? (d)   Continuously?

3.   Which is better: Bank X paying an 8.1% annual rate compounded annually, or Bank Y paying an 8% annual rate compounded monthly?

4.   Use a graph of $y = \left(1 + 0.07/x\right)^x$ to find the value of $\left(1 + 0.07/x\right)^x$ as $x \to \infty$. Confirm that the value you get is $e^{0.07}$.

5.   (a)   Find the effective annual yield for a 5% annual interest rate compounded
            (i) 1000 times/year,   (ii) 10,000 times/year,   (iii) 100,000 times/year.
     (b)   Look at the sequence of answers in part (a), and deduce the effective annual yield for a 5% annual rate compounded continuously.
     (c)   Compute $e^{0.05}$. How does this confirm your answer to part (b)?

6.   (a)   Find $\left(1 + 0.04/n\right)^n$ for $n = 10,000$, and $100,000$, and $1,000,000$. Use the results to deduce the effective annual yield of a 4% annual rate compounded continuously.
     (b)   Confirm your answer by computing $e^{0.04}$.

7.   (a)   The Banque Nationale du Zaïre pays 100% nominal interest on deposits, compounded monthly. You invest 1 million zaïre. (The "zaïre" is the unit of currency of the Republic of Zaïre.) How much money do you have after one year?
     (b)   How much money do you have after one year if you invest 1 million zaïre with interest compounded daily? Hourly? Each minute?
     (c)   Does this amount increase without bound as interest is compounded more and more often, or does it level off? If it levels off, provide a close "upper" estimate for the total after one year.

INDEX